Frontiers in Clinical Drug Research: Diabetes and Obesity
(Volume 1)

Editor

Atta-ur-Rahman, *FRS*
Kings College
University of Cambridge
Cambridge
UK

Bentham Science Publishers Ltd.
Executive Suite Y - 2
PO Box 7917, Saif Zone
Sharjah, U.A.E.
subscriptions@benthamscienc.org

CONTENTS

PREFACE

Diabetes is a major metabolic disease, and the number of patients is still growing in both industrialized and developing countries. Diabetes is attributed to an increased intake of glucose, beta cell fatigue in the pancreas and insulin resistance. Drugs to reduce blood glucose levels by several mechanisms have been launched in the past decades, and the options for treating hyperglycemia have been extended. Obesity is a condition characterized by an unhealthy increase in body mass due to fat deposits in the body. Obesity in individuals can cause a variety of other medical problems such as heart diseases, obstructive sleep apnea, type 2 diabetes, among others. In 2013, the American Medical Association classified Obesity as a disease. While the link between obesity and type 2 diabetes is clear – with a high percentage of obese patients also accounting for new type 2 obesity cases – the search for new drugs to treat these conditions is still ongoing.

This first volume of **Frontiers in Clinical Drug Research – Diabetes and Obesity** brings comprehensive reviews to readers interested in the recent advances in the development of pharmaceutical agents for the treatment of these two metabolic diseases. The scope of the eBook series covers a range of topics including the medicinal chemistry, pharmacology, molecular biology and biochemistry of natural and synthetic drugs affecting endocrine and metabolic processes linked with diabetes and obesity. Reviews in this series also include research on specific receptor targets and pre-clinical / clinical findings on novel pharmaceutical agents.

In the first chapter Onuigbo suggests a revised classification system for renoprotective agents: single-pathway-blockers (SPBs) that block single pathways of diabetic chronic kidney disease (CKD) progression, and multiple-pathway-blockers (MPBs) that simultaneously block several pathways of CKD progression. This article goes beyond the angiotensin receptor blocking paradigm of addressing diabetic CKD – which is a very common complication in diabetic patients. This chapter highlights a multi-modal approach to treating this condition.

In chapter 2, Poretsky *et al.* present new approaches to preventing and treating diabetes mellitus. They have focused on advanced glycation end products (AGE). The authors highlight the treatment approaches which include inhibiting AGE formation, breaking formed cross-links or blocking negative effects of AGE. A high ACE diet is believed to cause obesity and this chapter brings to light a novel way to approach diabetes treatment.

In chapter 3, Zetu *et al.,* discuss the present and future of liraglutide, a new Novo Nordisk therapy. liraglutide is the first human incretin mimetics GLP-1 analogue sharing 97% of its amino acid sequence with native human GLP-1. They have discussed the potential benefits for liraglutide, such as reducing fasting and postprandial glucose levels, decreasing HbA1c levels by up to 1, 75% at 18-month follow up, preventing weight gain and inducing weight loss with reduced risk of adverse symptoms. While this study shows promising phase 3 clinical trial results, more clinical trials are needed to evaluate long term outcomes.

Singh, in chapter 4, reviews the role of MicroRNA in the expression of genes that might possibly be involved with the onset of diabetes. The article demonstrates that miRNAs can facilitate signaling pathways involved in insulin resistance and obesity. This work offers novel ways of identification and validation of new targets of therapeutic intervention *viz.* agents controlling the differential expression of genes involved in hyperglycemia.

In chapter 5, Ramachandran *et al.* have reviewed the gene expression, nuclear receptors of the Peroxisome Proliferator Activated Receptors (PPARs) and their ligands. They have also discussed fibrates and thiazolidinediones - modulators of PPAR α and γ nuclear receptors respectively that are drugs in current use. This chapter points out the heterogeneity of these drugs and suggests that the clues to understand this phenomenon may lie in studying the molecular biology of PPARs. This is clinically significant, given that the use of thiazolidinediones in obesity patients has resulted in some adverse effects and is gradually declining,

In the last chapter and Sano and Sasase summarize drugs for three major diabetic complications currently in use and in clinical trial phases. They aim to give readers an understanding of the trends of current pharmacological approaches and the tide of next-generation drugs used to treat diabetic complications such as retinopathies, neuropathies and renal failure. Consequently, there are implications for improved quality of life for patients, especially for aging individuals.

I am grateful for the timely efforts made by the editorial personnel, especially Mahmood Alam (Director Publications), Maria Baig and Taimur --- at Bentham Science Publishers for their assistance.

Atta-ur-Rahman, FRS
Kings College
University of Cambridge
Cambridge
UK

List of Contributors

Agustin Busta

Division of Endocrinology, Department of Medicine, Beth Israel Medical Center and Albert Einstein College of Medicine, New York, NY 10003, USA

Andrew Hartland

Consultant Chemical Pathologist, Walsall Manor Hospital, UK

Ashutosh Pareek

Division of Endocrinology, Department of Medicine, Beth Israel Medical Center and Albert Einstein College of Medicine, New York, NY 10003, USA

Cornelia Zetu

National Institute of Diabetes, Nutrition and Metabolic Diseases "N.C. Paulescu", Bucharest, Romania

Cristian Serafinceanu

National Institute of Diabetes, Nutrition and Metabolic Diseases "N.C. Paulescu", Bucharest, Romania

Dan Mircea Cheța

National Institute of Diabetes, Nutrition and Metabolic Diseases "N.C. Paulescu", Bucharest, Romania

Emilia Liao

Division of Endocrinology, Department of Medicine, Beth Israel Medical Center and Albert Einstein College of Medicine, New York, NY 10003, USA

Kevin Stuart

Specialist Registrar in Chemical Pathology/Metabolic Medicine, Heart of England Foundation Trust, UK

Leonid Poretsky

Division of Endocrinology, Department of Medicine, Beth Israel Medical Center and Albert Einstein College of Medicine, New York, NY 10003, USA

Macaulay A.C. Onuigbo

College of Medicine, Mayo Clinic, Rochester, MN, Luther Midelfort Site, Mayo Health System Practice-Based Research Network (MHS PBRN), Department of Nephrology, Mayo Clinic Health System Eau Claire, 1221 Whipple Street, Eau Claire, WI 54702., MBA Healthcare Executive, USA

Mithun Bhartia

Specialist Registrar in Diabetes and Endocrinology, Sandwell Hospital, West Birmingham NHS Trust, UK

Puneetpal Singh

Department of Human Genetics, Punjabi University, Patiala, Punjab, India

Ryuhei Sano

Central Pharmaceutical Research Institute, Japan Tobacco Inc., Takatsuki, 569-1125 Osaka, Japan

Sudarshan Ramachandran

Department of Clinical Biochemistry, Good Hope Hospital, Heart of England NHS Foundation Trust, Rectory Road, Sutton Coldfield, West Midlands, UK

Tomohiko Sasase

Central Pharmaceutical Research Institute, Japan Tobacco Inc., Takatsuki, Osaka, Japan

Send Orders for Reprints to reprints@benthamscience.net

Frontiers in Clinical Drug Research: Diabetes and Obesity, Vol. 1, 2014, 3-32   **3**

CHAPTER 1

Angiotensin Blockers and Renoprotection in Diabetic Chronic Kidney Disease is a Failed Paradigm? – A Revisionist View of Renoprotection in Diabetic Chronic Kidney Disease and A Novel Classification Scheme for Renoprotective Agents

Macaulay A.C. Onuigbo[*]

College of Medicine, Mayo Clinic, Rochester, MN, Luther Midelfort Site, Mayo Health System Practice-Based Research Network (MHS PBRN), Department of Nephrology, Mayo Clinic Health System Eau Claire, 1221 Whipple Street, Eau Claire, WI 54702., MBA Healthcare Executive, USA

Abstract: The excellent innovative work of Squibb Laboratory scientists in the 1970's led to the discovery of the ACE inhibitors and the ARBs. Despite their widespread use, for renoprotection, we have continued to experience a global ESRD pandemic. Angiotensin blockade has proved inadequate. In 2010, we asked the rhetorical question: Is renoprotection with RAAS blockade a failed paradigm?

Keywords: Angiotensin receptor blockers, ACE inhibitors, chronic kidney disease, diabetes mellitus, novel classification scheme, revisionist view, renoprotection, renoprotective agents.

Our knowledge of mechanisms of CKD progression to ESRD is incomplete. There is accumulating evidence for roles for several putative, independent, and disparately dissimilar pathogenetic mechanisms including predisposing genetic abnormalities, and several non-genetic mechanisms of renal injury. It therefore is naively unscientific to surmise that angiotensin blockade is the magic bullet. It is time to re-strategize and come up with more innovation.

We recently suggested a new classification system for renoprotective agents: single-pathway-blockers (SPBs) that block single pathways of CKD progression,

*Address correspondence to Macaulay A.C. Onuigbo: College of Medicine, Mayo Clinic, Rochester, MN, Luther Midelfort Site, Mayo Health System Practice-Based Research Network (MHS PBRN), Department of Nephrology, Mayo Clinic Health System Eau Claire, 1221 Whipple Street, Eau Claire, WI 54702., MBA Healthcare Executive, USA; Tel: 715 838 3891; Fax: 715 838 1946; E-mail: onuigbo.macaulay@mayo.edu

and multiple-pathway-blockers (MPBs) that simultaneously block several pathways of CKD progression. Bardoxolone Methyl (inducer of the transcription factor, Nrf2) and pentoxifylline, (suppressor of factors of inflammatory response), would tentatively qualify as MBPs, if ongoing trials confirm both agents as efficacious renoprotective agents. We therefore support the ongoing investigative searches for more effective and safer renoprotective agents. Only time will tell if Bardoxolone Methyl, an MBP renoprotective agent, or other agents yet to be trialed, will live up to this hope or hype of the prevention of ESRD in CKD patients.

For now, the only renoprotective agents available to us are the RAAS blocking agents; all are clearly SBPs. We support their continued use as long as physicians continue to acknowledge their limitations, and more importantly, remain aware of the various modalities whereby these agents can in fact produce clinically significant nephrotoxicities, which are not always reversible following drug discontinuation.

THE INTRODUCTION OF ACEIs AND ARBs INTO CLINICAL PRACTICE

Squibb laboratory scientists in 1971 discovered the first inhibitor of the angiotensin converting enzyme (ACE), given intravenously, with blood pressure lowering properties [1]. The Squibb group subsequently developed the first orally effective ACE inhibitor, captopril [2]. Subsequently, many more active pharmacological moieties with ACE inhibition activity got introduced into the pharmacopeia, and afterward, the angiotensin receptor blockers (ARBs) were introduced into clinical medicine [3-11]. During the last three decades, with the discovery of similar chemically related pharmaceuticals, coupled with the completion of several large randomized multi-center placebo-controlled studies, the indications for ACE inhibitors and ARBs have expanded to include hypertension, heart failure, post-acute myocardial infarction, diabetic nephropathy and non-diabetic nephropathies, with class-specific claims of cardio-protection and renoprotection [12-35]. As a result, currently, nearly 80% of some US diabetic patients are receiving an ACE inhibitor, an ARB or a combination of both agents [36-38].

THE LIMITATIONS AND IMPERFECTIONS OF ACE INHIBITORS AND ARBs AS RENOPROTECTIVE AGENTS

Current renoprotection paradigms depend almost entirely on the use of ACE inhibitors and/or ARBs, but in total, these agents have proved to be imperfect. Worldwide, CKD to ESRD progression has apparently continued, unrelenting, despite the extensive use of these agents of angiotensin blockade [39-58].

Despite the extensive, widespread and continued use of the ACE inhibitors and ARBs both here in the US, and worldwide, over the last two decades, most estimates show that CKD to ESRD progression has continued, almost unabated, in CKD patients around the world [39-58]. Undeniably, recent epidemiological data from the USA, Canada, Europe, Asia and the Indian subcontinent, Africa, the Far East, South America, the Middle East and Eastern Europe all point to this increasing incidence of ESRD, tantamount to a global ESRD pandemic [39-58].

According to ESRD statistics recently released by the United Sates Renal Data System (USRDS) in October 2010, the prevalent ESRD population in the USA attained an all-time high of 547, 982 as of December ending 2008 [58]. This represented a 1.2 percent jump from 2007, compared to a 0.85 percent increase from 2006 to 2007 [58]. For the most part, arguments by some authorities that the prevalence of ESRD in the USA is rising but only due to increased survival of patients with ESRD is not convincing since the incident US ESRD population for 2007 of 108, 926 had increased to 112, 476 by 2008 representing an increase of 1.2% over the year before, an even greater rate of growth compared to the US ESRD population growth rate of 0.88% between 2006 and 2007 [58]. Furthermore, there is the argument that the increasing US ESRD population is related to the increasing diabetic epidemic. However, a very detailed and impartial analysis of the evidence base points very much to the contrary. Most impressively, in 2005, Jones *et al.* had reviewed the time trends of the incidence of diabetic ESRD in the United States of America, studying the United States Renal Data System database, the Diabetes Surveillance Program of the Centers for Disease Control and Prevention, as well as the diabetes literature in general, and had clearly demonstrated that the recent growth of the number of individuals with diabetes accounted for <10% of the increase in the number of diabetes-related

ESRD [44]. Instead, most of the observed increases in ESRD among US diabetic patients was due to a three-fold increase in risk of ESRD in people with diabetes and therefore qualified as an epidemic [44].

We have reported severally on the pitfalls and deficiencies of the evidence-base for the ACE inhibitors and ARBs for renoprotection, based on several large RCTs including the RENAAL and IDNT trials, and the reader is advised to review some of these reports [59-71]. Due to space constraints, we would only revisit some of the criticisms of the RENAAL trial in some more detail in the following paragraph.

The RENAAL trial which compared Losartan, an ARB, versus placebo, analyzed renal outcomes in 1513 diabetic CKD patients, mean baseline serum creatinine of 1.9 mg/dL, all with confirmed diabetic nephropathy with urinary albumin excretion >300 mg/day, and followed for a mean of 3.4 years, had demonstrated a 25% risk reduction in the doubling of serum creatinine (p=0.02), and a 28% risk reduction in ESRD (p=0.006) [24]. Going by these statistically significant risk reductions, as we had argued previously, and given the high cardiovascular mortality associated with ESRD, over a mean follow up of 3.4 years, one would have expected at least some mortality benefits in favor of Losartan compared to placebo in the RENAAL study [67]. In reality, however, to the contrary, our critical analysis of the RENAAL study database had demonstrated *higher* all-cause mortality rates in the Losartan group *vs.* the placebo group - 21% *vs.* 20.3%, albeit not statistically significant [67]. Besides, if indeed the ACEIs and ARBs as suggested by the RENAAL trial was able to reduce ESRD rates by as much as 28%, arguably, we would have observed much lower ESRD rates here in the US in the late 2000s, a picture that is not supported by recent USRDS data released in late 2010, and already discussed in a previous section here [48].

FREQUENTLY UNRECOGNIZED BUT SIGNIFICANT IATROGENIC RENAL FAILURE ASSOCIATED WITH ANGIOTENSIN BLOCKADE

Another concerning aspect of the widespread use of angiotensin blockade, especially in the older CKD patient, >65 years old, with or without proven renal artery stenosis, is the underdiagnosed and unrecognized degree of iatrogenic renal failure associated with and that can result from the concurrent use of these agents [59-71]. Given the critical role of angiotensin II in renal autoregulation, it is not surprising to note such a downside to their use, more so in the older >65 year old CKD patients [8, 59-71]. In 2005, we first described the syndrome of late-onset renal failure from angiotensin blockade (LORFFAB) in a prospective analysis of the association of concurrent angiotensin blockade with sometimes irreversible worsening azotemia in CKD patients [59]. We have since then published several series demonstrating this relationship even more strongly [59-72].

In recent years, apart from the well described experiences at our Mayo Clinic Health System, Eau Claire, Wisconsin in the USA, [59-71], there are new increasing reports of these associations of concurrent angiotensin blockade with worsening renal failure in general, and in association with specific clinical syndromes [73-89]. The clinical scenarios in which worsening renal failure has been described in association with concomitant angiotensin blockade include contrast-induced nephropathy [73-79], the post-operative state [80-82], during acute illness, [83, 84], and following bowel preparations for colonoscopy procedures [85-89]. Furthermore, Suissa *et al.* in a Canadian report, in 2006, demonstrated in a population-based historical cohort analysis of 6102 Canadian diabetic patients, mean age 66 years, an increased rate ratio of ESRD of 4.2 (95% CI: 2.0-9.0), after 3 years or longer of ACE inhibition [90].

Even more impressive was a recent report from the United Kingdom where El Nahas *et al.* from the Sheffield Kidney Institute, Sheffield, United Kingdom, had succinctly demonstrated sustained increases in eGFR from 16.4 to 26.6 ml/min/ 1.73 m^2 BSA, in 52 patients with advanced CKD who were receiving concurrent ACE inhibition and/or ARB therapy, mean age 73 years, following the discontinuation of ACE inhibitors and/or ARBs [91, 92].

We must acknowledge here that the commonly held speculative notion that any such worsening renal failure associated with angiotensin blockade is easily and always reversible with drug discontinuation is false, presumptuous and not backed by available evidence. Clearly, irreversible ESRD has been increasingly demonstrated in mostly older CKD patients despite subsequent discontinuation of concurrent angiotensin blockade with ACE inhibitors and/or ARBs, and this is without prejudice to the presence or absence of renal artery stenosis [83, 84, 90, 93-96]. Here again, we must recognize that in 2010, we described the first report of the previously unrecognized syndrome of rapid onset end-stage renal disease (SORO-ESRD), symbolized by the *de novo* development of irreversible ESRD rapidly following acute kidney injury (AKI), superimposed on CKD [94]. In that report, during 4 years prospective follow up of a cohort of 100 CKD patients, 15 of 17 (88%) patients who had progressed to ESRD, reached the status of ESRD rapidly and unpredictably, in an unanticipated manner, and this accelerated progression to ESRD was precipitated by acute medical and surgical events [94]. Pertinently, we had observed that of the 15 patients with SORO-ESRD, mean age 68 years, 9 of 15 (60%) were aged 65 years and older, and 6 of 15 (40%) were aged 80 years or older. In a significant number of these patients, 5 of 15 (33%), the AKI occurred in consonance with concurrent RAAS blockade [94]. Clearly from the foregoing, there is now increasing evidence of the strong association of concurrent angiotensin blockade with worsening AKI on CKD, and indeed sometimes irreversible ESRD. These observations demand further intense examination and study [67-71, 91-96].

THE NEW PARADIGM

The point remains that at this time, given the complexity of the issues, it is almost impossible to tease out all the confounding variables in order to make a clear determination of the number of cases of ESRD that can be directly attributable to concurrent RAAS blockade. Our approach is that this data, albeit impossible to get, is not even needed after all [59-67, 69-71]. We continue to insist that the treating physician who prescribes the ACEI and/or the ARB, for renoprotection, or any other indication for that matter, must remain watchful and cognizant of the potential of these agents to cause sometimes irreversible renal failure including ESRD, and to be ready to discontinue these agents at anytime, one day, one week, one month, one year or one decade after RAAS blockade initiation [59-67, 69-71]. This is the new paradigm.

In our experience, we have seen CKD patients followed closely by physicians for months to years, who had continued to demonstrate progressive increases in creatinine (acute renal failure on CKD) while on concurrent ACEI/ARB and the physicians had continued to prescribe the ACEI/ARB, purely on the notion that these pharmaceutical agents are renoprotective agents, and therefore can do no harm to the kidneys. This notion must stop since sometimes, belatedly and most unfortunately, some of these patients have gone on to directly and inexorably develop irreversible ESRD [59-67, 69-71]. The new paradigm that we are proposing is that pre-emptive drug discontinuation of the ACEI/ARB, in these circumstances, ought to be the norm, rather than the exception [59-67, 69-71]. This paradigm calls for a lifetime monitoring of kidney function by serum creatinine measurements, in our experience, at least 2-3 monthly after established drug dosing in these CKD patients on ACEI/ARB, more so in the older >65 year old CKD patients. It must be recognized, that at RAAS blockade initiation, and following dose increases of ACEIs and/or ARBs, serum creatinine testing frequency should be monitored more often, as often as after one week, again, more so in the older >65 year old CKD patients [59-67, 69-71].

This brings us to the issue of ecology of medical care as it relates to the question of the general applicability of results of clinical trials that are mostly conducted at our tertiary medical centers, to the general population, given that only a very small proportion of the patient population gets to be seen in these tertiary centers of medical research [97, 98].

THE ECOLOGY OF MEDICAL CARE REVISITED AS RELATED TO RESEARCH FINDINGS FROM TERTIARY MEDICAL CENTERS BEING THE PRIMARY SOURCE FOR EVIDENCE-BASED PRACTICE GUIDELINES

We would like to, once more, draw attention of the medical community to the issue of the ecology of medical care. The fact that most medical research occurs in our tertiary medical institutions is concerning, more so given that generally, practice-based guidelines and recommendations originate from research findings involving the limited percentage of the general patient population that get to be seen in these tertiary medical units [97, 98]. This potentially should limit the widespread applicability of findings of such research to the general patient population especially given the fact that the patients seen and managed in these tertiary medical centers represent only 0.1% of the patient population seen by physicians in the USA [97, 98]. White *et al.,* 50 years ago, in the article "The ecology of medical care" published in the New England Journal of Medicine, very eloquently demonstrated that the vast majority of research was conducted on the 0.1% of the population who were cared for in the academic tertiary hospitals [97, 98]. This fact must never be lost on us and must constantly remain a source of trepidation and concern to critical observers of the medical literature [71]. Therefore, we submit that for sustainability and good medical care and enhanced patient outcomes, old so-called established paradigms in medicine must and should always be open to revision, anytime, especially when such revisions are supported by emerging new convincing data [71, 98].

THE BARDOXOLONE STORY – A NEW WAY FORWARD?

In the June 2011 issue of the New England Journal of Medicine, the Bardoxolone Methyl Treatment: Renal Function in CKD/Type 2 Diabetes (BEAM) trial Investigators, rekindled new interest in the concept of renoprotection and the role of renoprotective agents when they reported significant increases in the mean (±SD) estimated GFR (eGFR) in diabetic CKD patients with enrollment eGFR of 20 to 45 ml per minute per 1.73 m^2 of body-surface area who received the trial drug, Bardoxolone Methyl versus placebo [99]. The report demonstrated that patients receiving Bardoxolone Methyl had significant increases in the mean (±SD) estimated GFR, as compared with placebo, at 24 weeks (with between-group differences per minute per 1.73 m^2 BSA of 8.2±1.5 ml in the 25-mg group, 11.4±1.5 ml in the 75-mg group, and 10.4±1.5 ml in the 150-mg group; P<0.001). The increases in eGFR were maintained through week 52, with significant differences per minute per 1.73 m^2 of 5.8±1.8 ml, 10.5±1.8 ml, and 9.3±1.9 ml, respectively [99]. This report prompted us to ask the loaded rhetoric question: Is Bardoxolone Methyl the right way forward in our quest for the perfect renoprotective agent? Bardoxolone Methyl is an antioxidant inflammation modulator (AIM) and the most potent known inducer of Nrf2 to enter clinical trials [100]. Nrf-2 (NF-E2-related factor 2) is a regulator of anti-oxidant, anti-inflammatory and detoxification pathways [101].

It must be conceded here that the detailed knowledge of the modalities of action of Bardoxolone Methyl as an anti-oxidant remain mostly conjectural [100, 101].

Nrf2 activates over 250 antioxidant and detoxification genes, improves endothelial function, and maintains kidney structure and function [100, 101]. It is hoped that the ongoing global phase III Bardoxolone Methyl trial, the BEACON (Bardoxolone methyl evaluation in patients with Chronic kidney disease and type 2 diabetes: the Occurrence of renal events) trial, that is anticipated to recruit approximately 1600 patients at 300 sites, worldwide, would yield even more impressive results. Again, it must be conceded here that the hopes of Bardoxolone Methyl turning out to be a potent and beneficial renoprotective agent remain to be validated by ongoing clinical trials.

STAND-ALONE *DE NOVO* ESRD RATES SHOULD BE THE "GOLD STANDARD" FOR MEASURING RENAL OUTCOMES IN CKD TRIALS

We had earlier reported on the pitfalls of the frequent use of so-called combination renal surrogates in measurement and estimation of renal outcomes in previously reported CKD trials [22-26, 65, 67-71, 102]. This is reflected in the over-reliance on the use of combination renal end points of so-called surrogates of renal function including the rates of doubling of serum creatinine, the degrees of reduction of proteinuria, with or without ESRD rates, as the basis for determining and defining renoprotection [69]. The BEAM study had shown that Bardoxolone Methyl in fact produced increased levels of albuminuria as measured by albumin creatinine ratios (ACR), yet another limitation of using changes in proteinuria as a measure of renoprotection [16, 99]. In addition, in our single center experiences, we have demonstrated temporal disassociations between kidney function and measured ACR in patients exhibiting clearly improved kidney function, following discontinuation of ACE inhibitors and/or ARBs, while at the same time, showing concurrently increased levels of ACR [65, 67, 69, 71, 102, 103]. Moreover, in an elegant set of observations, Tsalamandris *et al.* reported that of 40 patients with diabetes, followed over a period of 8-14 years, 15 developed progressive increase in albumin excretion rate (AER) with no decline in GFR, 13 had progressive increase in AER in association with decreasing GFR, and 12 (8 type 2) had decreasing GFR values without a significant increase in AER [104, 105]. Besides, El Nahas *et al.* from the Sheffield Kidney Institute, United Kingdom, followed 52 late stage CKD (4 & 5) patients, 24 (46%) with diabetes mellitus, mostly type 2, for at least 12 months after the ACEI/ARB was discontinued [91]. This study had demonstrated sustained improvements in eGFR following drug discontinuation, with no demonstrable increase in proteinuria, suggesting a significant dissociation between the level of proteinuria and renal function in late stage CKD patients, both diabetic and non-diabetic [91].

The over-reliance of the Nephrology literature on composite end-points in different combinations - doubling of serum creatinine, changes in levels of proteinuria, and end-stage renal disease (ESRD) - all at the same time, to determine statistical differences in renal outcomes between study groups, we have contested, is very troubling and unquestionably flawed [65, 67, 69, 71, 103]. One major drawback of the use of such composite aggregates of renal end-points is that the utilization of such potentially confounding variables, as if they represented strictly independent variables, does establish a threat to the internal validity of the statistical analyses [69, 71, 106-108]. This statistical anomaly or fallacy is referred to as the phenomenon of ecological fallacy or Simpson's Paradox [106-108]. Above all, the magnitude and even the direction of existing relationships between these competing variables, found by comparing group-level statistics, may not eventually be indicative of the true relationships tested with individual participant data [106-108]. Therefore, proof of association from such statistical analyses and deductions are error-prone and cannot be presumed to represent proof of causation [69, 71, 106-108]. We submit that for CKD trials, and in drug to drug comparisons, or drug to placebo comparisons, the only one hard renal outcome endpoint that is bereft of these encumbrances and statistical pitfalls and hazards, would be the exclusive use of incident *de novo* ESRD rates [67, 69, 71]. *De novo* ESRD rate is defined by the development during a drug trial of new irreversible and permanent renal failure requiring renal replacement therapy for 90 days or longer [71]. We therefore look forward to seeing the comparison of incident ESRD rates as primary stand-alone renal end-points in the ongoing global BEACON Bardoxolone Methyl phase III clinical trials.

There is one often cited major drawback of the potential use of ESRD rates to assess renoprotection trials in that it would take very long clinical trials, given the low rates of CKD-ESRD progression, to demonstrate any differences between treatments groups and that such trials are therefore untenable. We posit that if pharmaceutical companies refuse to fund such studies, government bodies such as the National Institutes of Health (NIH) should therefore champion such efforts. We would, on the contrary, argue that the alternative situation existing currently where we have a multitude of short trials with inconclusive endings is indeed worse. We would continue to argue that ESRD rates, without surrogate renal markers, must be the "gold standard" to help nephrologists resolve the current quagmire of the availability of many inconclusive trial reports. The argument for more practical suggestions in CKD research endpoints would only continue to give us what we currently have – inconclusive data. ESRD rate reductions, confirmed over time, between treatment groups, would represent hard-core evidence for renoprotection for future drug trials, we submit.

PUTATIVE PATHOGENETIC MECHANISMS FOR CKD PROGRESSION IN DIABETIC AND NON-DIABETIC NEPHROPATHY – CAN A SINGLE AGENT TRULY AND CONSISTENTLY DELIVER RENOPROTECTION?

Venkatachalam in 2010 declared that despite extensive research effort and time, it is still valid to posit that in 2010, we yet do not have a complete and unquestionable knowledge of the mechanisms that trigger and/or lead to CKD initiation and progression to ESRD, respectively [109]. Furthermore, even though current renoprotection paradigms depend wholly and entirely on the use of ACE inhibitors and/or ARBs [1-35], the overwhelming evidence-base suggests that a multitude of putative, different, independent and unrelated pathogenetic mechanisms drive (diabetic and non-diabetic) CKD initiation and progression to ESRD [110-136]. These putative mechanisms are herein summarized below. Also, even the exact anatomical site of the damage to the kidneys underlying the a priori initiation and subsequent progression of albuminuria in diabetic nephropathy and subsequent poor renal outcomes in diabetic nephropathy remains questionable, and is still under debate [137, 138]. As recently as 2009, Russo *et al.* reported new experimental evidence to suggest that in fact, tubular dysfunction might constitute the primary factor in the causation of early albuminuria from diabetic nephropathy as opposed to a glomerular pathology [137, 138]. What's more, Venkatachalam *et al.* in a study of the impact of acute kidney injury (AKI) and post-AKI repair mechanisms, concluded that the renal pathology that develops in regenerating tubules after AKI is characterized by failure of differentiation and persistently high signaling activity and that these post-AKI changes represent the proximate cause(s) that drive(s) downstream events in the interstitium - inflammation, capillary rarefaction and fibroblast proliferation [109].

The following factors have been suggested by various studies and evidence-lines to cause the initiation and subsequent propagation of renal disease progression in CKD patients:

-   Several predisposing genetic abnormalities including variations of the nonmuscle myosin heavy chain 9 gene (MYH9) on chromosome 22, and variants at chromosome 6q24-27 among African Americans [110, 111].

-   Oxidative stress combined with a paradoxic hypoxic renal environment conditioned by an underlying genetic predisposition (see above) [112-114].

-   The production of advanced glycosylation end-products and the interaction of these end-products on the multiligand receptor of the immunoglobulin superfamily receptor for advanced glycation end-products (RAGE) [115, 116].

-   Intrarenal angiotensin II and/or rennin production [117].

-   Inflammation [118].

-   Lipid toxicity [119-121].

-   Podocyte injury and apoptosis [122, 123].

-   Cytokine/chemokine/growth factor release causing renal injury [124, 125].

-   Asymmetric diethylarginine (ADMA) [126].

-   Uric acid in CKD progression continues to attract increasing global attention [127-132].

- APOL 1 gene variants in APOL 1 variants in the CKD pathogenesis and progression [133-136].

Suffice it to say that space limitations would not allow us to describe in any further detail, the implications of the above listed various putative mechanisms with regards to their involvement in CKD progression, as described variously in the current literature. Moreover, the potential interplay between these varying mechanisms, operational in the respective individual CKD patient, for now can only remain conjectural.

Keenly mindful of this array of potential mechanisms, and given the disparateness and dissimilarities between these suggested mechanisms of CKD propagation to ESRD, it would be naïve and simplistically unscientific and unsophisticated for practicing nephrologists to assume that single mechanism-blocking agents such as the ACE inhibitors and/or ARBs, which can only antagonize the angiotensin cascade, would successfully, consistently and unfailingly deliver perfect renoprotecton results in (diabetic and non-diabetic) CKD patients, all the time [65-67, 69-71, 103]. Such a presumption, it would appear, is simply unrealistic, unlikely, improbable and truly flies in the face of reason [65-67, 69-71, 103]. This "one-size fits all" approach to medicine, in general, must be questioned and challenged, and in our opinion ought to be discouraged especially in relation to CKD management [71, 139]. Bansal and Hsu, following a recent analysis of the disparate ESRD rates and mortality outcomes in various CKD populations, as reported by various studies in the literature, confirmed the complex heterogeneity of CKD populations [139]. These same authors had concluded that nephrologists must consider several factors for CKD prognostication, with a need to *individualize* CKD management for each individual CKD patient [139].

Consonant with the multiplicity of pathways of CKD progression, there has been research into the fact that the use of ACEIs may lead to the release of pharmacologically active kinins and that the use of ARBs could be potentially influenced by the presence of the various angiotensin II receptors, AT1-AT4 and beyond [7, 8]. Such ideas led to the concept of total RAAS blockade whereby the use of combination ACEI and ARB have been suggested to more than likely result in improved outcomes. Unfortunately, in general, these ideas have not been borne out by research evidence. Indeed, to the contrary, the findings from the ONTARGET study revealed that despite higher reductions of the level of proteinuria by combination therapy, an expected outcome, if anything, combination Ramipril-Telmisartan therapy led to higher incidence of renal failure, indeed more dialysis rates and of course more hyperkalemia, when compared to monotherapy with Telmisartan alone, or Ramipril alone [102].

We would end this review of the topic of renoprotection by discussing our novel proposed schema of classification of agents capable of renoprotection, first reported in a recent book publication [71]. This classification scheme is based on the number of putative mechanisms of CKD to ESRD progression that an individual pharmaceutical agent is capable of antagonizing or blocking, to produce the desired effects of renoprotection and therefore enhanced kidney survival [71].

SINGLE-PATHWAY BLOCKERS (SPBs) VERSUS MULTIPLE-PATHWAY BLOCKERS (MPBs)

The Renin Angiotensin Aldosterone System (RAAS) blocking drugs, the ACE inhibitors and the ARBs, only block a single putative pathogenetic pathway involved in CKD progression, the angiotensin cascade [1-35, 117]. We dubbed such renoprotective agents that specifically block or antagonize a single mechanism or pathway, single-pathway blockers (SPBs) [71]. These SPBs however would therefore have the obvious shortfall of an inability to attenuate or antagonize other potentially important putative, independent and unrelated pathogenetic pathways or mechanisms that also drive CKD to ESRD progression [71, 110-136]. They should therefore, not be expected to deliver perfect renoprotecton results in (diabetic and non-diabetic) CKD patients, all the time; from the foregoing review, they simply do not [65-67, 69-71, 103].

On the other hand, a renoprotective agent that is able to block or antagonize the effects of multiple pathogenetic mechanistic pathways or mechanisms, through its ability to simultaneously block, downstream, the effects of several upstream pathways or mechanisms of CKD to ESRD progression, would therefore concurrently interfere with several unrelated upstream pathways or mechanisms [71, 110-136]. Such agent(s) would arguably prove to be more effectual and more reliably successful renoprotective agents [71]. We dubbed this group of (novel) renoprotective agents, multiple-pathway blockers (MPBs) [71]. Recently, we proposed that these yet to be identified or developed agents – the MPBs – may potentially prove to be more effective useful renoprotective agents than the currently available SPB agents exemplified and typified by the ACE inhibitors and the ARBs [71]. Only time will tell if any new MPB renoprotective agent will live up to this hope or hype of preventing the progression of CKD to ESRD, as was proposed by the renowned nephrologist, Thomas Hostetter in 2001 [140].

A LAST WORD ON THE RELATIONSHIP BETWEEN PROTEINURIA, ANTIHYPERTENSIVE THERAPY, ANGIOTENSIN BLOCKADE AND RENAL OUTCOMES IN CKD TRIALS

For years, a whole plethora of nephrology literature has been built on the assessments of the degrees of reduction of proteinuria in CKD patients, both with and without diabetes mellitus, and the relationship of the impact of the various antihypertensive drug classes on this index and renal outcomes [14-20, 22-25, 28, 102, 141-146]. Indeed, the vast majority of the estimates and comparisons of renal outcomes in CKD trials have often used the degrees of achieved reductions in measured proteinuria as a measure of superiority of renoprotection [22-25, 102]. We have repeatedly raised concerns about the fallibility of such premises [65, 67-71].

In type I diabetes mellitus, the relationship between urinary protein excretion, diabetic nephropathy progression and the influence of angiotensin blockade is even more complicated, murkier and very unpredictable [147-149]. According to the recently published extended Diabetes Control and Complications Trial/Epidemiology of Diabetes Interventions and Complications Study Research Group (DCCT/EDIC) study, although RAAS inhibitors reduce urine albumin excretion and can induce regression, most regression to normoalbuminuria observed in this study was spontaneous, as previously observed in other studies [149]. The new DCCT/EDIC results, an extended median 13-year follow up of

1441 study participants also documented regression to normoalbuminuria after more than a decade of persistent microalbuminuria, mostly without RAAS inhibitor use, and generally in the setting of excellent control of glycemia and blood pressure [149]. From clinical anecdotes and this author's experiences in the last ten years at the Mayo Clinic Health System, Eau Claire, Wisconsin, USA, we have observed similar patterns of variability of measured proteinuria by urinary albumin creatinine ratios among type II diabetic CKD patients, on and off the influence of concomitant angiotensin blockade [65, 67-71]. Moreover, while we observed a tendency to increased proteinuria in our CKD patients following discontinuation of ACE inhibitors and/or ARBs [67], El Nahas *et al.* from Sheffield, the United Kingdom, who reported on the results of discontinuation of ACE inhibitors and/or ARBs in 52 patients (21 females and 31 males) with advanced CKD (stages 4 and 5), showed no change in proteinuria subsequent to the discontinuation of angiotensin blockade [91]. Baseline urine protein:creatinine ratio (PCR) was 77 ± 20 mg/mmol, and compared to end PCR values of 121.6 ± 33.6 mg/mmol, was not statistically significant [91].

We posit from the foregoing, that the use of degrees of measured proteinuria as a surrogate of renal outcomes, at best is only marginally useful, and must be discouraged. Again, as we had noted earlier, stand-alone *de novo* ESRD rates should henceforth serve as the only true measure of renal outcomes in drug to drug comparisons in CKD trials [71].

CONCLUSIONS

In a newly published book epilogue, we had raised the hope of a better tomorrow in the global delivery of nephrology care around the world [71]. Our hope in publishing the book was to rekindle new thinking and re-engineering in the ways we do things in medicine, especially in renal medicine or nephrology [71]. All healthcare providers must be ready, all the time to question guidelines and practice paradigms especially when we begin to observe events to the contrary [65-67, 68-71, 103]. No physician hesitates in stopping warfarin, and further to administer antidotes to warfarin such as vitamin K and fresh frozen plasma when a patient presents with life-threatening bleeding, irrespective of whatever the indication for anticoagulation. Similarly, every provider would promptly discontinue insulin or an oral hypoglycemic agent, and urgently and simultaneously administer intravenous glucose or parenteral glucagon when a diabetic presents with life-threatening hypoglycemia, to avert the potential demise of the patient. The primary physician's pledge is to do (the patient) no harm. Holding onto long held paradigms in the face of obvious deficiencies and shortcomings, or worse potential harm or injury to a patient, we posit, is unfair and "unphysicianly". Therefore, we would argue that given the preponderance of albeit circumstantial evidence linking worsening azotemia to concurrent RAAS blockade in certain clinical circumstances, every physician should strongly consider discontinuation of angiotensin blockade when faced with worsening renal failure of otherwise uncertain etiology. Significant renal salvage can often follow such prompt discontinuations [59-71, 97, 98]. Conversely, irreversible ESRD could result from late or no discontinuation of the suspected culprit ACE inhibitor and/or ARB [59-71, 93]. Physicians must not lose sight of the fact that physiologically speaking, angiotensin II in fact has several critically important physiological roles in renal-cardiovascular hemodynamics [8, 150, 151]. Angiotenisn II is involved in the regulation of glomerular filtration rate (GFR) and renal vascular resistance under conditions of volume depletion (autoregulation) and the regulation of the renal glomerular-macula densa apparatus that responds to a "baroreceptor" mechanism that is involved in the intricate balancing of systemic and intra-renal blood pressure homeostasis [8, 150]. Angiotensin II also regulates proximal renal tubular Na^+-HCO_3^- reabsorption *via* augmented luminal Na^+-H^+ antiporter activity and by increased basolateral Na^+-HCO_3^- co-transport as well as stimulation of proximal renal tubular Na^+-K^+ ATPase activity [8, 151]. Besides, there is accruing experimental evidence that angiotensin II, acting through several genetic pathways, plays crucial physiological roles in important tissue repair processes in the kidney, especially following renal injury [152, 153]. Therefore, whereas angiotensin blockade has some merit, there must be cognizance of a need to have some residual function of the RAAS especially in the older CKD patient, both with and without renal artery stenosis [60, 61, 63, 67, 69-71].

Finally, as noted by Renke *et al.* in 2010, despite the use of ACE inhibitors and ARBs in the management of patients with CKD, however, there is no universal therapy that is currently available to physicians that can stop progression of CKD [154]. Chronic kidney disease, previously called chronic kidney failure, and its progression to ESRD requiring renal replacement therapy, is a major health problem worldwide, accounting for gigantic and escalating healthcare costs globally, both in developed countries and in the poorer developing countries [39-58, 155-158]. Although current nephrology renoprotection paradigms are focused mainly on the blockade and antagonism of the renin-angiotensin system, we hypothesize that the search for new therapeutic modalities of renoprotection must be intensified, to bring relief to a straining global healthcare community [71, 155-158]. A positive result from this search is long overdue. This is even now more urgent and pressing, especially with the recent global financial recession and the cutbacks in government expenditures on health and social support services around the world, due to the dire economic recession and other economic stresses facing most of the world economies in 2011 [71, 159]. The time to act is now [71]. This critical mission is urgent, especially if we are to make any significant progress in our current efforts to slow down CKD to ESRD progression and to begin to retard the pace of the growing and costly global ESRD pandemic [71, 153-157].

EPILOGUE

It is our hope to enable and encourage the practitioner of the art of medicine, both nephrologists and non-nephrology physicians alike, to question long held paradigms particularly in the face of observed evidence to the contrary [71]. It is mandatory for physicians to strive to do no harm to the (CKD) patient, and for all of us to continue to endeavor to improve and elevate patient and renal outcomes in CKD management [71]. If we manage to attain these lofty CKD patient management goals addressed in this review, we would have achieved the objectives that we set out to accomplish with the publication of the book in the summer of 2011 [71].

Finally, we must state herein that we fully support the use of ACEIs and ARBs for all the various clinical indications, including renoprotection. However, the practicing physician must remain cognizant of the limitations of these agents and must remain vigilant to detect acutely worsening renal failure in CKD patients concurrently on RAAs blockade and to invariably include RAAS blockade as a potential cause of sometimes irreversible acute renal failure on CKD, if not discontinued promptly. Also, we suggest the pre-emptive temporary withholding of ACEIs and/or ARBs prior to elective major surgical procedures, during serious illness and before parenteral contrast exposure in CKD patients to further limit worsening renal failure in such patients – this is the concept of reno-prevention which we first enunciated in the Quarterly Journal of Medicine in 2009 [67]. The ACEI/ARB would be easily continued thereafter with stabilized or normalized baseline kidney function [67]. Substitution with vasodilators or calcium channel blockers for antihypertensive therapy during the withholding period is recommended and we have found this approach most useful in our Mayo Clinic Health System facility in Northwestern Wisconsin [67, 71].

We indeed remain very hopeful for the future of renoprotection.

ACKNOWLEDGEMENTS

I acknowledge the total support and encouragement of my wife, Nnonyelum Onuigbo MSc, our children, Mark Onuigbo B.S.E. Honors Cum Laude (Aerospace Engineering, University of Michigan, Ann Arbor, MI, May 2012), Victor Onuigbo (Stanford University) and Chimdi Onuigbo (Northwestern University).

This work is dedicated to the memory of a very dear friend, Ikechukwu Ojoko (Idejuogwugwu), who passed away back home in Port Harcourt, Nigeria, some years ago, after a reported brief illness. Idejuogwugwu, you are truly missed.

Finally, we dedicate this work to the memories of the 153 Nigerians who died in a fiery plane crash in Lagos, Nigeria, on June 3, 2012. The plane crash once again had brought to the fore, the several perennial and troubling ills such as corruption, greed and maladministration that have plagued this country for decades! May the souls of the 153 Nigerians rest in perfect peace.

CONFLICT OF INTEREST

There is no conflict of interest to be reported.

REFERENCES

[1]   Ondetti MA, Williams NJ, Sabo EF, Pluscec J, Weaver ER, Kocy O. Angiotensin-converting enzyme inhibitors from the venom of Bothrops jararaca. Isolation, elucidation of structure, and synthesis. Biochemistry 1971; 10(22): 4033-9.

[2]   Ondetti MA, Rubin B, Cushman DW. Design of specific inhibitors of angiotensin-converting enzyme: new class of orally active antihypertensive agents. Science 1977; 196(4288): 441-4.

[3]   Brunner F, Kukovetz WR. Postischemic antiarrhythmic effects of angiotensin-converting enzyme inhibitors. Role of suppression of endogenous endothelin secretion. Circulation 1996; 94(7): 1752-61.

[4]   Ondetti MA, Rubin B, Cushman DW. Design of specific inhibitors of angiotensin-converting enzyme: new class of orally active antihypertensive agents. Science 1977; 196(4288): 441-4.

[5]   Rizzoni D, Muiesan ML, Porteri E, Castellano M, Zulli R, Bettoni G, Salvetti M, Monteduro C, Agabiti-Rosei E. Effects of long-term antihypertensive treatment with lisinopril on resistance arteries in hypertensive patients with left ventricular hypertrophy. J Hypertens 1997; 15(2): 197-204.

[6]   Taddei S, Virdis A, Ghiadoni L, Mattei P, Salvetti A. Effects of angiotensin converting enzyme inhibition on endothelium-dependent vasodilatation in essential hypertensive patients. J Hypertens 1998; 16(4): 447-56.

[7]   Onuigbo M. Angiotensin II receptor antagonists: what future? South Med J. 1998; 91(8): 794-6. (Editorial).

[8]   Ballermann BJ, Onuigbo MAC (2000) Angiotensins. In: Handbook of Physiology, edited by Fray, JCS: Section 7, The Endocrine System, Vol III. Endocrine Regulation of water and Electrolyte Balance. New York, NY: Oxford University Press; pp 104-55.

[9]   Brilla CG, Funck RC, Rupp H. Lisinopril-mediated regression of myocardial fibrosis in patients with hypertensive heart disease. Circulation 2000; 102(12): 1388-93.

[10]   Goto K, Fujii K, Onaka U, Abe I, Fujishima M. Angiotensin-converting enzyme inhibitor prevents age-related endothelial dysfunction. Hypertension 2000; 36(4): 581-7.

[11]   van Ampting JM, Hijmering ML, Beutler JJ, *et al.* Vascular Effects of ACE Inhibition Independent of the Renin-Angiotensin System in Hypertensive Renovascular Disease : A Randomized, Double-Blind, Crossover Trial. Hypertension 2001; 37(1): 40-45.

[12]   Pollare T, Lithell H, Berne C. A comparison of the effects of hydrochlorothiazide and captopril on glucose and lipid metabolism in patients with hypertension. N Engl J Med 1989; 321(13): 868-73.

[13]   Effect of enalapril on mortality and the development of heart failure in asymptomatic patients with reduced left ventricular ejection fractions. The SOLVD Investigators. N Engl J Med 1992; 327(10): 685-691. Erratum in: N Engl J Med 1992; 327(24): 1768.

[14]   Lewis EJ, Hunsicker LG, Bain RP, Rohde RD. The effect of angiotensin-converting-enzyme inhibition on diabetic nephropathy. The Collaborative Study Group. N Engl J Med 1993; 329(20): 1456-1462. Erratum in: N Engl J Med 1993; 330(2): 152.

[15]   Laffel LM, McGill JB, Gans DJ. The beneficial effect of angiotensin-converting enzyme inhibition with captopril on diabetic nephropathy in normotensive IDDM patients with microalbuminuria. North American Microalbuminuria Study Group. Am J Med 1995; 99(5): 497-504.

[16]   Ruggenenti P, Perna A, Gherardi G, *et al.* Renoprotective properties of ACE-inhibition in non-diabetic nephropathies with non-nephrotic proteinuria. Lancet 1999; 354(9176): 359-364. (The REIN trial).

[17]   Packer M, Poole-Wilson PA, Armstrong PW, *et al*. Comparative effects of low and high doses of the angiotensin-converting enzyme inhibitor, lisinopril, on morbidity and mortality in chronic heart failure. ATLAS Study Group. Circulation 1999; 100(23): 2312-2318.

[18]   Mathiesen ER, Hommel E, Hansen HP, Smidt UM, Parving HH. Randomised controlled trial of long term efficacy of captopril on preservation of kidney function in normotensive patients with insulin dependent diabetes and microalbuminuria. BMJ 1999; 319(7201): 24-5.

[19]   Mogensen CE, Neldam S, Tikkanen I, *et al*. Randomised controlled trial of dual blockade of renin-angiotensin system in patients with hypertension, microalbuminuria, and non-insulin dependent diabetes: the candesartan and lisinopril microalbuminuria (CALM) study. BMJ 2000; 321(7274): 1440-1444.

[20]   Effects of ramipril on cardiovascular and microvascular outcomes in people with diabetes mellitus: results of the HOPE study and MICRO-HOPE substudy. Heart Outcomes Prevention Evaluation Study Investigators. Lancet 2000; 355(9200): 253-9. Erratum in: Lancet 2000; 356(9232): 860.

[21]   Pitt B, Poole-Wilson PA, Segal R, *et al*. Effect of losartan compared with captopril on mortality in patients with symptomatic heart failure: randomised trial--the Losartan Heart Failure Survival Study ELITE II. Lancet 2000; 355(9215): 1582-1587.

[22]   Jafar TH, Stark PC, Schmid CH, *et al*; AIPRD Study Group. Angiotensin-Converting Enzymne Inhibition and Progression of Renal Disease. Proteinuria as a modifiable risk factor for the progression of non-diabetic renal disease. Kidney Int 2001; 60(3): 1131-40.

[23]   Parving HH, Lehnert H, Brochner-Mortensen J, Gomis R, Andersen S, Arner P: The effect of irbesartan on the development of diabetic nephropathy in patients with type 2 diabetes. N Engl J Med 2001; 345: 870–878.

[24]   Brenner BM, Cooper ME, de Zeeuw D, *et al*. Effects of losartan on renal and cardiovascular outcomes in patients with type 2 diabetes and nephropathy. N Engl J Med 2001; 345: 861–869.

[25]   Lewis EJ, Hunsicker LG, Clarke WR, *et al*. Renoprotective effect of the angiotensin-receptor antagonist irbesartan in patients with nephropathy due to type 2 diabetes. N Engl J Med 2001; 345: 851–860.

[26]   Jafar TH, Schmid CH, Landa M, *et al*. Angiotensin-converting enzyme inhibitors and progression of nondiabetic renal disease. A meta-analysis of patient-level data. Ann Intern Med 2001; 135(2): 73-87. Erratum in: Ann Intern Med 2002 Aug 20; 137(4): 299.

[27]   Luño J, Barrio V, Goicoechea MA, *et al*. Effects of dual blockade of the renin-angiotensin system in primary proteinuric nephropathies. Kidney Int Suppl 2002; (82): S47-52.

[28]   Rossing K, Jacobsen P, Pietraszek L, Parving HH. Renoprotective effects of adding angiotensin II receptor blocker to maximal recommended doses of ACE inhibitor in diabetic nephropathy: a randomized double-blind crossover trial. Diabetes Care 2003; 26(8): 2268-74.

[29]   Pfeffer MA, McMurray JJ, Velazquez EJ *et al*. Valsartan, captopril, or both in myocardial infarction complicated by heart failure, left ventricular dysfunction, or both. N Engl J Med 2003; 349(20): 1893–1906.

[30]   Onuigbo M, Weir MR. Evidence-based treatment of hypertension in patients with diabetes mellitus. Diabetes Obes Metab 2003; 5(1): 13-26.

[31]   MacKinnon M, Shurraw S, Akbari A, Knoll GA, Jaffey J, Clark HD. Combination therapy with an angiotensin receptor blocker and an ACE inhibitor in proteinuric renal disease: a systematic review of the efficacy and safety data. Am J Kidney Dis 2006; 48(1): 8-20.

[32]   Jennings DL, Kalus JS, Coleman CI, Manierski C, Yee J. Combination therapy with an ACE inhibitor and an angiotensin receptor blocker for diabetic nephropathy: a meta-analysis. Diabet Med 2007; 24(5): 486-93. Epub 2007 Mar 15.

[33]   Bakris GL, Weir MR. Comparison of dual RAAS blockade and higher-dose RAAS inhibition on nephropathy progression. Postgrad Med 2008; 120(1): 33-42.

[34]   Weir MR, Bakris GL. Combination therapy with Renin-Angiotensin-aldosterone receptor blockers for hypertension: how far have we come? J Clin Hypertens (Greenwich) 2008; 10(2): 146-52.

[35]   Ruilope LM. Angiotensin receptor blockers: RAAS blockade and renoprotection. Curr Med Res Opin 2008; 24(5): 1285-93. Epub 2008 Mar 25.

[36]   Carter BL, Malone DC, Ellis SL, Dombrowski RC. Antihypertensive Drug Utilization in Hypertensive Veterans With Complex Medication Profiles. J Clin Hypertens (Greenwich) 2000; 2(3): 172-180.

[37]   Nelson CR, Knapp DA.Trends in antihypertensive drug therapy of ambulatory patients by US office-based physicians. Hypertension 2000; 36(4): 600-603.

[38]   Scarsi KK, Bjornson DC. The use of ACE inhibitors as renoprotective agents in Medicaid patients with diabetes. Ann Pharmacother 2000; 34(9): 1002-1006.

[39]   Ursea N, Mircescu G, Constantinovici N, Verzan C. Nephrology and renal replacement therapy in Romania. Nephrol Dial Transplant 1997; 12(4): 684-90.

[40]   Hsu CY, Vittinghoff E, Lin F, Shlipak MG. The incidence of end-stage renal disease is increasing faster than the prevalence of chronic renal insufficiency. Ann Intern Med 2004; 141(2): 95-101.

[41]   Szczech LA, Lazar IL. Projecting the United States ESRD population: issues regarding treatment of patients with ESRD. Kidney Int Suppl 2004; 90: S3-S7.

[42]   Afifi A, El Setouhy M, El Sharkawy M, *et al.* Diabetic nephropathy as a cause of end-stage renal disease in Egypt: a six-year study. East Mediterr Health J 2004; 10(4-5): 620-626.

[43]   Jha V. End-stage renal care in developing countries: the India experience. Ren Fail 2004; 26(3): 201-8.

[44]   Jones CA, Krolewski AS, Rogus J, Xue JL, Collins A, Warram JH Epidemic of end-stage renal disease in people with diabetes in the United States population: Do we know the cause? Kidney Int 2005; 67(5): 1684-91.

[45]   Perico N, Plata R, Anabaya A, Codreanu I, Schieppati A, Ruggenenti P, Remuzzi G. Strategies for national health care systems in emerging countries: the case of screening and prevention of renal disease progression in Bolivia. Kidney Int Suppl 2005; (97): S87-94.

[46]   Badmus TA, Arogundade FA, Sanusi AA, Akinsola WA, Adesunkanmi AR, Agbakwuru AO, Salako AB, Faponle AF, Oyebamiji EO, Adetiloye VA, Famurewa OC, Oladimeji BY, Fatoye FO. Kidney transplantation in a developing economy: challenges and initial report of three cases at Ile Ife. Cent Afr J Med 2005; 51(9-10): 102-6.

[47]   Liu WJ, Hooi LS. Patients with end stage renal disease: a registry at Sultanah Aminah Hospital, Johor Bahru, Malaysia. Med J Malaysia 2007; 62(3): 197-200.

[48]   US Renal Data System: USRDS 2007 Annual Data Report: Atlas of End-Stage Renal Disease in the United States. Bethesda, National Institutes of Health, National Institute of

Diabetes   and   Digestive   and   Kidney   Disease,   2007.   Available   at http//:www.usrds.org/adr.htm.

[49]   Stewart JH, McCredie MR, Williams SM, Jager KJ, Trpeski L, McDonald SP; ESRD Incidence Study Group. Trends in incidence of treated end-stage renal disease, overall and by primary renal disease, in persons aged 20-64 years in Europe, Canada and the Asia-Pacific region, 1998-2002. Nephrology (Carlton) 2007; 12(5): 520-527.

[50]   Imai E, Horio M, Iseki K, *et al.* Prevalence of chronic kidney disease (CKD) in the Japanese general population predicted by the MDRD equation modified by a Japanese coefficient. Clin Exp Nephrol 2007; 11(2): 156-163. Epub 2007 Jun 28.

[51]   Chiu YL, Chien KL, Lin SL, Chen YM, Tsai TJ, Wu KD. Outcomes of stage 3-5 chronic kidney disease before end-stage renal disease at a single center in Taiwan. Nephron Clin Pract 2008; 109(3): c109-c118. Epub 2008 Jul 25.

[52]   Collins AJ, Foley RN, Herzog C, *et al.* United States Renal Data System 2008 Annual Data Report. Am J Kidney Dis 2009 Jan; 53(1 Suppl): S1-S374.

[53]   Hada R, Khakurel S, Agrawal RK, Kafle RK, Bajracharya SB, Raut KB. Incidence of end stage renal disease on renal replacement therapy in Nepal. Kathmandu Univ Med J (KUMJ) 2009 Jul-Sep; 7(27): 301-5.

[54]   Canadian Institute for Health Information, Treatment of End-Stage Organ Failure in Canada, 1999 to 2008—CORR 2010 Annual Report (Ottawa, Ont.: CIHI, 2010).

[55]   Abraham G, Jayaseelan T, Matthew M, Padma P, Saravanan AK, Lesley N, Reddy YN, Saravanan S, Reddy YN. Resource settings have a major influence on the outcome of maintenance hemodialysis patients in South India. Hemodial Int 2010; 14(2): 211-7.

[56]   Ulasi II, Ijoma CK. The enormity of chronic kidney disease in Nigeria: the situation in a teaching hospital in South-East Nigeria. J Trop Med 2010; 2010: 501957. Epub 2010 Jun 2.

[57]   Farrington K, Hodsman A, Casula A, Ansell D, Feehally J. UK Renal Registry 11th Annual Report (December 2008): Chapter 4 ESRD prevalent rates in 2007 in the UK: national and centre-specific analyses. Nephron Clin Pract 2009; 111 Suppl 1: c43-68. Epub 2009 Mar 26.

[58]   U.S. Renal Data System, USRDS 2010 Annual Data Report: Atlas of Chronic Kidney Disease and End-Stage Renal Disease in the United States, National Institutes of Health, National Institute of Diabetes and Digestive and Kidney Diseases, Bethesda, MD, 2010. http://www.usrds.org/2010/ADR_booklet_2010_lowres.pdf, http://www.usrds.org/2009/pdf/V2_02_INC_PREV_09.PDF accessed December 20, 2010.

[59]   Onuigbo MA, Onuigbo NT. Late onset renal failure from angiotensin blockade (LORFFAB): a prospective thirty-month Mayo Health System clinic experience. Med Sci Monit 2005; 11(10): CR462-9. Epub 2005 Sep 26.

[60]   Onuigbo MA, Onuigbo NT. Late-onset renal failure from angiotensin blockade (LORFFAB) in 100 CKD patients. Int Urol Nephrol 2008; 40(1): 233-9. Epub 2008 Jan 15.

[61]   Onuigbo MA, Onuigbo NT. Does renin-angiotensin aldosterone system blockade exacerbate contrast-induced nephropathy in patients with chronic kidney disease? A prospective 50-month Mayo Clinic study. Ren Fail 2008; 30(1): 67-72.

[62]   Onuigbo MA, Onuigbo NT. Late onset azotemia from RAAS blockade in CKD patients with normal renal arteries and no precipitating risk factors. Ren Fail 2008; 30(1): 73-80.

[63]   Onuigbo MA, Onuigbo NT. Worsening renal failure in older chronic kidney disease patients with renal artery stenosis concurrently on renin angiotensin aldosterone system

blockade: a prospective 50-month Mayo-Health-System clinic analysis. QJM 2008; 101(7): 519-27. Epub 2008 Mar 28.

[64]   Onuigbo MA, Onuigbo NT. Renal failure and concurrent RAAS blockade in older CKD patients with renal artery stenosis: an extended Mayo Clinic prospective 63-month experience. Ren Fail 2008; 30(4): 363-71.

[65]   Onuigbo MAC. An analytical review of the evidence-base for reno-protection from the large RAAS blockade trials after ONTARGET. Re-visitation of the potential for iatrogenic renal failure with RAAS blockade ? A call for caution. Nephron Clin Pract 2009; 113(2): c63-9, Epub 2009 Jul 14.

[66]   Onuigbo MA. The natural history of chronic kidney disease revisited--a 72-month Mayo Health System Hypertension Clinic practice-based research network prospective report on end-stage renal disease and death rates in 100 high-risk chronic kidney disease patients: a call for circumspection. Adv Perit Dial 2009; 25: 85-88.

[67]   Onuigbo MAC: Reno-prevention *vs.* renoprotection: a critical re-appraisal of the evidence-base from the large RAAS blockade trials after ONTARGET – a call for more circumspection. QJM 2009; 102: 155–167.

[68]   Onuigbo MA. Relation between kidney function, proteinuria, and adverse outcomes - A critical look at the application of medical statistics in the Nephrology literature. QJM 2010 Jul; 103(7): 537-8. Epub 2010 Apr 11.

[69]   Onuigbo MA. Is renoprotection with RAAS blockade a failed paradigm? Have we learnt any lessons so far? Int J Clin Pract 2010; 64(10): 1341-6.

[70]   Onuigbo MA. Can ACE inhibitors and angiotensin receptor blockers be detrimental in CKD patients? Nephron Clin Pract 2011; 118: c407-c419.

[71]   Onuigbo MA, Onuigbo N. In "Chronic Kidney Disease and RAAS Blockade: A New View of Renoprotection". Lambert Academic Publishing GmbH & Co. KG. London. 2011. ISBN: 978-3-8454-1523-9.

[72]   Onuigbo MA. Radiographic contrast-induced nephropathy and patient mortality. Mayo Clin Proc 2008; 83(12): 1412-3.

[73]   Louis BM, Hoch BS, Hernandez C, *et al.* Protection from the nephrotoxicity of contrast dye. Renal Fail 1996; 18(4): 639-646.

[74]   Kini AS, Mitre CA, Kamran M, *et al.* Changing trends in incidence and predictors of radiographic contrast nephropathy after percuatneous coronary intervention with use of fenoldopam. Am J Cardiol 2002; 89: 999-1002.

[75]   Toprak O, Cirit M, Bayata S, Yesil M, Aslan SL. The effect of pre-procedural captopril on contrast-induced nephropathy in patients who underwent coronary angiography. Anadolu Kardiyol Derg 2003; 3(2): 98-103.

[76]   Cirit M, Toprak O, Yesil M, *et al.* Angiotensin-Converting Enzyme Inhibitors as a risk factor for contrast-induced nephropathy. Nephron Clin Pract 2006; 104: c20-c27.

[77]   Komenda P, Zalunardo N, Burnett S, *et al.* Conservative outpatient renoprotective protocol in patients with low GFR undergoing contrast angiography: a case series. Clin Exp Nephrol 2007; 11(3): 209-13. Epub 2007 Sep 28.

[78]   From AM, Bartholmai BJ, Williams AW, Cha SS, McDonald FS. Mortality associated with nephropathy after radiographic contrast exposure. Mayo Clin Proc 2008; 83(10): 1095-100.

[79]   Kiski D, Stepper W, Brand E, Breithardt G, Reinecke H. Impact of renin-angiotensin-aldosterone blockade by angiotensin-converting enzyme inhibitors or AT-1 blockers on frequency of contrast medium-induced nephropathy: a post-hoc analysis from the Dialysis-

versus-Diuresis (DVD) trial. Nephrol Dial Transplant 2010; 25(3): 759-64. Epub 2009 Nov 9.

[80]   Thakar CV, Kharat V, Blanck S, Leonard AC. Acute kidney injury after gastric bypass surgery. Clin J Am Soc Nephrol 2007; 2(3): 426-30. Epub 2007 Mar 14.

[81]   Arora P, Rajagopalam S, Ranjan R, *et al.* Preoperative use of angiotensin-converting enzyme inhibitors/angiotensin receptor blockers is associated with increased risk for acute kidney injury after cardiovascular surgery. Clin J Am Soc Nephrol 2008; 3(5): 1266-73. Epub 2008 Jul 30.

[82]   Metz LI, LeBeau ME, Zlabek JA, Mathiason MA. Acute renal failure in patients undergoing cardiothoracic surgery in a community hospital. WMJ 2009; 108(2): 109-14.

[83]   Al-Azzam SI, Al-Husein BA, Abu-Dahoud EY, Dawoud TH, Al-Momany EM. Etiologies of acute renal failure in a sample of hospitalized Jordanian patients. Ren Fail 2008; 30(4): 373-6.

[84]   Ozturk S, Arpaci D, Yazici H, Taymez DG, Aysuna N, Yildiz A, Sever MS. Outcomes of acute renal failure patients requiring intermittent hemodialysis. Ren Fail 2007; 29(8): 991-6.

[85]   Markowitz GS, Nasr SH, Klein P, *et al.* Renal failure due to acute nephrocalcinosis following oral sodium phosphate bowel cleansing. Hum Pathol 2004; 35(6): 675-84.

[86]   Rose M, Karlstadt RG, Walker K. Renal failure following bowel cleansing with a sodium phosphate purgative. Nephrol Dial Transplant 2005; 20(7): 1518-9. Epub 2005 May 26.

[87]   Russmann S, Lamerato L, Marfatia A, *et al.* Risk of impaired renal function after colonoscopy: a cohort study in patients receiving either oral sodium phosphate or polyethylene glycol. Am J Gastroenterol 2007; 102(12): 2655-63. Epub 2007 Oct 26.

[88]   Khurana A, McLean L, Atkinson S, Foulks CJ. The effect of oral sodium phosphate drug products on renal function in adults undergoing bowel endoscopy. Arch Intern Med 2008; 168(6): 593-7.

[89]   Pálmadóttir VK, Gudmundsson H, Hardarson S, Arnadóttir M, Magnússon T, Andrésdóttir MB. Incidence and outcome of acute phosphate nephropathy in Iceland. PLoS One 2010; 5(10): e13484.

[90]   Suissa S, Hutchinson T, Brophy JM, Kezouh A. ACE-inhibitor use and the long-term risk of renal failure in diabetes. Kidney Int 2006; 69: 913-919.

[91]   Ahmed AK, Kamath NS, El Kossi M *et al.* The impact of stopping inhibitors of the renin-angiotensin system in patients with advanced chronic kidney disease. Nephrol Dial Transplant 2009 doi:10.1093/ ndt/gfp511 [Epub ahead of print].

[92]   Onuigbo MA. The impact of stopping inhibitors of the renin-angiotensin system in patients with advanced chronic kidney disease. Nephrol Dial Transplant. 2010; 25(4): 1344-5. Epub 2009 Dec 27.

[93]   Devoy MAB, Tomson CR, Edmunds ME, Feehally J, Walls J. Deterioration in renal function associated with angiotensin converting enzyme inhibitor therapy is not always reversible. J Intern Med 1992; 232(6): 493-498.

[94]   Onuigbo MA. Syndrome of rapid-onset end-stage renal disease: a new unrecognized pattern of CKD progression to ESRD. Ren Fail 2010; 32(8): 954-8.

[95]   Onuigbo MAC, Onuigbo N. Syndrome of rapid onset end-stage renal disease revisited – Observations from two chronic kidney disease populations in two continents. US Nephrology 2010; 5(2): 81-5.

[96]   Onuigbo M, Onuigbo N. Syndrome of rapid onset end-stage renal disease (SORO-ESRD): A new unrecognized pattern of progression of CKD to ESRD. The International Society of Nephrology (ISN). World Congress of Nephrology 2011, Vancouver, Canada. (Abstract).

[97] White KL, Williams TF, Greenberg BG. The ecology of medical care. N Engl J Med 1961;265:885-892.

[98] Green LA, Fryer GE Jr, Yawn BP, Lanier D, Dovey SM. The ecology of medical care revisited. N Engl J Med 2001; 344: 2021-2025.

[99] Pergola PE, Raskin P, Toto RD, Meyer CJ, Huff JW, *et al.* Bardoxolone Methyl and Kidney Function in CKD with type 2 diabets. N Eng J Med 2011; June 24: 1-10. (10.1056/NEJMoa1105351).

[100] Pergola PE, Krauth M, Huff JW, Ferguson DA, Ruiz S, Meyer CJ, Warnock DG. Effect of Bardoxolone Methyl on Kidney Function in Patients with T2D and Stage 3b-4 CKD. Am J Nephrol 2011; 33(5): 469-476. [Epub ahead of print].

[101] Thomas MC. Bardoxolone: augmenting the Yin in chronic kidney disease. Diab Vasc Dis Res 2011; 8(4): 303-4. doi: 10.1177/1479164111421034.

[102] Mann JF, Schmieder RE, McQueen M, *et al*; ONTARGET investigators. Renal outcomes with telmisartan, ramipril, or both, in people at high vascular risk (the ONTARGET study): a multicentre, randomised, double-blind, controlled trial. Lancet 2008; 372(9638): 547-53.

[103] Onuigbo MAC, Onuigbo NTC. Chapter 1: Angiotensin Converting Enzyme Inhibitors. In: DeBrue, AN. Angiotensin Converting Enzyme Inhibitors New York: Nova Biomedical Books. Published by Nova Science Publishers; 2009. p. 1-41.

[104] Tsalamandris C, Allen TJ, Gilbert RE, *et al.* Progressive decline in renal function in diabetic patients with and without albuminuria. Diabetes 1994; 43: 649-655.pmid:8168641.

[105] Onuigbo MA. Causes of renal failure in patients with type 2 diabetes mellitus. JAMA 2003; 290(14): 1855; author reply 1855-6. PMID:14532312. DOI:10.1001/jama.290.14.1855-b. (Letter).

[106] Riley RD, Lambert PC, Staessen JA, *et al.* Meta-analysis of continuous outcomes combining individual patient data and aggregate data. Statist. Med 2008; 27: 1870-1893.

[107] Ameringer S, Serlin RC, Ward S. Simpson's Paradox and experimental research. Nurs Res 2009; 58(2): 123-127.

[108] Cooper H, Patall EA. The relative benefits of meta-analysis conducted with individual participant data versus aggregated data. Psychol Methods 2009; 14(2): 165-176.

[109] Venkatachalam MA, Griffin KA, Lan R, Geng H, Saikumar P, Bidani AK. Acute kidney injury: a springboard for progression in chronic kidney disease. Am J Physiol Renal Physiol 2010 Mar 3. [Epub ahead of print].

[110] Freedman BI, Hicks PJ, Bostrom MA, Comeau ME, Divers J, Bleyer AJ, Kopp JB, Winkler CA, Nelson GW, Langefeld CD, Bowden DW: Non-muscle myosin heavy chain 9 gene MYH9 associations in African Americans with clinically diagnosed type 2 diabetes mellitus-associated ESRD. Nephrol Dial Transplant 2009; 24: 3366–3371.

[111] Leak TS, Mychaleckyj JC, Smith SG, Keene KL, Gordon CJ, Hicks PJ, Freedman BI, Bowden DW, Sale MM: Evaluation of a SNP map of 6q24–27 confirms diabetic nephropathy loci and identifies novel associations in type 2 diabetes patients with nephropathy from an African-American population. Hum Genet 2008; 124: 63–71.

[112] Forbes JM, Coughlan MT, Cooper ME: Oxidative stress as a major culprit in kidney disease in diabetes. Diabetes 2008; 57: 1446–1454.

[113] Singh DK, Winocour P, Farrington K: Mechanisms of disease: The hypoxic tubular hypothesis of diabetic nephropathy. Nat Clin Pract Nephrol 2008; 4: 216–226.

[114] Swaminathan S, Shah SV: Novel approaches targeted toward oxidative stress for the treatment of chronic kidney disease. Curr Opin Nephrol Hypertens 2008; 17: 143–148.

[115] Goh SY, Cooper ME: Clinical review: The role of advanced glycation end products in progression and complications of diabetes. J Clin Endocrinol Metab 2008; 93: 1143–1152.

[116] Flyvbjerg A, Denner L, Schrijvers BF, Tilton RG, Mogensen TH, Paludan SR, Rasch R. Long-term renal effects of a neutralizing RAGE antibody in obese type 2 diabetic mice. Diabetes 2004; 53(1): 166-72.

[117] Park S, Bivona BJ, Kobori H, Seth DM, Chappell MC, Lazartigues E, Harrison-Bernard LM: Major role for ACE-independent intrarenal ANG II formation in type II diabetes. Am J Physiol Renal Physiol 2010; 298: F37–F48.

[118] Matavelli LC, Huang J, Siragy HM. (Pro)renin receptor contributes to diabetic nephropathy by enhancing renal inflammation. Clin Exp Pharmacol Physiol 2010; 37(3): 277-82. Epub 2009 Sep 21.

[119] Sasaki H, Kamijo-Ikemori A, Sugaya T, Yamashita K, Yokoyama T, Koike J, Sato T, Yasuda T, Kimura K: Urinary fatty acids and liver-type fatty acid binding protein in diabetic nephropathy. Nephron Clin Pract 2009; 112: c148–c156.

[120] Lennon R, Pons D, Sabin MA, Wei C, Shield JP, Coward RJ, Tavare JM, Mathieson PW, Saleem MA, Welsh GI: Saturated fatty acids induce insulin resistance in human podocytes: Implications for diabetic nephropathy. Nephrol Dial Transplant 2009; 24: 3288–3296.

[121] An WS, Kim HJ, Cho KH, Vaziri ND. Omega-3 fatty acid supplementation attenuates oxidative stress, inflammation, and tubulointerstitial fibrosis in the remnant kidney. Am J Physiol Renal Physiol 2009; 297(4): F895-903. Epub 2009 Aug 5.

[122] Yilmaz MI, Saglam M, Qureshi AR, Carrero JJ, Caglar K, Eyileten T, Sonmez A, Cakir E, Oguz Y, Vural A, Yenicesu M, Stenvinkel 172 Nephrology Self-Assessment Program - Vol 9, No 3, May 2010 P, Lindholm B, Axelsson J: Endothelial dysfunction in type-2 diabetics with early diabetic nephropathy is associated with low circulating adiponectin. Nephrol Dial Transplant 2008; 23: 1621–1627.

[123] Yamaguchi Y, Iwano M, Suzuki D, Nakatani K, Kimura K, Harada K, Kubo A, Akai Y, Toyoda M, Kanauchi M, Neilson EG, Saito Y: Epithelial-mesenchymal transition as a potential explanation for podocyte depletion in diabetic nephropathy. Am J Kidney Dis 2009; 54: 653–664.

[124] Yokoi H, Mukoyama M, Mori K, Kasahara M, Suganami T, Sawai K, Yoshioka T, Saito Y, Ogawa Y, Kuwabara T, Sugawara A, Nakao K: Overexpression of connective tissue growth factor in podocytes worsens diabetic nephropathy in mice. Kidney Int 2008; 73: 446–455.

[125] Jaffa AA, Usinger WR, McHenry MB, Jaffa MA, Lipstiz SR, Lackland D, Lopes-Virella M, Luttrell LM, Wilson PW: Connective tissue growth factor and susceptibility to renal and vascular disease risk in type 1 diabetes. J Clin Endocrinol Metab 2008; 93: 1893–1900.

[126] Fujimi-Hayashida A, Ueda S, Yamagishi S, Kaida Y, Ando R, Nakayama Y, Fukami K, Okuda S. Association of asymmetric dimethylarginine with severity of kidney injury and decline in kidney function in IgA nephropathy. Am J Nephrol 2011; 33(1): 1-6. Epub 2010 Nov 26.

[127] Chonchol M, Shlipak MG, Katz R, Sarnak MJ, Newman AB, Siscovick DS, Kestenbaum B, Carney JK, Fried LF. Relationship of uric acid with progression of kidney disease. Am J Kidney Dis 2007 Aug;50(2):239-47.

[128] Yen CJ, Chiang CK, Ho LC, Hsu SH, Hung KY, Wu KD, Tsai TJ. Hyperuricemia associated with rapid renal function decline in elderly Taiwanese subjects. J Formos Med Assoc 2009; 108(12): 921-8.

[129] Feig DI. Uric acid: a novel mediator and marker of risk in chronic kidney disease? Curr Opin Nephrol Hypertens 2009; 18(6): 526-30.

[130] Kuo CF, Luo SF, See LC, Ko YS, Chen YM, Hwang JS, Chou IJ, Chang HC, Chen HW, Yu KH. Hyperuricaemia and accelerated reduction in renal function. Scand J Rheumatol 2011; 40(2): 116-21. Epub 2010 Sep 26.

[131] Bellomo G, Venanzi S, Verdura C, Saronio P, Esposito A, Timio M. Association of uric acid with change in kidney function in healthy normotensive individuals. Am J Kidney Dis 2010; 56(2): 264-72. Epub 2010 Apr 10.

[132] Badve SV, Brown F, Hawley CM, Johnson DW, Kanellis J, Rangan GK, Perkovic V. Challenges of conducting a trial of uric-acid-lowering therapy in CKD. Nat Rev Nephrol 2011; 7(5): 295-300. Epub 2011 Feb 15.

[133] Tzur S, Rosset S, Shemer R, Yudkovsky G, Selig S, Tarekegn A, Bekele E, Bradman N, Wasser WG, Behar DM, Skorecki K. Missense mutations in the APOL1 gene are highly associated with end stage kidney disease risk previously attributed to the MYH9 gene. Hum Genet 2010; 128(3): 345-50. Epub 2010 Jul 16.

[134] Behar DM, Kedem E, Rosset S, Haileselassie Y, Tzur S, Kra-Oz Z, Wasser WG, Shenhar Y, Shahar E, Hassoun G, Maor C, Wolday D, Pollack S, Skorecki K. Absence of APOL1 Risk Variants Protects against HIV-Associated Nephropathy in the Ethiopian Population. Am J Nephrol 2011; 34(5): 452-9. Epub 2011 Oct 3.

[135] Kopp JB, Nelson GW, Sampath K, Johnson RC, Genovese G, An P, Friedman D, Briggs W, Dart R, Korbet S, Mokrzycki MH, Kimmel PL, Limou S, Ahuja TS, Berns JS, Fryc J, Simon EE, Smith MC, Trachtman H, Michel DM, Schelling JR, Vlahov D, Pollak M, Winkler CA. APOL1 Genetic Variants in Focal Segmental Glomerulosclerosis and HIV-Associated Nephropathy. J Am Soc Nephrol 2011; 22(11): 2129-37. Epub 2011 Oct 13.

[136] Papeta N, Kiryluk K, Patel A, Sterken R, Kacak N, Snyder HJ, Imus PH, Mhatre AN, Lawani AK, Julian BA, Wyatt RJ, Novak J, Wyatt CM, Ross MJ, Winston JA, Klotman ME, Cohen DJ, Appel GB, D'Agati VD, Klotman PE, Gharavi AG. APOL1 Variants Increase Risk for FSGS and HIVAN but Not IgA Nephropathy. J Am Soc Nephrol 2011; 22(11): 1991-6. Epub 2011 Oct 13.

[137] Comper WD, Hilliard LM, Nikolic-Paterson DJ, Russo LM. Disease-dependent mechanisms of albuminuria. Am J Physiol Renal Physiol 2008; 295(6): F1589-600. Epub 2008 Jun 25.

[138] Russo LM, Sandoval RM, Campos SB, Molitoris BA, Comper WD, Brown D: Impaired tubular uptake explains albuminuria in early diabetic nephropathy. J Am Soc Nephrol 2009; 20: 489–494.

[139] Bansal N, Hsu CY. Long-term outcomes of patients with chronic kidney disease. Nat Clin Pract Nephrol 2008; 4: 532–3.

[140] Hostetter TH. Prevention of end-stage renal disease due to type 2 diabetes. N Engl J Med 2001; 345(12): 910-2.

[141] De Jong PE, Navis G, de Zeeuw D. Renoprotective therapy: titration against urinary protein excretion. Lancet 1999; 354(9176): 352-3.

[142] de Zeeuw D, Remuzzi G, Parving HH, Keane WF, Zhang Z, Shahinfar S, Snapinn S, Cooper ME, Mitch WE, Brenner BM. Proteinuria, a target for renoprotection in patients with type 2 diabetic nephropathy: lessons from RENAAL. Kidney Int 2004; 65(6): 2309-20.

[143] Lorenzo V, Saracho R, Zamora J, Rufino M, Torres A. Similar renal decline in diabetic and non-diabetic patients with comparable levels of albuminuria. Nephrol Dial Transplant 2010; 25: 835–841

[144]   Astor BC, Matsushita K, Gansevoort RT, van der Velde M, Woodward M, Levey AS, *et al.* Lower estimated glomerular filtration rate and higher albuminuria are associated with mortality and end-stage renal disease. A collaborative meta-analysis of kidney disease population cohorts. Kidney Int 2011; 79(12): 1331-40. Epub 2011 Feb 2.

[145]   de Zeeuw D, Parekh R, Soman S. CKD treatment: time to alter the focus to albuminuria? Adv Chronic Kidney Dis 2011; 18(4): 222-3.

[146]   Packham DK, Alves TP, Dwyer JP, Atkins R, de Zeeuw D, Cooper M, Shahinfar S, Lewis JB, Lambers Heerspink HJ. Relative Incidence of ESRD Versus Cardiovascular Mortality in Proteinuric Type 2 Diabetes and Nephropathy: Results From the DIAMETRIC (Diabetes Mellitus Treatment for Renal Insufficiency Consortium) Database. Am J Kidney Dis 2012; 59(1): 75-83. Epub 2011 Nov 3.

[147]   Perkins BA, Ficociello LH, Silva KH, Finkelstein DM, Warram JH, Krolewski AS. Regression of microalbuminuria in type 1 diabetes. N Engl J Med 2003; 348 (23): 2285-2293.

[148]   Steinke JM, Sinaiko AR, Kramer MS, Suissa S, Chavers BM, Mauer M; International Diabetic Nephopathy Study Group. The early natural history of nephropathy in type 1 diabetes, III: predictors of 5-year urinary albumin excretion rate patterns in initially normoalbuminuric patients. Diabetes 2005; 54(7): 2164-2171.

[149]   de Boer IH, Rue TC, Cleary PA; *et al.* Diabetes Control and Complications Trial/Epidemiology of Diabetes Interventions and Complications Study Research Group. Long-term renal outcomes of patients with type 1 diabetes mellitus and microalbuminuria: an analysis of the Diabetes Control and Complications Trial/Epidemiology of Diabetes Interventions and Complications cohort. Arch Intern Med 2011; 171(5): 412-420.

[150]   Rose BD. Clinical Physiology of Acid Base And Electrolyte disorders, 5th Ed., NY, McGrawHill, 2001; Primer of Kidney Disease.

[151]   Geibel J, Giebisch G, Boron WF. Angiotensin II stimulates both Na(+)-H+ exchange and Na+/HCO3- cotransport in the rabbit proximal tubule. Proc Natl Acad Sci USA 1990; 87(20): 7917-20.

[152]   Zhang SL, Guo J, Moini B, Ingelfinger JR. Angiotensin II stimulates Pax-2 in rat kidney proximal tubular cells: impact on proliferation and apoptosis. Kidney Int 2004; 66(6): 2181-92.

[153]   Chen CO, Park MH, Forbes MS, *et al.* Angiotensin Converting Enzyme inhibition aggravates renal interstitial injury resulting from partial unilateral ureteral obstruction in the neonatal rat. Am J Physiol Renal Physiol 2007; 292(3): F946-55. Epub 2006 Nov 14.

[154]   Renke M, Tylicki L, Rutkowski P, Knap N, Zietkiewicz M, Neuwelt A, Aleksandrowicz E, Łysiak-Szydłowska W, Woźniak M, Rutkowski B Effect of pentoxifylline on proteinuria, markers of tubular injury and oxidative stress in non-diabetic patients with chronic kidney disease - placebo controlled, randomized, cross-over study. Acta Biochim Pol 2010; 57(1): 119-23. Epub 2010 Mar 22.

[155]   Chugh KS, Jha V. Differences in the care of ESRD patients worldwide: required resources and future outlook. Kidney Int Suppl 1995; 50: S7-13.

[156]   Lysaght MJ. Maintenance dialysis population dynamics: current trends and long-term implications. J Am Soc Nephrol 2002; 13 Suppl 1: S37-40.

[157]   Jha V. Current status of end-stage renal disease care in South Asia. Ethn Dis 2009; 19(1 Suppl 1): S1-27-32.

[158]   Shah SV. Progress toward novel treatments for chronic kidney disease. J Ren Nutr 2010; 20(5 Suppl): S122-6.

[159] Onuigbo M. Healthcare expenditure in the United States of America in the last year of life: where ethics, medicine and economics collide? Int J Clin Pract 2012; 66(2): 226-7. doi: 10.1111/j.1742-1241.2011.02846.

Send Orders for Reprints to reprints@benthamscience.net

Frontiers in Clinical Drug Research: Diabetes and Obesity, Vol. 1, 2014, 33-66

CHAPTER 2

New Approach to Preventing and Treating Diabetes Mellitus: Focus on Advanced Glycation End Products (AGE)

Emilia Liao, Agustin Busta, Ashutosh Pareek and Leonid Poretsky[*]

Division of Endocrinology, Department of Medicine, Beth Israel Medical Center and Albert Einstein College of Medicine, New York, NY 10003, USA

Abstract: In this chapter, we discuss the role of advanced glycation end products (AGE) in the pathogenesis of diabetes and its complications, as well as potential AGE-focused treatment strategies. AGE are compounds that form as a result of chemical reactions from the glycation of proteins and lipids. Glycation of proteins alters their structure and function (extracellular effects), and AGE also bind to receptors which promote inflammatory pathways (intracellular effects). AGE are formed endogenously through normal metabolism and aging, and AGE formation is accelerated in certain pathologic states such as diabetes (due to hyperglycemia) and oxidative stress (via reactive oxygen species). AGE accumulation and deposition are reported in chronic inflammatory diseases such as atherosclerosis and diabetes. AGE deposits form cross-links with matrix proteins resulting in increased stiffness of collagen, decreased elasticity of vasculature, increased muscle stiffness and reduction in function. AGE can also be formed through prolonged heating of food (such as frying and grilling). Tobacco smoke is another source of exogenous AGE. High AGE diet is associated with increasing weight gain, adiposity, and insulin resistance. Treatment approaches include inhibiting AGE formation, breaking formed cross-links, or blocking negative effects of AGE. Treatment of underlying pathologic states that accelerate AGE formation (hyperglycemia, oxidative stress and hypoxia) is also important. Reducing exogenous intake of AGE is a lifestyle modification in the treatment and prevention of diabetes and obesity that will lower the risk of developing complications, and probably also lower the risk of developing diabetes itself. The pathogenesis of type 2 diabetes is not yet fully elucidated. AGE are implicated in the development of diabetes and its complications, and anti-AGE therapies represent a novel category of therapeutic interventions against diabetes.

Keywords: Diabetes, advanced glycation end products (AGE), retinopathy, neuropathy, nephropathy, diabetes complications, receptor for AGE (RAGE), methylglyoxal.

---

[*]**Address correspondence to Leonid Poretsky:** Division of Endocrinology, Beth Israel Medical Center, 317 East 17th Street, New York, NY 10003, USA; Tel: 212-420-4666; Fax: 212-420-2224; E-mail: lporetsk@chpnet.org

INTRODUCTION

According to the Centers for Disease Control (CDC), 25.8 million people (or 8.3% of the population) living in the United States suffers from diabetes, with 18.8 million diagnosed and 7 million yet to be diagnosed [1]. Furthermore, 79 million Americans are estimated to be prediabetic. According to the International Diabetes Federation (IDF), diabetes affects approximately 366 million people worldwide. This number is projected to rise to 522 million by 2030 [2]. Coupled with the rise of prevalence of diabetes mellitus is the cost associated with the disease. In the United States, the estimated total cost of diabetes was 174 billion dollars in 2007 [1]. Globally, in 2011, diabetes was responsible for at least 464 billion dollars in healthcare expenditures [2].

Diabetes mellitus, according to the American Diabetes Association (ADA), can be diagnosed in several ways: a fasting plasma glucose level greater than or equal to 126 mg/dl [3], a two hour plasma glucose of greater than 200 mg/dl following a 75 gram oral glucose load, or a hemoglobin A1c (HbA1c) value of 6.5% or greater. Symptoms of hyperglycemia such as excessive thirst, polyuria, and weight loss, in combination with a random plasma glucose of >200 mg/dl also diagnose diabetes (Table **1**). The diagnostic criteria for diabetes are established because the risk of retinopathy is increased at these levels of glycemia.

Table 1: American diabetes association diagnostic criteria for diabetes mellitus

Criteria	*Threshold*
Fasting plasma glucose	$\geq$ 126 mg/dl
2 hour plasma glucose during OGTT	$\geq$ 200 mg/dl
Symptoms (*e.g.,* excessive thirst, polyuria, blurry vision) plus a random plasma glucose of $\geq$ 200mg/dl	$\geq$ 200 mg/dl
Hemoglobin A1C	$\geq$ 6.5%

Impaired fasting glucose (IFG) and impaired glucose tolerance (IGT) are hyperglycemic states that place a patient at increased risk of developing diabetes, and both fall under the umbrella term of prediabetes. IFG is defined as glucose between 100 mg/dl to 125 mg/dl after an overnight fast. IGT is defined as a glucose falling in the range of 140 mg/dl to 199 mg/dl, two hours after the administration of a 75 gram oral glucose load. Prediabetes can also be diagnosed when glycated hemoglobin is between 5.7 to 6.4 %.

In addition to causing morbidity, microvascular complications of diabetes mellitus—retinopathy, nephropathy, and neuropathy—contribute significantly to the rising cost of treating diabetes. The Diabetes Control and Complications Trial (DCCT), conducted from 1983 to 1993, demonstrated unequivocally that strict glucose control (HbA1c of <7%) delayed the onset and slowed the progression of microvascular complications of type 1 diabetes [4]. The United Kingdom Prospective Diabetes Study (UKPDS) later confirmed this conclusion to be valid for type 2 diabetes as well [5].

THE POTENTIAL ROLE OF AGE IN CAUSING DIABETES

Traditionally, type 2 diabetes mellitus is best thought of as a continuum that begins with the development of insulin resistance and sustained normoglycemia, but progresses to overt hyperglycemia and impairment of both insulin sensitivity and insulin secretion. Data reveal that at the time of diagnosis, most patients with type 2 diabetes have lost up to 80-90% of their pancreatic beta cell function [6]. Observations that suggest a genetic influence on the development of type 2 diabetes include the following: high prevalence of diabetes in certain ethnic groups (*e.g.,* Pima Indians); one third of patients with type 2 diabetes with a first degree relative that shares the same diagnosis [7]; and the lifetime risk for developing diabetes in a first-degree relative of a patient with type 2 diabetes being 5 to 10 times higher than that of age- and weight-matched subjects without a family history of diabetes [7].

Though the contribution of genetics to diabetes is important, it is unlikely that genetics of mankind has changed significantly in recent decades. It is generally accepted that overnutrition plays a pivotal role in the development of diabetes and the metabolic syndrome. Humans convert an excess of consumed energy into adipose tissue which acts as an active endocrine organ [8] and contributes to the development of insulin resistance. The mechanisms by which excessive adipose tissue accumulation, particularly in the intra-abdominal space, leads to insulin resistance are not completely understood.

As an endocrine organ, adipose tissue produces several compounds that participate in the regulation of appetite and energy homeostasis, lipid metabolism, insulin sensitivity, immunity, and angiogenesis [9]. Key components in this process include pro-inflammatory cytokines such as tumor necrosis factor alpha (TNF-α) and interleukin-6 (IL-6). TNF-α and IL-6 promote inflammation and induce insulin resistance *via* suppression of insulin receptor signal transduction in hepatocytes [9]. Conversely, adiponectin, an insulin-sensitizing cytokine secreted by adipose tissue, suppresses inflammation, promotes beta cell function and survival, increases insulin sensitivity, and decreases hepatic glucose output [9].

While the traditional theory of the pathogenesis of diabetes focuses on overnutrition as the main environmental factor, a new hypothesis focusing on the role of advanced glycation end products (AGE) is under investigation (Fig. **1**).

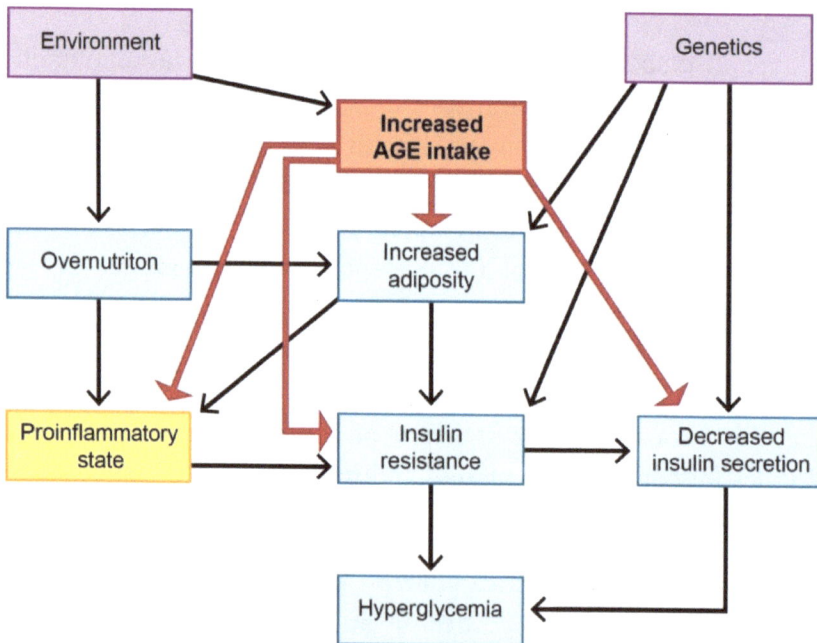

Figure 1: Traditional view (purple) of the pathogenesis of diabetes mellitus describes a combination of genetic and environmental factors leading to obesity, insulin resistance, reduced insulin secretion, and hyperglycemia. A more recent approach includes a proinflammatory state contributing to insulin resistance (yellow). The current hypothesis also includes the effects of AGEs (red), which induce a proinflammatory state, increase adiposity, induce insulin resistance, and inhibit insulin secretion (details and references are provided in main text). Adapted from Poretsky L. Looking beyond overnutrition for causes of epidemic metabolic disease. PNAS, 109:39, 15537-15538, 2012.

Although AGEs are formed under normal physiologic conditions *via* spontaneous, non-enzymatic processes, their generation is enhanced in hyperglycemic and insulin resistant states.

On a cellular level, advanced glycation results in irreversible cross-linking of proteins, loss of protein structure and function, and eventual apoptosis. The endoplasmic reticulum (ER), the primary regulatory organelle of protein synthesis and transport, counteracts the effects of AGE. ER function is increased in cells with elevated AGE [10]. AGE effects on the cell are mediated by AGE binding to the receptor RAGE (receptor for advanced glycation endproducts) that is present on cell surface. A cascade of cellular reactions is initiated and results in increased levels of several proinflammatory molecules, for example, nitric oxide and TNF-α [10, 11].

Individuals who consume an AGE-rich diet have higher C-reactive protein, fibrinogen, and other pro-inflammatory markers [12]. In addition to the intake of thermally processed foods, intake of a high-fat diet can induce both AGE formation and oxidative stress [11]. In diabetic animal models, AGE restriction prevents renal and cardiovascular complications [13]. In mice models, researchers have demonstrated that AGE are responsible for inducing an inflammatory state, which leads to insulin resistance and hyperglycemia [14].

In a recent experiment, Cai *et al.* fed a diet enriched in AGE to mice that were maintained on isocaloric diets similar to control animals. After three generations, the mice fed an AGE-enriched diet developed a higher level of adiposity and insulin resistance. Abnormalities in the insulin receptor signaling cascade (tyrosine-phosphorylated insulin receptor, IRS-1, IRS-2, and Akt) were observed, the amount of skeletal muscle was reduced, and macrophages and adipocytes shifted to a proinflammatory phenotype. Mice fed an AGE-rich diet demonstrated reduced insulin-stimulated 2-deoxyglucose uptake by adipose tissue and skeletal muscle [15].

Recent data and ongoing clinical studies are attempting to determine whether these or similar processes occur in the diabetic population. Diabetic patients are at increased risk for chronic kidney disease (CKD). Patients with diabetes and high levels of AGE have been shown to have a seven-fold increase risk of developing CKD [16]. Since a strict dietary regimen aimed at reducing AGE is difficult for most people to follow, sequestering AGE in the gut, before they are absorbed, is a new avenue of research. A pilot randomized clinical trial in twenty diabetic patients has tested the hypothesis whether sevelamer carbonate, a phosphate binder that acts in the human gut, can bind AGE and help improve outcomes in diabetes [13]. In this study, the use of sevelamer in diabetic patients with CKD stage 2-4 reduced glycated hemoglobin and improved lipid abnormalities. These findings need to be replicated on a larger scale.

CHEMISTRY AND SOURCES OF AGE

AGE are formed as a result of nonenzymatic reactions between intracellular glucose-derived dicarbonyl precursors (glyoxal, methylglyoxal, 3-deoxyglucosone) and the amino groups of both intracellular and extracellular proteins. AGE formation is accelerated in the presence of increased levels of glucose.

AGE are formed through three known pathways: (1) The Maillard reaction, which begins with the aldehyde group of sugars reacting with amino groups of proteins, lipids or nucleic acids; (2) oxidative stress, which involves auto-oxidation of glucose and the peroxidation of lipids [17]; and (3) the polyol pathway, with the fructose metabolite (fructose-3-phosphate) converting to alpha-oxaldehydes and reacting with monoacids to form AGE [18].

The Maillard Reaction is a non-enzymatic reaction between glucose and protein. Initially, it forms reversible intermediate products within hours. However, when AGE have been formed for several weeks, the process becomes irreversible and the newly formed AGE can cross-link proteins and will remain in the tissues for prolonged periods (Fig. **2**). The initial (reversible) reaction results in formation of Schiff base and Amadori product (for example, HbA1c), which subsequently undergo numerous reactions of condensation, rearrangement, fragmentation and oxidative modification leading to the formation of irreversible AGE. AGE are very stable, and accumulates inside and outside cells, interfering with protein function [19, 20].

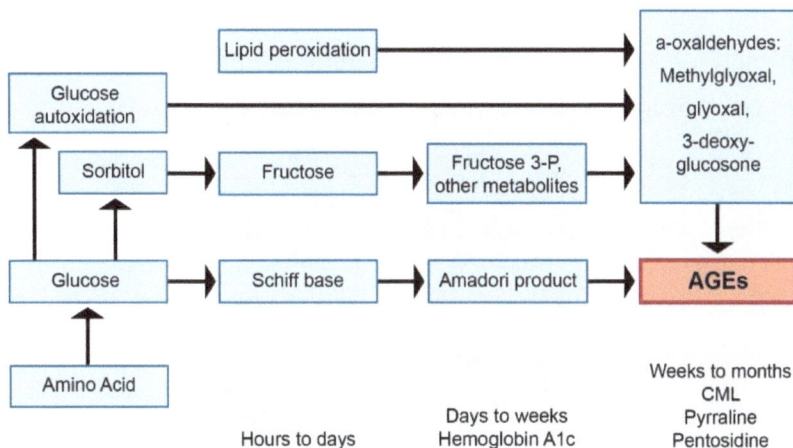

Figure 2: Formation of AGE, Maillard reaction. Adapted from Luevano-Contreras C, Chapman-Novakofski K. Dietary advanced glycation end products and aging. Nutrients, 2:1247-1265, 2010.

The best known AGE (Table **2** and Fig. **3**) include N-carboxymethyl-lysine (CML), N-carboxyethy-llysine (CEL), the cross-linked pentosidine, methylglyoxal lysine dimer (MOLD); and glyoxal lysine dimer (GOLD) [21].

Table 2: Some widely distributed AGEs

Glyoxal-derived lysine dimer (GOLD)
Methylglyoxal-derived lysine dimer (MOLD)
3-deoxyglucosone-derived lysine dimer (DOLD)
Pentosidine
N-carboxymethyl-lysine (CML)
N-carboxyethyl-lysine (CEL)
N-5-hydro-5-methyl-4-imidazolon-2-yl (MG-H1)
5-(2-amino-hydro-5-methyl-4-dydroe-4-imidazolon-1-tl (MG-H2)
5-(2-amino-4-hydro-4methyl-5-imidazolon-1-yl (MG-H3)
Pyrraline

As mentioned above, AGE bind to a specific receptor (RAGE), which is expressed on many types of cells such as macrophages, T-cells, endothelium, and vascular smooth muscle. Through the receptor-mediated AGE/RAGE pathway, AGE can affect the vascular compartment by releasing pro-inflammatory cytokines [22] and growth factors from macrophages; by generating reactive oxygen species in endothelial cells; by increasing procoagulant activity in endothelial cells and macrophages; and by enhancing the proliferation of vascular smooth muscle cells and the synthesis of extracellular matrix.

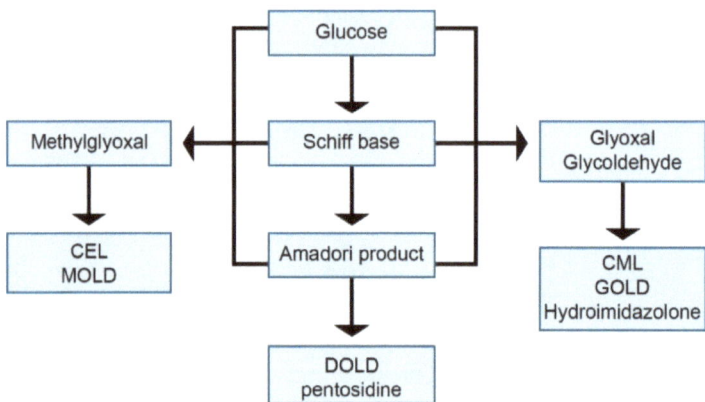

Figure 3: Maillard reaction and the formation of AGEs. CEL: N-carbosyethyl-lysine; CML: N-carboxymethyl-lysine; MOLD: methylglyoxal lysine dimer; DOLD: 3-deoxyglucosone-derived lysine dimer; GOLD: glyoxal lysine dimer.

AGE can also directly cross-link extracellular matrix proteins. Cross-linking of collagen type 1 molecule in large vessels decreases their elasticity [23, 24], leading to shear stress and endothelial injury. Similarly, AGE-induced cross-linking of type IV collagen in basement membrane decreases endothelial cell adhesion and increases extravasation of fluid. Proteins cross-linked by AGE are resistant to proteolytic enzyme action.

Normally, oxidants are generated during the course of cellular metabolism and include superoxide, hydrogen peroxide, peroxyl radicals, and hydroxyl radicals. These are collectively referred to as "reactive oxygen species" (ROS). ROS react with proteins, nucleic acids, lipids and other molecules altering their structure and damaging tissues. Potential toxic oxygen species can be referred to as pro-oxidants. Anti-oxidants, on the other hand, suppress their formation or oppose their action. In a normal cell, there is an appropriate pro-oxidant: antioxidant balance (ratio) which can be shifted to pro-oxidant when the production of oxygen species is greatly increased or when there is a decrease of antioxidants. This condition is called "oxidative stress" [25]. Oxidative stress leading to increase ROS accelerates AGE formation [11].

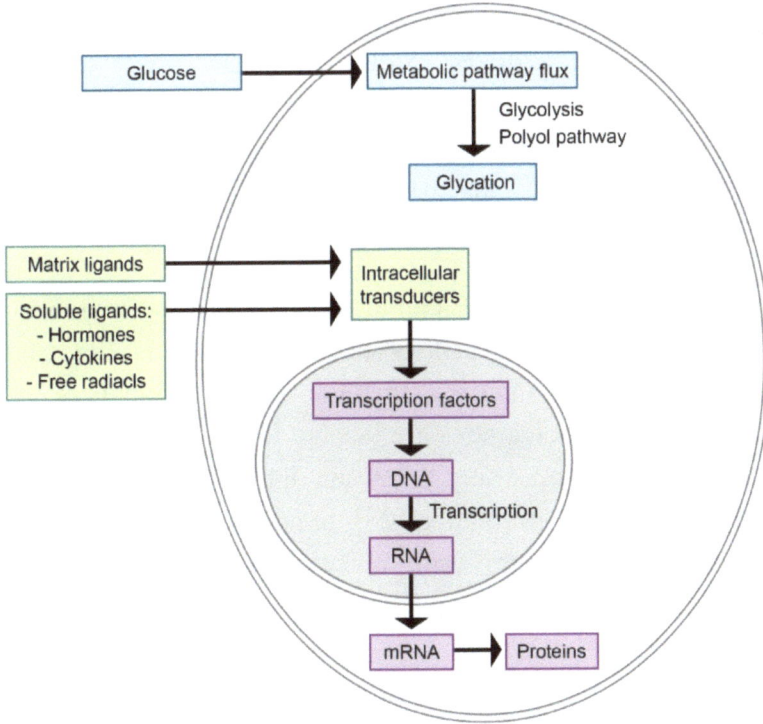

Figure 4: Schematic cell showing three general mechanisms by which AGEs may cause pathological changes in diabetes. Adapted, with permission, from Brownlee M, Glycation and diabetic complications. Diabetes, 43, 1994.

Intracellular production of AGE precursors damages target cells by three general mechanisms (Fig. **4**): (1) intracellular proteins modified by AGE have altered function; (2) extracellular matrix components modified by AGE precursors interact abnormally with other matrix components and with matrix receptors (integrins) on cells; and (3) plasma proteins modified by AGE precursors bind to AGE receptors on cells such as macrophages, inducing receptor-mediated ROS production. This AGE-receptor interaction activates the pleiotropic transcription factor nuclear factor-κB (NFκB), causing pathologic changes in gene expression [20], ultimately causing inflammation and endothelial dysfunction. Extracellular effects of AGE include alteration of structural proteins and deposition in arterial walls, resulting in reduced elasticity of vasculature [12].

AGE come from endogenous and exogenous sources. In the body, hyperglycemia promotes metabolic activity and is the best known pathway by which levels are increased in cells. Hyperglycemia leads to further oxidative stress which then perpetuates oxidation of sugars/lipids, leading to further production of AGE. AGE-modified proteins and lipids, as well as nucleic acids, can be removed from the body by proteolytic digestion. However, these compounds tend to be resistant to degradation due to the cross-links that are formed between AGE and their target tissues.

In addition to *in vivo* production, AGE can also be found in cigarette smoke and food. The curing of tobacco leaves has been proposed as the source for compounds that can readily increase *in vivo* AGE. Cermai *et al.* have demonstrated that toxins from cigarettes are inhaled into the alveoli, and then are transported to the bloodstream where they can interact with other glycation products and contribute to AGE formation [26]. In addition, many studies suggest that the modern diet is also a significant source of AGE. With the increase in popularity of processed foods, soft drinks, and red meats, the source of dietary oxidants has grown exponentially.

For centuries heat has been used for treatment of food to improve its safety and taste. Applying heat to foods has been found to increase the amount of AGE. Heat treatment in some foods results in promotion of the Maillard reaction, which adds desirable flavor, color, and aroma [27]. The Maillard reaction is also applied to industrialized products such as sodas and juices. Taking this into consideration, there is growing evidence that the average Western diet is a plentiful source of exogenous AGE. The amount of AGE in food can be reduced by changing the way the food is prepared. For instance, a fried potato contains almost one hundred times more AGE than a potato that is boiled.

The amount of AGE in our diets has been measured by researchers. N-carboxymethyl-lysine (CML), the most readily measurable AGE in the human body, is used as a marker of AGE levels in foods. In one study, Uribarri *et al.* estimated the amount of AGEs from food records and found a significant correlation between ingested AGE and plasma CML levels. In contrast, a subgroups of patients fed an AGE-deplete diet were found to have decreased levels of CML [28].

In summary, AGE come from two main sources: endogenous and the diet. AGE react with tissues and blood vessels to create crosslinks that predispose tissue to vascular injury and atherosclerosis. AGE in diet are due mainly to the thermal processing of foods. Thus, approaches that allows reduction of AGE content of food warrant further investigation.

DIABETES COMPLICATIONS AND THE ROLE OF AGE

Diabetes is a multiorgan disease with both microvascular and macrovascular complications. Microvascular complications include retinopathy, nephropathy and neuropathy, while macrovascular complications include coronary artery disease (CAD) and peripheral arterial disease (PAD). Individuals with diabetes are at greater risk for developing CAD, myocardial infarction and stroke. In fact, diabetes is considered a CAD equivalent. As mentioned above, both the DCCT and UKPDS trials in patients with type 1 and type 2 diabetes showed 35-75% higher incidence of various complications with higher hemoglobin A1c [4, 29].

The principal factor contributing to the development of diabetes complications is hyperglycemia. Other contributing factors include hypertension and hyperlipidemia. Chronic hyperglycemia leads to formation of AGE, and AGE accumulation has been shown in sites of diabetes complications, including kidney, retina, peripheral nerves and arterial walls.

Retinopathy

Diabetes is the leading cause of blindness in the United States [1]. Diabetic retinopathy is characterized by thickening of the capillary basement membrane, microaneurysm formation, loss of pericytes, capillary acellularity and neovascularization. Microaneurysms and hard exudates (lipid transudates) are present in mild, nonproliferative retinopathy. In more severe disease, cotton wool spots and intraretinal microvascular abnormalities (vessel tortuosity and neovascularization) develop. Increased vascular permeability and breakdown of the blood-retinal barrier lead to macular edema, which is the most common cause of vision loss in nonproliferative retinopathy. Retinal ischemia induces vascular endothelial growth factor (VEGF) overexpression, which is thought to play a key role in the development of neovascularization [30].

Retinal pericytes accumulate AGE [31] and AGE-RAGE interaction induces apoptosis of pericytes [32]. AGE also stimulate VEGF expression in pericytes [33]. In addition to promoting development of proliferative retinopathy, VEGF levels are associated with the breakdown of the blood-retinal barrier and retinal vascular hyperpermeability, possibly by inducing intracellular adhesion molecule-1 (ICAM-1) expression [34].

AGE promote vascular inflammation and thrombogenesis, leading to retinal ischemia, through multiple mechanisms. First, AGE induce expression of pro-atherogenic genes ICAM-1 and monocyte chemoattractant protein-1 (MCP-1) in microvascular endothelial cells and also decrease nitric oxide activity in endothelial cells, by inactivating nitric oxide and by reducing nitric oxide synthase transcription [35]. AGE may also augment platelet aggregation. Correlation between serotonin-induced platelet aggregation and serum AGE level has been reported in diabetic patients, and animal studies showed enhancement of platelet aggregating reactivity induced by AGE proteins [36]. AGE also induce plasminogen activating inhibitor-1 (PAI-1) *via* RAGE [37]. AGE- RAGE interactions also lead to stimulation of growth and tube formations (which are needed for new vessel formation) of microvascular endothelial cells [38].

The AGE-RAGE system may be working in concert with the renin-angiotensin system, which contributes to the development of diabetic retinopathy. Angiotensin II (Ang II) stimulates intracellcular ROS in retinal pericytes and upregulates VEGF in pericytes. Ang II levels correlate with VEGF levels, and both are higher in patients with proliferative retinopathy than in patients with background retinopathy [39]. Yamagishi *et al.* have found that Ang II induces RAGE overexpression in retinal pericytes and augments AGE-induced pericyte apoptosis [40]. The angiotensin receptor blocker (ARB) telmisartan blocked AngII-induced RAGE expression [41] and ROS generation, possibly by down-regulating RAGE expression *via* its unique PPAR-γ modulating ability [42]. Troglitazone, a PPAR-γ agonist, also down-regulated RAGE expression [43]. Interestingly, candesartan, a different ARB, did not reduce RAGE expression [43].

Vitreous levels of AGE and VEGF are significantly higher in diabetic patients than in controls and correlate with severity of retinopathy. There is significant correlation between AGE and VEGF levels [44, 45]. AGE and VEGF were found to be inversely correlated to total antioxidant status, measured by enzyme-linked immunosorbent assay (ELISA) [46].

Blocking AGE-RAGE interaction or inhibiting downstream signaling (to reduce ROS generation) are potential therapeutic strategies in the treatment of retinopathy. Anti-oxidant agents may attenuate the effects of AGE. Anti-VEGF therapies are already used as a modality for retinopathy.

Nephropathy

Diabetes is the leading cause of renal failure in the United States [1]. Diabetic kidney disease is characterized by glomerular basement membrane thickening, expansion of mesangium and hyaline accumulation. Progressive proteinuria results from increased permeability of the glomerular filter membrane. Hyperglycemia contributes to nephropathy by increasing hyperfiltration of excess glucose, inducing activation of protein kinase C, and increasing ROS [47].

AGE deposited in the glomerular basement membrane cause thickening and expansion of the extracellular membrane. As discussed above, AGE also exert their deleterious effects *via* RAGE. RAGE is expressed in kidneys and blood vessels of patients with nephropathy. Inflammatory cytokines, such as IL-6 and transforming growth factor-beta (TGF-β), are increased in the setting of hyperglycemia and consequently increased AGE formation, and contribute to progression of kidney disease [48].

Wendt showed that diabetic RAGE knockout mice did not develop mesangial matrix expansion or glomerular basement membrane thickening [49]. Myint *et al.* reported significant suppression of kidney function abnormalities, such as elevated creatinine, increased albuminuria, kidney enlargement and glomerulosclerosis, in a RAGE null mouse model of diabetic nephropathy [50]. Diabetic mice with RAGE overexpression demonstrated progressive glomerulosclerosis and renal dysfunction compared to diabetic RAGE-null mice [51]. Activation of RAGE in podocytes may contribute to VEGF expression and promote inflammation in glomeruli, leading to albuminuria and glomerulosclerosis [49].

Neuropathy

Diabetic neuropathy is the most common complication of long-term diabetes and affects sensory, motor and autonomic functions. The most common diabetic neuropathy is distal symmetrical sensorimotor polyneuropathy, which is characterized by symmetrical loss of distal small cutaneous nerve fibers resulting in stocking-and-glove distribution of affected nerves [52]. Loss of small myelinated Aδ fibers and unmyelinated C fibers results in altered mechanical, thermal and pain sensations. Progression of diabetic neuropathy leads to impairment of large myelinated Ab fibers which affect vibration, proprioception and balance [53]. It is not known why some patients develop painful neuropathy, while others predominately experience numbness.

The DCCT showed that intensive therapy reduced the risk of neuropathy by 64% compared to conventional therapy. The lower rate of neuropathy in the intensive group persisted after the study concluded, even though at that point, both groups showed similar glucose control, supporting the benefit of early glucose control and harm of prolonged hyperglycemia ("metabolic memory") [54]. Accumulation of AGE has been shown in peripheral nerves of diabetic patients [55]. Measurement of AGE by skin autofluorescence correlated with slowed sensory nerve conduction velocity and presence of microvascular complications [56].

AGE contribute to neuropathy through multiple mechanisms. RAGE is present in dorsal root ganglia, peripheral nerves, Schwann cells and epidermal fibers in rodent models [57]. First, AGE binding to RAGE leads to activation of PI-3K, Ki-Ras, and MAPK pathways that ultimately leads to increased NF-κB, which, in turn, results in increased expression of pro-inflammatory cytokines. NF-κB also upregulates RAGE and stimulates NADPH oxidase, leading to excess ROS formation [58]. Upregulation of AGE and ROS generation both contribute to neuronal damage. Diabetic RAGE knockout mice, as well as diabetic mice treated with RAGE antibody, showed incomplete suppression of NF-κB activation [59]. Pain perception and other peripheral nervous system deficits were reduced in RAGE knockout mice [57].

Second, glycation of structural and cellular proteins also may contribute to neuropathy, by causing dysfunction of neurons. Glycation of mitochondrial proteins is particularly damaging, since neurons have high metabolic demands and depend on mitochondria to generate ATP [60].

Third, differences in the glyoxalase system activity, which prevents formation of AGE by metabolizing reactive dicarbonyls to α-hydroxyacids, may make an individual more or less susceptible to neuropathy. The glyoxalase system consists of two enzymes, glyoxalase I and glyoxalase II. Glyoxalase I is the key enzyme and rate limiting step in the glyoxalase pathway (Fig. **5**). Glyoxalase I is ubiquitously expressed in the cytosol of all cells [61] and expressed at high abundance in small unmyelinated peptidergic neurons [62], which mediate pain, thermal and mechanical sensitivity. Diabetic mice with reduced glyoxalase I levels showed greater development of insensate loss of epidermal fibers and reduced components of mitochondrial oxidative phophorlyation. Conversely, diabetic BALB/cByJ mice with higher glyoxalase I levels were protected from these indices of neuropathy [63]. Overexpession of glyoxalase I has also been shown to reduce markers of other diabetes complications. In a mouse model, overexpession of glyoxalase I in the lens protected against hyperglycemia-induced protein modification [64]. In retinal capillary pericytes, glyoxalase I protected against apoptosis in hyperglycemic conditions [65]. Diabetic transgenic rats overexpressing glyoxalase I demonstrated reduced levels of markers of oxidative damage in the kidneys [66]. Thus, some patients may be protected from development of neuropathy and other diabetes complications through genetic differences relating to the glyoxalase pathway.

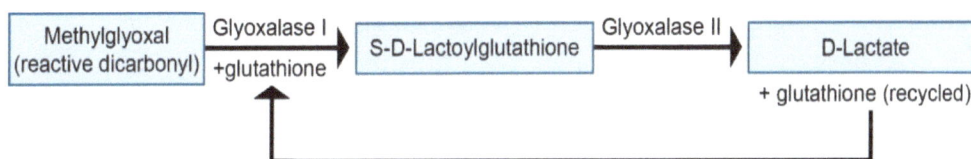

Figure 5: Glyoxalase system. Glyoxalase system detoxifies reactive dicarbonyls prior to AGE formation. Glyoxalase I is the key enzyme that converts reactive dicarbonyls (such as methylglyoxal) to a-hydroxyacids. Methylglyoxal reacts with glutathione, which is recycled after conversion to D-lactate. Overexpression of glyoxalase I has protective effects against diabetes complications. Adapted, with permission, from Jack, Wright. Transl Res 2012; 159: 355-365.

Macrovascular Disease

The DCCT showed approximately 50% reduction in CVD events in patients with type 1 diabetes, 11 years after study conclusion [67]. Similar to microvascular disease, AGE contribute to atherosclerosis in multiple ways.

AGE-RAGE interaction leads to activation of ROS and NF-κB in vascular wall cells, which promotes expression of pro-inflammatory genes such as ICAM-1, vascular cell adhesion molecule-1 (VCAM-1), MCP-1, PAI-1 and VEGF. NF-κB also upregulates RAGE, creating a positive cycle [68]. AGE-RAGE binding potentiates Ang II-induced smooth muscle proliferation and activation [69]. Homozygous RAGE null mice demonstrated suppressed smooth muscle proliferation and reduced neointimal expansion compared to wild type mice [70].

Deposition of AGE in arterial walls causes structural modification of collagen by AGE and results in decreased elasticity of vasculature. AGE also inactivate nitric oxide, causing defective vasodilatation [71]. LDL molecules modified by AGE have impaired clearance, contributing to higher LDL levels in diabetic patients [72]. AGE have been detected within atherosclerotic lesions [73], and RAGE overexpression is associated with enhanced inflammatory reaction and matrix metalloproteinase (MMP) expression in plaque macrophages in diabetic patients [74], thus contributing to plaque destabilization.

TREATMENT

Therapies targeted to reduce AGE formation and action work through several mechanisms. First, AGE production can be decreased by controlling factors that accelerate AGE formation, such as glycemic control, oxidative stress, and exogenous AGE exposure through diet modification and smoking cessation. Second, inhibitors of AGE formation act by inhibiting glycation reactions or by preventing formation of cross-links to proteins. RAGE is another therapeutic target, either *via* RAGE antibody or soluble RAGE (sRAGE). Fourth, blocking downstream signaling or decreasing RAGE expression will reduce RAGE activity. Once AGE are formed, agents that break cross link formation can decrease RAGE levels (Fig. **6**).

A. INHIBITORS OF AGE FORMATION

Aminoguanidine

One of the most well-studied inhibitors of AGE formation is aminoguanidine. Aminoguanidine reacts with Amadori intermediates and dicarbonyl compounds (methylglyoxal, glyoxal, 3-deoxyglucosone) preventing their conversion into AGEs. The first report of aminoguanidine intervention demonstrated the prevention of arterial wall protein cross linking in diabetic animals [75]. In other animal studies, aminoguanidine was shown to increase vascular elasticity and decrease vascular permeability. Aminoguanidine also preserved myocardial compliance and prevented cardiac hypertrophy [76, 77]. In cholesterol-fed rabbits, aminoguanidine reduced vascular RAGE expression and plaque formation in aorta [78]. Aminoguanidine prevented oxidative modification of LDL, and administration of aminoguanidine to diabetic patients reduced LDL by 28% [72].

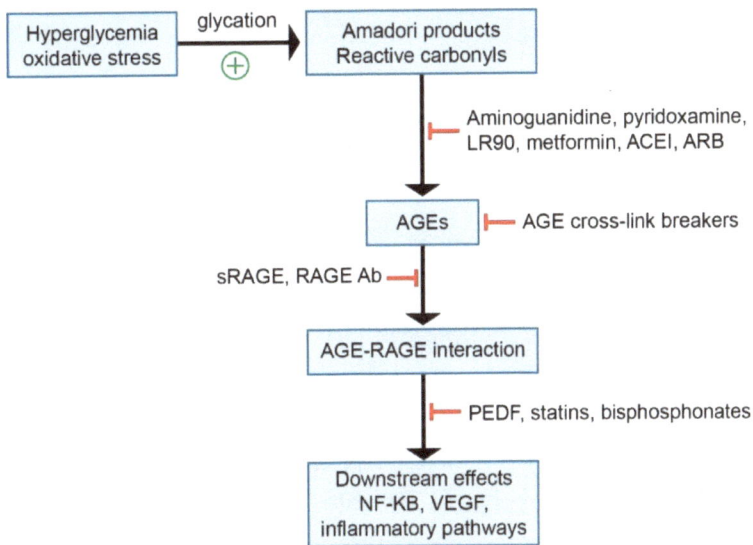

Figure 6: Target mechanisms of AGE therapies. See text for discussion.

In diabetic rats, aminoguanidine prevented AGE accumulation in pre-capillary arterioles, preventing abnormal endothelial cell proliferation and pericyte loss [79]. Aminoguanidine-treated diabetic rats showed improved motor nerve conduction velocity despite lack of changes in body weight, blood glucose, and HbA1c levels. Aminoguanidine also inhibited an accumulation of fluorescent AGE in diabetic nerves [80]. In other rat studies, aminoguanidine normalized impaired nerve blood flow [81] and improved endoneurial microvasculature [82]. However, baboons with type 1 diabetes treated with aminoguanidine for 3 years did not show improvements in nerve conduction velocity or myelinated fiber pathology. It was suggested that because the baboons had type 1 diabetes for less than 5 years, AGE accumulation contributing to nerve damage may not have yet occurred [83]. The largest clinical trial with aminoguanidine was the ACTION (A Clinical Trial in Overt Nephropathy of Type 1 Diabetes), involving 690 patients with type 1 diabetes. Patients treated with aminoguanidine for 2-4 years showed reduced 24 hour proteinuria and slower decreases in GFR. Progression of retinopathy was also slowed. However, the primary end point of doubling of serum creatinine did not reach statistical significance [84]. The ACTION II trial enrolled 599 patients with type 2 diabetes, and like the ACTION trial, was designed to assess the safety and efficacy of aminoguanidine in slowing progression to overt nephropathy. ACTION II was terminated prematurely due to side effects, which included gastrointestinal disturbances, liver function abnormalities, flu-like symptoms, development of anti-nuclear antibody (ANA), and rarely, an anti-neutrophil cytoplasmic antibody (ANCA)-associated vasculitis [85].

Even though further clinical studies were not pursued, the aminoguanidine studies produced proof of concept that inhibition of AGE formation can prevent or attenuate AGE-mediated development of diabetic complications.

Pyridoxamine

Pyridoxamine is a derivative of pyridoxine (vitamin B6) that can inhibit AGE formation at three different levels: 1) by blocking oxidative degradation of Amardori intermediates; 2) by scavenging toxic carbonyl products of glucose and lipid degradation; and 3) by trapping ROS [86]. In streptozotocin-induced diabetic rats, pyridoxamine inhibited early renal disease and reduced hyperlipidemia [87]. Pyridoxamine also prevented diabetes-induced retinal vascular lesions [88]. In obese Zucker rats, pyridoxamine inhibited development of renal and vascular disease [89]. Clinical trials with pyridoxamine involving patients with type 1 and type 2 diabetes showed slower rise in creatinine in treated subjects compared to placebo, but no change in urine albumin excretion was seen. Urinary TGF-β, along with CML and CEL (AGE measured in the study), were decreased in the pyridoxamine groups and increased in placebo groups [90]. Pyridoxamine also antagonized Ang II-induced increases in AGE, prevented renal hypertrophy and reduced salt retention [91].

Thiamine and Benfotiamine

Thiamine (vitamin B1) is an essential cofactor, and its active metabolite, thiamine diphosphate, (TDP, which is also known as thiamine pyrophosphate), is required in glycolysis, Krebs cycle and pentose phosphate pathway. TDP is required for transketolase, an enzyme which shifts excess fructose-6-phosphate and glyceraldehyde-3-phosphate for glycolysis to the pentose phosphate pathway, reducing substrate available for AGE formation. Thiamine is also important in neuronal transmission, and thiamine deficiency is known to cause neurologic disorders (such as beriberi and Wernicke-Korsakoff syndrome).

Benfotiamine is a lipid soluble derivative of thiamine, which was developed for improved bioavailability. It is more easily absorbed and oral administration results in five times greater plasma concentration then the equivalent dose of thiamine [92]. Like thiamine, benfotiamine increases TDP levels and enhances transketolase activity. Both thiamine and benfotiamine may also have anti-oxidant effects by increasing glutathione levels [93].

Streptozotocin-induced diabetic rats treated with benfotiamine and thiamine showed reversal of increased AGE levels and absence of tissue accumulation of AGE [94]. Diabetic rats treated with benfotiamine, but not thiamine, showed improvement in nerve conduction velocity [95]. Benfotiamine also prevented formation of AGE in peripheral nerve tissue [95]. In another animal study, benfotiamine prevented development of retinopathy by inhibiting AGE formation, increasing transketolase activity in the retina and preventing pericyte apoptosis [96]. While animal studies using thiamine and benfotiamine showed promising results, clinical studies in humans are scarce. Short term (3-6 weeks) studies with benfotiamine show improved neuropathy symptoms scores and pain reduction [97, 98].

LR-90

LR-90 is a new AGE inhibitor that has action similar to aminoguanidine and pyridoxamine. It inhibits formation of glycoxidative AGE and their interactions with ROS [99]. In streptozotocin-induced diabetic rats, LR-90 prevented progression of diabetic nephropathy. LR-90 also prevented a cellular capillaries and pericyte dropout in experimental retinopathy in rats [100]. Furthermore, LR-90 exhibited anti-inflammatory effects in human monocytes by inhibiting expression of RAGE and other pro-inflammatory genes, such as MCP-1 and cyclooxygenase-2, thus possibly conferring additional vascular benefit [101].

Metformin

Metformin is an oral diabetes agent is a biguanide, similar in structure to aminoguanidine. Metformin has the ability to trap methylglyoxal (MG), a reactive carbonyl species, and form a more stable compound, triazepinone, thereby reducing MG levels. Metformin also chelates metal ions which are cofactors of oxidation processes in the Maillard reaction, and enhances the glyoxylase pathway, which (as mentioned earlier) detoxifies reactive carbonyl species to D-lactate [102]. Studies in patients with diabetes show significantly reduced MG levels and increased triazepinone levels in patients treated with metformin. [103]. D-lactate levels are also higher in patients taking metformin [102].

In vitro studies by Ishibashi showed that metformin blocked AGE-induced apoptotic cell death of tubular cells in a dose-dependent manner [104]. Metformin also reduced RAGE expression in cultured endothelial cells [105] and prevented glycation of LDL or formation of foam cells [106]. Structural alterations in diabetic myocardium related to the effects of glycation were prevented with metformin treatment [107]. In the UKPDS trial, metformin reduced macrovascular complications such as stroke, while other diabetes treatments regimens (including sulfonylureas and insulin) did not [108]. Metformin may offer additional benefits beyond glucose lowering, probably by blocking formation of AGE and their downstream effects.

B. AGE BREAKERS

PTB and Alagebrium

Accumulation of AGE leads to formation of cross-links that make normally flexible proteins rigid. Long-lived proteins with slow turnover rates, such as extracellular matrix proteins, connective tissue matrix, and basement membrane components like collagen, are particularly susceptible [109]. The prototype AGE cross-link breaker was PTB (N-phenacylthiazolium bromide), which was shown to decrease AGE accumulation in tissue [110] but did not reduce proteinuria in diabetic rats [111]. PTB is very unstable and has a very short half-life, and therefore, a more stable derivative was developed. It is called ALT-711 or alagebrium, and it breaks carbon-carbon bonds between carbonyls [112]. In diabetic rats, treatment with alagebrium slowed progression of albuminuria [113]. In db/db mice treated with alagebrium, urinary albumin was decreased, and changes in glomerular hypertrophy, mesangial expansion and glomerular basement thickening were less severe [114]. In streptozotocin diabetic apoE knockout mice, alagebrium treatment resulted in reduced albuminuria and renal structural injury, and decreased renal RAGE expression [115]. Alagebrium also reduced cardiac AGE levels, reduced left ventricular weight, reduced cardiac brain natriuretic peptide (BNP) and increased collage solubility [116], suggesting that it may attenuate diabetes-associated (AGE-mediated) cardiac abnormalities. While initial trials showed promise, large clinical trials did not meet primary endpoints. In the SAPPHIRE trial (Systolic and Pulse Pressure Hemodynamic Improvement by Restoring Elasticity), 768 patients were randomized to placebo or 4 different doses of alagebrium for six months. All arms, including placebo, showed reduction in systolic blood pressure. Lower dose groups showed 4mmHg lower blood pressure than placebo. Alagebrium was safe and well-tolerated, but evidence of efficacy was lacking [117]. Combination trials of alagebrium with ACE inhibitor are planned to evaluate whether the addition of alagebrium is superior to treatment with ACE inhibitor alone.

C. INHIBITING AGE-RAGE INTERACTION

Soluble RAGE (sRAGE) and Endogenous Secretory RAGE (esRAGE)

Endogenous soluble RAGE (sRAGE) are isoforms of the RAGE protein that are present in plasma and may be generated from cleavage of full length RAGE or from novel splice variant of RAGE. sRAGE lacks the COOH terminal and transmembrane and signaling domains. sRAGE can bind extracellular ligands, including RAGE ligands, thereby acting as a competitive inhibitor to prevent AGE-RAGE interaction. Administration of recombinant form of sRAGE suppressed atherosclerotic changes and stabilized established atherosclerosis in diabetic apoE-null mice [118, 119]. sRAGE also inhibited diabetic leukostasis and blood-retinal barrier breakdown in diabetic C57/BJ6 and RAGE transgenic mice

[120]. This was accompanied by reductions in VEGF and ICAM-12 expression in the retina [120].

Surprisingly, multiple studies have shown that total sRAGE levels are increased, not decreased, in patients with type 1 and type 2 diabetes [121, 122], and AGE levels are positively associated with sRAGE [123]. In type 2 diabetic patients, positive correlation of sRAGE and urine albumin excretion [124] and sRAGE and inflammatory markers TNF-α and MCP-1, [125] are reported. It has been postulated that sRAGE reflects tissue RAGE expression and rises in parallel to serum AGE levels as a counter response [126]. However, not all studies confirm this conclusion [127].

Endogenous secretory RAGE (esRAGE) is a type of sRAGE and represents less than half of total sRAGE in human plasma [128]. In contrast to total sRAGE, esRAGE is reduced in patients with type 1 and type 2 diabetes [129, 130], diabetic retinopathy [129, 130], and atherosclerosis [131], suggesting separate roles of sRAGE and esRAGE in diabetes [126]. esRAGE levels are approximately 3-4 folds lower than total sRAGE levels. sRAGE, but not esRAGE, was positively associated with urine albumin excretion in type 2 diabetic individuals [124]. However, levels of both sRAGE and esRAGE are increased in chronic kidney disease, including end stage renal disease, and renal transplantation reduces sRAGE levels. It is not known whether the increase sRAGE and esRAGE is due to reduced clearance or upregulation to protect against toxic RAGE ligands [128]. At this time, it is not known whether esRAGE and sRAGE have distinct roles in different clinical situations.

In streptozotocin-treated diabetic rats, ramipril decreased circulating and renal AGE. Renal upregulation of RAGE was observed, as well as increase in sRAGE. sRAGE and AGE complexes were identified in plasma of rats and humans. Similarly, perindopril treatment significantly increased sRAGE in patients with type 1 diabetes. These studies suggest that angiotensin converting enzyme (ACE) inhibition reduces accumulation of AGE, in part by increasing production and secretion of sRAGE [132]. Telmisartan blocked Ang II-induced increase in RAGE and sRAGE in cultured endothelial cells, and also decreased sRAGE levels in patients with hypertension [133]. The study shows that sRAGE is secreted by Ang II-exposed endothelial cells and that sRAGE reflects endothelial RAGE expression. It also indicates interaction between AGE-RAGE and RAS systems.

D. REDUCTION OF DOWNSTREAM EFFECTS OF AGE

Nifedipine

The dihydropyridine calcium channel blocker nifedipine inhibits RAGE expression in AGE-exposed endothelial cells by suppressing ROS generation [134]. Nifedipine also blocked AGE-induced MCP-1 expression in human cultured mesangial cells, possibly by suppressing NADPH oxidase activity [135].

Pigment-Epithelium Derived Factor (PEDF)

PEDF is a glycoprotein belonging to the superfamily of serine protease inhibitors possessing multiple protective, beneficial properties. PEDF levels in aqueous humor or vitreous humor are reduced in patients with diabetes, especially those with proliferative retinopathy [136]. There is a positive correlation between vitreous AGE and VEGF levels in diabetic patients and inverse correlation with PEDF and antioxidant status [137]. Intravenous administration of AGE to normal rats decreased retinal PEDF levels and increased retinal vascular permeability through upregulation of VEGF expression [138]. PEDF inhibits AGE-induced vascular hyperpermeability by suppressing NADPH oxidase activity [138] and subsequent NF-κB-dependent VEGF gene induction.

PEDF inhibits AGE-induced ROS generation and subsequently prevents pericyte apoptosis [139] by down-regulating expression of bcl-2, an anti-apoptotic factor. PEDF also blocks AGE-induced upregulation of ICAM-1, VEGF and MCP-1 and AGE-induced nitric oxide suppression in endothelial cells [140]. Further, PEDF inhibits diabetes or AGE-induced RAGE expression by blocking NF-κB activation [140]. PEDF was also reported to inhibit VEGF binding to VEGF receptor 2 in retinal endothelial cells [141].

Thus, PEDF possesses anti-inflammatory and antioxidant properties that are protective against the effects of AGE and VEGF in diabetic retinopathy.

Statins and Bisphosphonates

HMG CoA reductase inhibitors, also known as statins, block cholesterol synthesis in the liver. HMG CoA reductase pathway (Fig. **7**). leads to formation of mevalonate pyrophosphate compounds (including farnesyl pyrophosphate) and ultimately, cholesterol. From farnesyl pyrophosphate, prenylated proteins are formed, and protein prenylation has been found to be important for AGE-RAGE signaling in endothelial cells [142]. Prenylation of small G proteins, specifically GTPases, RAS, Rho and Rac, is important for cellular functions, such as cell growth, cell differentiation, and cell adhesion. Overactive RAS signaling has been implicated in the development of cancer, and one of the RAS oncogenes (KRAS, HRAS, NRAS) is mutated in approximately 20% of all human tumors [143].

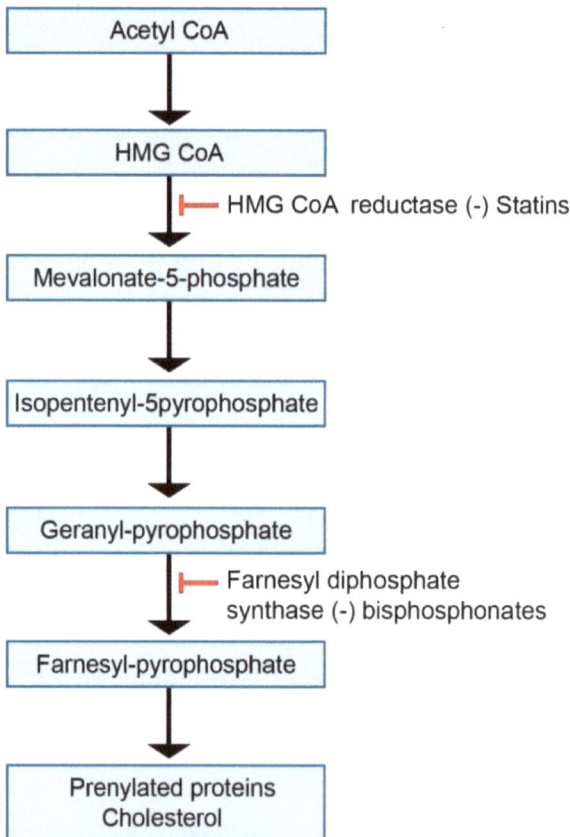

Figure 7: Mevalonate pathway. Inhibition of protein prenylation, either with HMG CoA reductase inhibitors (statins) or by blocking farnesyl diphosphate synthase (nitrogen containing bisphosphonates) may attenuate or block AGE signaling.

Cerivastatin inhibited AGE-induced increase in NF-κB activity and VEGF expression in endothelial cells [144]. Atorvastatin also inhibited AGE-induced ROS generation in Hep3B cells in a dose-dependent manner. Atorvastatin, as well as N-acetylcysteine (an antioxidant), suppressed C-reactive protein expression in AGE exposed Hep3B cells [145]. Mevalonate blocks inhibitory effects of cerivastatin on AGE exposed endothelial cells, and a farnesyl transferase inhibitor mimicked cerivastatin effects, suggesting that cerivastatin may block AGE-RAGE signaling through suppression of protein prenylation [144].

Bisphosphonates are potent inhibitors of bone resorption and are the most widely used agents to treat osteoporosis [146]. Their structure is similar to pyrophosphate, and bisphosphonates inhibit enzymes that utilize pyrophosphate. The two phosphonate groups have high binding affinity to calcium, and thus are able to accumulate in high concentrations in bone. Nitrogen-containing bisphosphonates (such as pamidronate, alendronate, risedronate) block farnesyl disphosphate synthase in the HMG CoA reductase pathway, and thus also block protein prenylation. Incardronate sodium was found to inhibit AGE-induced upregulation in NF-κB activity and VEGF expression in endothelial cells [146]. Minodronate, a novel bisphosphonate, inhibited AGE-induced NF-κB activation and suppressed VCAM-1 expression by reducing ROS generation in endothelial cells. These effects were reversed by geranylgeranyl pyrophosphate [147]. Taken together, these data suggest that blocking prenylation inhibits downstream inflammatory effects of AGE.

CONCLUSIONS

Advanced glycation end products are pro-oxidant metabolic derivatives of nonenzymatic reactions between reducing sugars and free amines of proteins and nucleic acids that have been shown to increase appetite, induce inflammation and worsen metabolic disease. They are found in processed food and their amount in the food can be increased by applying heat to food and decreased by using less heat and more water (stewing *vs.* frying). AGE may participate in the development of diabetes mellitus and advance its complications. Multiple approaches to the reduction of AGE are feasible. Therapies targeting this system will be complementary to existing diabetes therapies and have potential to prevent development of diabetes complications—a novel strategy going beyond just controlling glucose. However, except for management of glycemic control, these approaches remain experimental.

ACKNOWLEDGEMENTS

Declare none.

CONFLICT OF INTEREST

The authors do not have any conflicts of interest to report.

REFERENCES

[1]   Centers for Disease Control and Prevention. National Diabetes Fact Sheet: national estimates and general information on diabetes and prediabetes in the United States, 2011. Atlanta, GA: US Department of Health and Human Services, Centers for Disease Control and Prevention, 2011.

[2]   International Diabetes Federation. 2011 Diabetes Atlas, 5th Ed. Available from: www.idf.org

[3]   American Diabetes Association. Diagnosis and classification of diabetes mellitus. Diabetes Care 2010;33:Suppl 1:S62-S69. Erratum in Diabetes Care 2010;33(4):e57

[4]   The Diabetes Control and Complications Trial Research Group (DCCT). The Effect of Intensive Treatment of Diabetes on the Development and Progression of Long-Term Complications in Insulin-Dependent Diabetes Mellitus. N Engl J Med 1993; 329:977-986.

[5]   UKPDS Group. Intensive blood-glucose control with sulphonylureas or insulin compared with conventional treatment and risk of complications in patients with type 2 diabetes (UKPDS 33). Lancet 1998; 352: 837–53.

[6]   DeFronzo A. From the Triumvirate to the Ominous Octet: A New Paradigm for the Treatment of Type 2 Diabetes Mellitus. Diabetes April 2009; 58(4): 773-795.

[7]   McCulloch DK, Robertson RP. Pathogenesis of type 2 diabetes mellitus. In: D Nathan, J Mulder (Eds.), UpToDate. Retrieved from: http://www.uptodate.com.

[8]   Harford KA, Reynolds CM, McGillicuddy FC, Roche HM. Fats, inflammation and insulin resistance: Insights to the role of macrophage and T-cell acuumulation in adipose tissue. Proc Nutr Soc 2011; 70:408-417.

[9]   Odrowaz-Sypniewska G. Markers of pro-inflammatory and pro-thrombotic state in the diagnosis of metabolic syndrome. Adv Med Sci. 2007;52:246-50.

[10]   Hotamisligil GS. Endoplasmic reticulum stress and the inflammatory basis of metabolic disease. Cell 2006; 140:900-917.

[11]   Piperi C, Adamopoulos C, Dalagiorgou G, *et al.* Crosstalk between advanced glycation and endoplasmic reticulum stress: emerging therapeutic targeting for metabolic diseases. J Clin Endocrinol Metab 2012; 97(7): 2231-2242.

[12]   Vlassara H. AGE restriction in diabetes mellitus: a paradigm shift. Nat Rev Endocrinol. 2011 7:526-539.

[13]   Vlassara H, Uribarri J, Cai W, *et al*. Effects of sevelamer on HbA1C, inflammation, and advanced glycation end products in diabetic kidney disease. Clin J Am Soc Nephrol 2012; 7(6):934-42.

[14]   Cai W, Ramdas M, Zhu L, *et al*. Oral advanced glycation endproducts (AGEs) promote insulin resistance and diabetes by depleting antioxidant defenses AGE receptor-1 and sirtuin 1. Proc Natl Acad Sci USA 2012; 109(39): 15888-93.

[15]   Poretsky L. Looking beyond overnutrition for causes of epidemic metabolic disease. Proc Natl Acad Sci USA. 2012; Sep 25;109(39):15537-8.

[16]   Sun JK, Keenan HA, Cavallerano JD, *et al*. Protection from reinopathy and other complcations in patients with type 1 diabetes of extreme duration: the Joslin 50-year medalist study. Diabetes Care 2011; 34:968-974.

[17]   Uribarri J, Tuttle KR. Advanced glycation end products and nephrotoxicity of high-protein diets. Clin J Am Soc Nephrol 2006; 1: 1293-1299.

[18]   Lorenzi M. The polyol pathway as a mechanism for diabetic retinopathy: attractive, elusive and resilient. Exp Diabetes Res 2007; 2007: 61038.

[19]   Brownlee M, Vlassara H, Cerami A. Nonenzymatic glycosylation and the pathogenesis of diabetes complications. Ann Intern Med 1984; 101: 527-37.

[20]   Ahmed N. Advanced glycation endproducts—role in pathology of diabetic complications. Diabetes Res Clin Pract 2005; 67: 3-21.

[21]   Thomas MC. Advanced glycation end products. In: "Diabetes and the Kidney", pp 66-74, Lai KN, Tang SCW, eds., Karger Publishers, Basel, Switzerland, 2011.

[22]   Harja E, Bu DX, Hudson BI, *et al*. Vascular and inflammatory stresses mediate atherosclerosis *via* RAGE and its ligands in apoE-/mice. J Clin Invest 2008; 118(1): 183-94.

[23]   Brownlee M. Advanced protein glycosylation in diabetes and aging. Annu Rev Med. 1995; 46: 223-234.

[24]   Tanaka S, Avigad G, Brodsky B, Eikenberry EF. Glycation induces expansion of the molecular packing of collagen. J Mol Biol. 1988; 203: 295-505.

[25]   Murray RK. Red and white blood cells. In: Bender DA, Botham KM, Weil PA, Kennelly PJ, Murray RK, Rodwell VW, eds. Harper's Illustrated Biochemistry. Chapter 52, 29e, McGraw Hill, 2011.

[26]   Cerami C, Founds H, Nicholl I, *et al*. A. Tobacco smoke is a source of toxic reactive glycation products. Proc Natl Acad Sci 1997; 94(25): 13915-13920.

[27]   Luevano-Contreras C, Chapman-Novakofski K. Dietary advanced glycation end products and aging. Nutrients 2010; 2: 1247-65.

[28]   Uribarri J, Cai W, Peppa M, *et al*. Circulating glycotoxins and dietary advanced glycation endproducts: two links to inflammatory response, oxidative stress, and aging. J Gerontol A Biol Sci Med Sci 2007; 62(4): 427-33.

[29]   American Diabetes Association. Implications of the United Kingdom Prospective Diabetes Study. Diab Care 2002;25: S28-32.

[30]   Adamis AP, Miller JW, Bernal MT, *et al*. Increased VEGF levels in the vitreious of eyes with proliferative DR. Am J Ophthamol 1994; 118: 445-50.

[31]   Stitt AW, Li YM, Gardiner TA, *et al*. Advanced glycation end products co-localize with AGE receptors in the retinal vasculature of diabetic and AGE-infused rats. Am J Pathol 1997; 150: 523-31.

[32] Yamagishi S, Hsu CC, Taniguchi M, *et al.* Receptor-mediated toxicity to pericytes of advanced glycosylation end products: a possible mechanism of pericyte loss in diabetic microangiopathy. Biochem Biophys Res Commun 1995; 213: 681-7.

[33] Yamagishi S, Amano S, Inagaki Y, *et al.* Advanced glycation end products-induced apoptosis and overexpression of vascular endothelial growth factor in bovine retinal pericytes. Biochem Biophys Res Commun 2002; 296:877-82.

[34] Lu M, Perez VL, Ma N, *et al.* VEGF increases retinal vascular ICAM-1 expression *in vivo*. Invest Ophthalmol Vic Sci 1999; 40: 1808-12.

[35] Joussen AM, Poulaki V, Qin W, *et al.* Retinal vascular endothelial growth factor induces intercellular adhesion molecule-1 and endothelial nitric oxide synthase expression and initiates early diabetic retinal leukocyte adhesion *in vivo*. Am J Pathol 2002; 160: 501-9.

[36] Hasegawa Y, Suchiro A, Higasa S, *et al.* Enhancing effect of advanced glycation end products on serotonin-induced platelet aggregation in patients with diabetes mellitus. Thromb Res 2002; 107:319-23.

[37] Yamigishi S, Yonekura H, Yamamoto Y, *et al.* Vascular endothelial growth factor acts as a pericyte mitogen under hypoxic conditions. Lab Invest 1998; 79: 501-9.

[38] Yamagishi S, Yonekura H, Yamamoto Y, *et al.* Advanced glycation end products-driven angiogenesis *in vitro*. Induction of the growth and tube formation of human microvascular endothelial cells through autocrine vascular endothelial growth factor. J Biol Chem 1997; 272: 8723-30.

[39] Funatsu H, Yamashita H, Nakanishi Y, *et al.* Angiotensin II and vascular endothelial growth factor in the vitreous fluid of patients with proliferative diabetic retinopathy. Br J Ophthalmol 2002; 86: 311-5.

[40] Yamagishi S, Takeuchi M, Matsui T, *et al.* Angiotensin II augments advanced glycation end products-induced pericyte apoptosis through RAGE overexpression. FEBS 2005; 579: 4265-70.

[41] Nakamura K, Yamagishi S, Nakamura Y, *et al.* Telmisartan inhibits expression of a receptor for advanced glycation end products in angiotensin II-exposed endothelial cells and decreases serum levels of soluble RAGE in patients with essential hypertension. Microvasc Res 2005; 70: 137-41.

[42] Yamagishi S, Nakamura K, Matsui T. Potential utility of telmisartan, an angiotensin II type 1 receptor blocker with peroxisome proliferator-activated receptor gamma modulating activity for the treatment of cardiometabolic disorders. Curr Mol Med 2007; 7: 463-9.

[43] Yoshida T, Yamagishi S, Nakamura K, *et al.* Telmisartan inhibits AGE-induced C-reactive protein production through downregulation of the receptor for AGE *via* peroxisome proliferator-activated receptor gamma activation. Diabetologia 2006; 49: 3094-9.

[44] Miura J, Yamagishi S, Uchigata Y, *et al.* Serum levels of non-carboxymethyllysine advanced glycation end products are correlated to severity of microvascular complications in patients with type 1 diabetes. J Diabetes Complications 2003; 17: 16-21.

[45] Koga K, Yamagishi S, Okamoto T, *et al.* Serum levels of glucose-derived advanced glycation end products are associated with the severity of diabetic retinopathy in type 2 diabetic patients without renal dysfunction. Int J Clin Pharmacol Res 2002; 22: 13-7.

[46] Yokoi M, Yamagishi S, takeuchi M, *et al.* Elevations of AGE and vascular endothelial growth factor with decreased total antioxidant status in the vitreous fluid of diabetic patients with retinopathy. Br J Ophthalmol 2005; 89:673-5.

[47]   Najafian B, Alpers CE, Fogo AB. Pathology of human diabetic nephropathy. Contrib Nephrol 2011; 170: 36-47.

[48]   D'Aagati V, Schmidt AM. RAGE and the pathogenesis of chronic kidney disease. Nat Rev Nephrol 2010; 6: 352-60.

[49]   Wendt TM, Tanji N, Guo J, *et al.* RAGE drives the development of glomerulosclerosis and implicates podocyte activation in the pathogenesis of diabetic nephropathy. Am J Pathol 2003; 162: 1123-37.

[50]   Myint KM, Yamamoto Y, Doi T, *et al.* RAGE control of diabetic nephropathy in a mouse model: effects of RAGE gene disruption and administration of low-molecular weight heparin. Diabetes 2006; 55:2510-22.

[51]   Yamamoto Y, Kato I, Doi T, *et al.* Development and prevention of advanced diabetic nephropathy in RAGE-overexpressing mice. J Clin Invest 2001; 108:261-8.

[52]   Said G. Diabetic neuropathy—a review. Nat Clin Pract Neurol 2007; 3:331-40.

[53]   Yagihashi S. Yamagishi S, Wada R. Pathology and pathogenetic mechanisms of diabetic neuropathy: correlation with clinical signs and symptoms. Diabetes Res Clin Pract 2007; 99:S184-9.

[54]   The effect of intensive diabetes therapy on the development and progression of neuopathy. The Diabetes Control and Complications Trial Research Group. Ann Intern Med 1995; 122:561-8.

[55]   Misur I, Zarkovic K, Barada A, *et al.* Advanced glycation end products in peripheral nerve in type 2 diabetes with neuropathy. Acta Diabetol 2004; 41:158-66.

[56]   Meerwaldt R, Links TP, Graaff R, *et al.* Increased accumulation of skin advanced glycation endo products precedes and correlates with clinical manifestation of diabetic neuropathy. Diabetologia 2005; 48: 1637-44.

[57]   Toth C, Rong Ll, Yang C, *et al.* Receptor for advanced glycation end products and experimental diabetic neuropathy. Diabetes2007; 57: 1002-17.

[58]   Vincent AM, perrone L, Sullivan KA, *et al.* Receptor for advanced glycation end products activation injures primary sensory neurons *via* oxidative stress. Endocrinology 2007; 148: 548-58.

[59]   Bierhaus A, Haslbeck KM, Humpert PM, *et al.* Loss of pain perception in diabetes is dependent on a receptor of the immunoglobulin superfamily. J Clin Invest 2004; 114: 1741-51.

[60]   Rosca MG, Mustata TG, Kinter MT, *et al.* Glycation of mitochondrial proteins from diabetic rat kidneys is associated with excess superoxide formation. Am J Physiolo Renal Physiol 2005; 289: F420-30.

[61]   Rabbani N, Thornalley PJ. Glyoxalase in diabetes, obesity and related disorders. Semin Cell Dev Biol 2011; 22:309-17.

[62]   Jack MM, Ryals JM, Wright DE. Characterisation of glyoxalase I in a streptozocin-induced mouse model of diabetes with painful and insensate neuropathy. Diabetologia 2011; 54: 2174-82.

[63]   Velez L, Sokoloff G, Miczek KA, *et al*l. Differences in aggressive behavior and DNA copy number variatns between BALB/cJ and BALB/cByJ substrains. Behav Genet 2010; 40: 201-10.

[64]   Gangadhariah MH, Mailankot M, Reneker L, *et al.* Inhibition of methylglyoxal-mediated protein modification in glyoxalase I overexpressing mouse lenses. J Ophthalmol 2010: 274317.

[65] Miller AG, Smith DG, Bhat M, *et al*. Glyoxalase I is critical for human retinal capillary pericyte survival under hyperglycemic conditions. J Biol Chem 2006; 281: 11864-71.

[66] Brouwers O, Neissen PM, Ferreira I, *et al*. Overexpression of glyoxalase I reduces hyperglycemia-induced levels of advanced glycation end products and oxidative stress in diabetic rats. J Biol Chem 2010; 286: 1374-80.

[67] Nathan DM, Cleary PA, Backlund JY, *et al*. Intensive diabetes treatment and cardiovascular disease in patients with type 1 diabetes. N Engl J Med 1988; 318: 1315-21.

[68] Yamagishi S, Nakamura K, Matsui T, *et al*. Receptor for advanced glycation end products: a novel therapeutic target for diabetic vascular complications. Curr Pharm Des 2008; 14: 487-95.

[69] Shaw SS, Schmidt AM, Banes AK, *et al*. S100B-RAGE mediated augmentation of angiotensin II-induced activation of JAK2 in vascular smooth muscle cells is dependent on PLD2. Diabetes 2003; 52: 2381-8.

[70] Sakaguchi T, Yan SF, Yan SD, *et al*. Central role of RAGE-dependent neointimal expansion in arterial restenosis. J Clin Invest 2003; 162: 1123-37.

[71] Bucala R, Tracey KJ, Cerami A. Advanced glycosylation products quench nitric oxide and mediate defective endothelium-dependent vasodilatation in experimental diabetes. J Clin Invest 1991; 87: 432-8.

[72] Bucala R, Makita Z, Vega G, *et al*. Modification of low density lipoprotein by advanced glycation end products contributes to the dyslipidemia of diabetes and renal insufficiency. Proc Natl Acad Sci USA 1994; 91: 9441-5.

[73] Nakamura Y, Horri Y, Nishino T, *et al*. Immunohistochemical localization of advanced glycosylation end products in coronary atheroma and cardiac tissue in diabetes mellitus. Am J Pathol 1993; 43: 1649-56.

[74] Cipollone F, Iezzi A, Fazia M, *et al*. The receptor RAGE as a progression factor amplifying arachidonate-dependent inflammatory and proteolytic response in human atherosclerotic plaques: role of glycemic control. Circulation 2003; 108: 1070-7.

[75] Brownlee M, Vlassara H, Kooney A, *et al*. Aminoguanidine prevents diabetes-induced arterial wall protein cross-linking. Science 1986; 232: 1629-32.

[76] Norton GR, Candy G, Woodiwiss AJ. Aminoguanidine prevents the decreased myocardial compliance produced by streptozotocin-induced diabetes mellitus in rats. Circulation 1996; 93: 1905-12.

[77] Chang KC, Tseng CD, Wu MS, *et al*. Aminoguanidine prevents arterial stiffening in a new rat model type 2 diabetes. Eur J Clin Invest 2006; 36: 528-35.

[78] Panagiotopoulos S, O'Brien RC, Bucala R, *et al*. Aminoguanidine has an anti-atherogenic effect in the cholesterol-fed rabbit. Atherosclerosis 1998; 136-125-31.

[79] Hammes HP, Martin S, Federlin K, *et al*. Aminoguanidine treatment inhibits the development of experimental diabetic retinopathy. Proc Natl Acad Sci USA 1991; 88-11555-8.

[80] Yagihashi S, Kamijo M, Baba M, *et al*. Effect of aminoguanidine on functional and structural abnormalities in peripheral nerve of STZ-induced diabetic rats. Diabetes 1992; 41: 47-52.

[81] Kihara M, Schmelzer JD, Poduslo JF, *et al*. Aminoguanidine effects on nerve blood flow, vascular permeability, electrophysiology, and oxygen free radicals. Proc Natl Acad Sci USA 1991; 88: 6107-11.

[82]   Sugimoto K, Yagihashi S. Effects of aminoguanidine on structural alterations of microvessels in peripheral nerve of streptozotocin diabetic rats. Microvasc Res 1997; 53: 105-12.

[83]   Birrell AM, Heffernan SJ, Ansselin AD, *et al.* Functional and structural abnormailities in the nerves of type 1 diabetic baboons: aminoguanidine treatment does not improve nerve function. Diabetologia 2000; 43: 110-6.

[84]   Bolton WK, Cattran DC, Williams ME, *et al.* Randomized trial of an inhibitor of formation of advanced glycation end products in diabetic nephropathy. Am J Nephrol 2004; 24: 32-40.

[85]   Freedman BI, Wuerth JP, Cartwright K, *et al.* Design and baseline characteristics for the aminoguanidine clinical trial in overt type 2 diabetic nephropathy (ACTION II). Control Clin Trials 1999; 20;493-510.

[86]   Voziyan PA, Metz TO, Baynes JW, *et al.* A post-Amadori inhibitor pyridoxamine also inhibits chemical modification of proteins by scavenging carbonyl intermediates of carbohydrate and lipid degradation. J Biol Chem 2002; 277: 3397-403.

[87]   Degenhardt TP, Alderson NL, Arrington DD, *et al.* Pyridoxamine inhibits early renal disease and dyslipidemia in the streptozotocin-diabetic rat. Kidney Int 2002; 61: 939-50.

[88]   Stitt A, Gardiner TA, Alderson NL, *et al.* The AGE inhibitor pyridoxamine inhibits development of retinopathy in experimental diabetes. Diabetes 2002; 51: 2826-32.

[89]   Alderson Nl, Chachich MN, Youssef NN, *et al.* The AGE inhibitor pyridoxamine inhibits lipemia and development of renal and vascular disease in Zucker obese rats. Kidney Int 2003; 63: 2123-3.

[90]   Williams ME, Bolton WK, Khalifah RG, *et al.* Effects of pyridoxamine in combined phase 2 studies of patients with type 1 and type 2 diabetes and overt nephropathy. Am J Nephrol 2007; 27:605-14.

[91]   Thomas MC, Tikellis C, Burns WM, *et al.* Interactions between renin and angiotensin system and advanced glycation in the kidney. J Am Soc Nephrol 2005; 16: 2976-84.

[92]   Balakumar P, Rohilla A, Krishan P, *et al.* The multifaceted therapeutic potential of benfotiamine. Pharmacol Res 2010; 61: 482-8.

[93]   Bakker SJ, Heine RJ, Gans RO. Thiamine may indirectly act as an antioxidant. Diabetologia 1997; 40: 741-2.

[94]   Karachalias N, Babaei-Jadidi R, Rabbani N, *et al.* Increased protein damage in renal glomeruli, retina, nerve, plasma and urine and its prevention by thiamine and benfotiamine therapy in a rat model of diabetes. Diabetologia 2010; 53: 1506-16.

[95]   Stracke H, Hammes HP, Werkmann D, *et al.* Efficacy of benfotiamine *vs.* thiamine on function and glycation products of peripheral nerves in diabetic rats. Exp Clin Endocrinol Diabetes 2001; 109: 330-6.

[96]   Beltramo E, Berrone E, Tarallo S, *et al.* Different apoptotic responses of human and bovine pericytes to fluctuating glucose levels and protective role of thamine. Diab Metab Res Dev 2009; 25: 566-76.

[97]   Stracke H, Gaus W, Achenbach U, *et al.* Benfotiamine in diabetic polyneuropathy (BENDIP): results of a randomixed, double blind, placebo controlled study. Exp Clin Endocrinol Diabetes 2008; 116: 600-5.

[98]   Haupt E, Ledermann H, Kopcke W, *et al.* Benfotiamine in the treatment of diabetic polyneuropathy—a three week randomized, controlled pilot study. Int J Clin Pharmacol Ther 2005; 43: 71-7.

[99]   Figarola JL, Scott S, Loera S, *et al*. LR-90, a new advanced glycation endproduct inhibitor prevents progression of diabetic nephropathy in sterptozotocin-diabetic rats. Diabetologia 2003; 46: 1140-52.

[100]   Bhatwadekar A, Glann JV, Figarola JL, *et al*. A new advanced glyaction inhibitor, LR-90, prevents experimental diabetic retinopathy in rats. Br J Ophthalmol 2008; 92: 545-7.

[101]   Figarola JL, Shanmugam N, Natarajan R, *et al*. Anti-inflammatory effects of the advanced glycation end product inhibitor LR-90 in human monocytes. Diabetes 2007; 56: 647-55.

[102]   Beisswenger P, Ruggiero-Lopez D. Metformin inhibition of glycation processes. Diabetes Metab 2003; 29: 6S95-103.

[103]   Ruggerio-Lopez D, Howell SK, Szwergold BS, *et al*. Metformin reduces methylglyoxal levels for formation of a stable condensation product (trazepinone). Diabetes 2000; 49: A124.

[104]   Ishibashi Y, Matsui T, Takeuchi M, *et al*. Metformin inhibits advanced glycation end products(AGEs)-induced renal tubular cell injury by suppressing reactive oxygen species generation *via* reducing receptor for AGEs expression. Horm Metab Res 2012; 44: 891-5.

[105]   Ouslimani N, Mahrouf M, Peynet J, *et al*. Metfomin reduces endothelial cell expression of both the receptor for advanced glycation end products and lectin-like oxidized receptor 1. Metabolism 2007; 56: 308-13.

[106]   Brown B, Mahroof FM, Cook NL, *et al*. Hydrazine compounds inhibit glycation of low-density lipoproteins and prevent the *in vitro* formation of model foam cells from glycoaldehyde-modified low-density lipoproteins. Diabetologia 2006; 49: 775-83.

[107]   Regan TJ, Jyothirmayi GN, Laharm C, *et al*. Left ventricular diastolic dysfunction in diabetic or hypertensive subjects: role of collagen alterations. Adv Exp Med Bio.1 2001; 298: 127-32.

[108]   UK Prospective Diabetes Study Group. Effect of intensive blood glucose control with metformin on complications in overweight patients with type 2 diabetes. Lancet 1998; 352: 854-65.

[109]   Vlassara H. Advanced glycation end products and atherosclerosis. Ann Med 1996; 28: 419-26.

[110]   Cooper ME, Thallas V, Forbes J, *et al*. The cross-link breaker, N-phenacylthiazolium bromide prevents vascular advanced glycation end product accumulation. Diabetologia 2000; 43: 660-4.

[111]   Schwedler SB, Verbeke P, Bakala H, *et al*. N-phenacylthiazolium bromide decreases renal and increases urinary advanced glycation end products excretion without ameliorating diabetic nephropathy in C57Bl/6 mice. Diabetes Obes Metab 2001; 3:230-9.

[112]   Coughlan MT, Forbes JM, Cooper ME. Role of the AGE crosslink breaker, alagebrium, as a renoprotective agent in diabetes. Kidney Int 2007; 72: S54-60.

[113]   Forbes JM, Thallas V, Thomas MC, *et al*. the breakdown of preexisting advanced glycation end products is associated with reduced renal fibrosis in experimental diabetes. FASEB 2003; 17: 1762-4.

[114]   Peppa M, Brem H, Cai W, *et al*. Prevention and reversal of diabetic nephropathy in db/db mice treated with alagebrium (ALT-711). Am J Nephrol 2006; 26: 430-6.

[115]   Lassila M, Seah KK, Allen TJ, *et al*. Accelerated nephropathy in diabetic apolipoprotein e-knockout mouse: role of advanced glycation end products. J Am Soc Nephrol 2004; 15: 2125-38.

[116] Candido R, Forbes JM, Thomas MC, *et al.* A breaker of advanced glycation end products attenuates diabetes-induced myocardial structural changes. Circ Res 2003; 92: 785-92.

[117] Bakris GL, Bank AJ, Kass DA, *et al.* Advanced glycation end product crosslink breakers. A novel approach to cardiovascular pathologies related to the aging process. Am J Hypertens 2004; 17: 23S-30S.

[118] Park L, Raman KG, Lee KJ, *et al.* Suppression of accelerated diabetic atherosclerosis by the soluble receptor for advanced glycation end products. Nat Med 1998; 4: 1025-31.

[119] Bucciarelli L, Wendt T, Qu W, *et al.* RAGE blockade stabilized established atherosclerosis in diabetic apolipoprotein E-null mice. Circulation 2002: 106: 2827-35.

[120] Kaji Y, Usui T, Ishida S, *et al.* Inhibition of diabetic leukostasis and blood-retinal barrier breakdown with a soluble form of a receptor for advanced glycation end products. Invest Ophthalmol Vis Sci 2007; 48: 858-65.

[121] Challier M, Jacqueminet S, Badbdesslam O, *et al.* Increased serum concentrations of soluble receptor for advanced glycation end products in patients with type 1 diabetes. Clin Chem 2005; 51: 1749-50.

[122] Tan KC, Shiu SW, Chow WS, *et al.* Association between serum levels of soluble receptor for advanced glycation end products and circulating advanced glycation end products in type 2 diabetes. Diabetologia 2006; 49: 2756-62.

[123] Yamagishi S, Adachi H,Nakamura K, *et al.* Positive association between serums levels of advanced glycation end products and the soluble form of receptor for advanced glycation end products in nondiabetic subjects. Metabolism 2006; 55: 1227-31.

[124] Humpert PM, Kopf S, Djuirc Z, *et al.* Plasma sRAGE is independently associated with urinary albumin excretion in type 2 diabetes. Diabetes Care 2006; 29: 1111-13.

[125] Nakamura K, Yamagishi S, Adachi H, *et al.* Serums levels of sRAGE, the soluble form of receptor for advanced glycation end products, are associated with inflammatory markers in patients with type 2 diabetes. Mol Med 2007; 13: 185-9.

[126] Yamagishi S, Imaizumi T. Serum levels of soluble form of receptor for advanced glycation end products may reflects tissue RAGE expression in diabetes. Arterioscler Thromb Vasc Biol 2007: 27: e32.

[127] Basta G, Sironi AM, Lazzerini G, *et al.* Circulating soluble receptor for advanced glycation end products is inversely associated with glycemic control and S100A12 protein. J Clin Endocrinol Metab 2006; 91: 4628-34.

[128] Koyama H, Yamamoto H, Nishizawa Y. Endogenous secretory RAGE as a novel biomarker for metabolic syndrome and cardiovascular diseases. Biomarker Insights 2007; 2: 331-9.

[129] Katakami N, Matsuhisa M, Kaneto H, *et al.* Decreased endogenouse secretory advanced glycation end product receptor in type 1 diabetic patients: its possible association with diabetic vascular complications. Diabetes Care 2005; 28: 2716-21.

[130] Sakurai S, Yamamoto Y, Tamei H, *et al.* Development of an ELISA for esRAGE and its application to type 1 diabetic patients. Diabetes Res Clin Pract 2006; 73: 158-65.

[131] Koyama H, Shoji T, Yokoyama H, *et al.* Plasma level of endogenous secretory RAGE is associated with components of the metabolic syndrome and atherosclerosis. Arterioscler Thromb Vasc Biol 2005; 25:2587-93.

[132] Forbes JM, Thrope SR, Thallas-Bonke V, *et al.* Modulation of soluble receptor for advanced glycation end products by angiotensin-converting enzyme-1 inhibition in diabetic nephropathy. J Am Soc Nephrol 2005; 16: 2363-72.

[133] Nakamura K, Yamagishi S, Nakamura Y, *et al.* Telmisartan inhibits expression of a receptor for advanced glycation end products in angiotensin-II-exposed endothelial cells and decreases serum levels of soluble RAGE in patients with essential hypertension. Microvascl Res 2005; 70: 137-41.

[134] Yamagishi S, Takeuchi M. Nifedipine inhibits gene expression of receptor for advanced glycation end product in endothelial cells by suppressing reactive oxygen species generation. Drugs Exp Clin Rex 2004; 30: 169-75.

[135] Matusi T, Yamagishi S, Nakamura K, *et al.* Nifedipine, a calcium-channel blocker, inhibits advanced glycation end product-induced expression of monocyte chemoattractant protein-1 in human cultured mesangial cells. J Int Med Res 2007; 35: 107-12.

[136] Spranger J, Osterhoff M, Reimann M, *et al.* Loss of the antiangiogenic pigment epithelium-derived factor in patients with angiogenic eye disease. Diabetes 2002; 50: 2641-5.

[137] Yokoi M, Yamagishi S, Saito A, *et al.* Positive association of pigment epithelium-derived factor with total antioxidant capacity in the vitreous fluid of patients with proliferative diabetic retinopathy. Br J Ophthalmol 2007; 91: 885-7.

[138] Yamagishi S, Nakamura K, Matsiu T, *et al.* Pigment epithelium-derived factor inhibits advanced glycation end product-induced retinal vascular hyperpermability by blocking reactive oxygen species-mediated vascular endothelial growth factor expression. J Biol Chem 2006; 281: 20213-20.

[139] Yamagishi S, Inagaki Y, Amano S, *et al.* Pigment epithelium-derived factor protects cultured retinal pericytes from advanced glycation end product-induced injury through its antioxidative properties. Biochem Biophys Res Commun 2002; 296: 877-82.

[140] Inagaki Y, Yamagishi S, Okamoto T, *et al.* Pigment epithelium-derived factor prevents advanced glycation end prodecuts-induced monocyte chemoattractant protein-1 production in microvascular endothelial cells by suppressing intracellular reactive oxygen species generation. Diabetologia 2003; 46: 284-7.

[141] Zhang SX, Wang JJ, Gao G, *et al.* Pigment epithelium-derived factor downregulates vascaulr endothelial growth factor (VEGF) expression and inhibits VEGF-VEGF receptor 2 binding in diabetic retinopathy. J Mol Endocrinol 2006; 37: 1-12.

[142] Okamoto T, Yamagishi S, Inagaki Y, *et al.* Angiogenesis induced by advanced glycation end products and its prevention by cerivastatin. FASEB J 2002; 16: 1928-30.

[143] Downward J. Finding the weakness in cancer. N Engl J Med 2009; 361: 922-4.

[144] Yoshida T, Yamagishi S, Nakamura K, *et al.* Atorvastatin inhibits advanced glycation end products-induced C-reactive protein expression in hepatoma cells by suppressing reactive oxygen species generation. Vasc Dis Prev 2007; 4: 213-6

[145] National Osteoporosis Society. Drug treatment. UK National Osteoporosis Society. August 2012.

[146] Okamoto T, Yamagishi S, Inagaki Y, *et al.* Incadronate disodium inhibits advanced glycation end products-induced angiogenesis *in vitro*. Biochem Biophys Res Commun 2002; 297: 419-24.

[147] Yamagishi S. Matsui T, Nakamura K, *et al.* Minodronate, a nitrogen-containing bisphosphonate, inhibits advanced glycation end product-induced vascular cell adhesion molecule-1 expression in endothelial cells by suppressing reactive oxygen species generation. Int J Tissue React 2005; 27: 189-95.

CHAPTER 3

Liraglutide: Present and Future

Dan Mircea Cheța, Cristian Serafinceanu and Cornelia Zetu*

National Institute of Diabetes, Nutrition and Metabolic Diseases "N.C. Paulescu", Bucharest, Romania

Abstract: The conventional treatment of type 2 diabetes cannot stop the declining of beta-cell function. Weight gain and hypoglycemia represent frequent side effects. These important limitations justify the search for new therapeutic options, one of them being incretin-mimetics GIP and GLP-1 analogues.The incretin phenomenon comprises up to 60% of the postprandial insulin secretion *via* the stimulation of the pancreatic islet cells. The first incretin identified, glucose-dependent insulinotropic polypeptide (GIP), exhibits weak effects on gastric acid secretion but more potent insulinotropic actions. Its release is stimulated by both fat and glucose ingestion. The second incretin hormone, glucagon-like peptide 1 (GLP-1) stimulates insulin and suppresses glucagon secretion, inhibits gastric emptying and reduces appetite and food intake. Both GLP-1 and GIP receptor activation lead to a rapid insulin release in a glucose dependent manner, and also promote resistance to apoptosis as well as enhance beta-cell survival. Most pharmaceutical efforts related to incretin action in the treatment of type 2 diabetes were focused on GLP-1 agonists, due to considerable loss of GIP activity at these patients. Both ADA (American Diabetes Association)/EASD (European Association for the Study of Diabetes) and AACE (American Association of Clinical Endocrinologists) guidelines recognize GLP-1 analogues as efficient agents for patients who do not achieve metabolic control through lifestyle changes and selected oral drugs. Liraglutide, a new Novo Nordisk therapy, is the first human GLP-1 analogue sharing 97% of its amino acid sequence with native human GLP-1. Furthermore, the prolonged half-life of liraglutide makes it the first human GLP-1 analogue suitable for once-daily administration. An extensive series of preclinical and clinical studies have revealed considerable potential benefits for liraglutide, *i.e.* reducing fasting and postprandial glucose, decreasing HbA1c levels by up to 1, 75% at 18-month follow up, preventing weight gain and inducing weight loss. Moreover, this agent showed positive influence both on beta-cell function and mass preservation, and cardiac function by increasing cardiomyocyte survival and reducing systolic blood pressure. The phase 3 clinical trial program (LEAD studies) has demonstrated that the glycemic benefits of liraglutide are not associated with an increased risk of hypoglycemia, or with an increased risk of pancreatitis. In terms of tolerability, gastrointestinal adverse effects were less persistent with liraglutide when compared with an active comparator in a large head-to-head study. Practitioners can consider liraglutide as a very promising option for type 2 diabetes. However, new large trials are still necessary to evaluate its long-term clinical outcomes.

***Address correspondence to Cornelia Zetu:** National Institute of Diabetes, Nutrition and Metabolic Disease "N. C. Paulescu", Bucharest, Romania; Tel: 004 0724 30 53 63; Fax: 004 021 771 45 86; Email: corapnc@yahoo.com

Keywords: Agonists, beta-cell function, blood pressure, DPP-4 inhibitors, fasting plasma glucose, gastric emptying, glucagon, glucagon-like peptide 1 (GLP-1), HbA1C, hypoglycemia, incretin phenomenon, incretin-mimetics, LEAD trials, liraglutide, nausea, Novo Nordisk A/S, postprandial plasma glucose, type 2 diabetes, Victoza, weight loss.

TYPE 2 DIABETES MELLITUS: DIFFICULTIES IN THERAPY AND PREVENTION

The main problems discussed in the first part of our work are: the increase of diabetes mellitus and its consequences; the modern therapy of diabetes mellitus; difficulties and hindrances; the prevention of type 2 diabetes is still hard riddled with unknowns.

Increase of Diabetes Mellitus and Its Consequences

On the occasion of the World Diabetes Congress (Dubai, 4-8 December 2011), International Diabetes Federation (IDF) published the fifth edition of the IDF Diabetes Atlas. This important document presents the current situation of the *diabetes epidemic,* with the most significant data and concepts [1, 2].

The total number of people with diabetes in 2011 was 366 million and the estimate for 2030 is 552 million. Fifty percent of the total number (183 million people) still remained undiagnosed. Diabetes caused 4.6 million deaths in 2011. All countries are suffering under the burden of this disease, but particularly affected are those who are socially and economically disadvantaged: 80% of people with diabetes live in low–and middle–income countries. The healthcare expenditures occasioned by diabetes in 2011 represent 11% of total healthcare expenditures in adults (at least USD 465 billion dollars).

Diabetes mellitus can lead to severe complications, such as: cardiovascular disease, kidney disease, eye disease, nerve damage, diabetic foot and others. These complications represent a major cause of disability, reduced quality of life and death [2].

Eighty-five to ninety-five percent of all diabetes or an even higher percentage in low-and middle-income countries is made up of type 2 diabetes [2]. This is regarded as a progressive disorder, including beta-cell dysfunction, insulin resistance and hyperglycemia, amongst others. Several clinical trials proved that intensive glycemic control can significantly reduce the development of chronic complications in patients with type 2 diabetes. To obtain good metabolic control, such patients often require an association of oral drugs and insulin in addition to lifestyle changes [3].

The Modern Therapy of Type 2 Diabetes

American Diabetes Association (ADA) and other scientific authorities recommend the intervention at the beginning of the treatment with metformin and lifestyle changes (medical nutrition therapy and physical exercise) and then the augmentation with additional agents (including early insulin therapy) as a mean of achieving and maintaining metabolic targets (HbA1c <7%). The intensification of treatment is achieved by the utilization of another class. Non-insulin agents are listed below [4]:

- Biguanides: Metformin

- Sulfonylureas (2nd generation): Glibenclamide/Glyburide, Glipizide, Gliclazide, Glimepiride

- Meglitinides: Repaglinide, Nateglinide

- Thiazolidinediones (Glitazones): Pioglitazone, Rosiglitazone

- α-Glucosidase inhibitors: Acarbose, Miglitol

- GLP-1 receptor agonist (incretin-mimetics): Exenatide, Liraglutide

- DPP-4 inhibitors (incretin enhancers): Sitagliptin, Vildagliptin, Saxagliptin, Linagliptin

- Bile acid sequestrants: Colesevelam

- Dopamine-2 agonists: Bromocriptine

Difficulties and Hindrances [4, 5]

Metformin. In the first part of the treatment 5-20% of patients presented gastro-intestinal side effects: diarrhoea, nausea, flatulence, abdominal cramping and pain. In few cases there were hematological disorders due to vitamin B12 deficiency. Lactic acidosis was rare.

Sulfonylureas. The most common disadvantage of sulfonylureas is hypoglycemia. This is particularly dangerous for the aged and for those with hepatic and/or renal disorders. Sometimes hypoglycemic episodes can result in serious neurologic consequences or death.

Weight gain could be another side effect. This possibility is reduced by the association of sulfonylureas with metformin and acarbose.

There is a controversy with respect to the influence of sulfonylureas on cardiovascular risk; they may "blunt" myocardial ischemic preconditioning.

The durability of the positive effects of sulfonylureas is not very long.

Meglitinides can also induce hypoglycemia and weight gain.

Thiazolidinediones have a longer list of side effects and disadvantages. Their cost is high. They may cause weight gain, fluid retention, heart failure, increased cardiovascular events, bone fractures and others.

α-Glucosidase inhibitors cause gastrointestinal disturbances (diarrhoea, flatulence *etc.*).

GLP-1 receptors agonists can induce gastrointestinal disorders (nausea, vomiting, diarrhoea), acute pancreatitis and others. Their price is high. Details about liraglutine will be discussed later.

DPP-4 inhibitors are also expensive. Occasionally they can provoke urticaria and pancreatitis.

Bile acid sequestrants have a high cost. They can cause constipation and increased level of triglycerides.

Dopamine-2 agonists can produce dizziness, nausea, fatigue, rhinitis [4].

Several papers tried to accomplish a global vision on the disadvantages of the oral antidiabetes drugs and to justify the search for new and better compounds [6-13].

The progressive nature of type 2 diabetes mellitus is generally accepted, as well as the central role of the beta-cell in its pathogenesis. The gradual decline in beta-cell function is considered to be the major reason why oral antidiabetes drugs, particularly sulfonylureas, lose their effectiveness in patients with type 2 diabetes mellitus [9].

Insulin resistance has also an important contribution to the pathophysiology of the disease. These defects lead to chronic hyperglycemia and an increased risk of complications. The clinicians are well aware that a large proportion of diabetic patients have suboptimal or poor glycemic control. "Oral drug failure" is a reality for many patients [5].

An extensive explanation of the pathogenesis was offered by De Fronzo (the "ominous octet"), with multiple implications for the modern therapy of type 2 diabetes [14].

The difficulty of achieving good metabolic control is often associated with the side effects of the drugs, particularly weight gain and hypoglycemia. Unfortunately, weight gain is an independent risk factor for cardiovascular disease, reduces patients ' quality of life and can hamper the adherence to the treatment regimen [10, 12]. On the other hand, severe hypoglycemia can lead to loss of consciousness, brain damage or death. Control of the adverse reactions is especially difficult when treatment intensification or a combination therapy is needed [10, 13].

The optional therapy for type 2 diabetes mellitus should offer a number of advantages, such as:

- improvement of beta-cell function;

- good metabolic control, without hypoglycemia;

- weight loss;

- control of the additional risk factors accompanying type 2 diabetes (dyslipidemia, hypertension *etc.*);

- simple and flexible therapeutic regimen [10].

Therefore, we need new therapeutic agents that can facilitate the attainment of metabolic targets and weight reduction while addressing other comorbidities associated with type 2 diabetes mellitus, and with a favorable safety profile. The capacity to slow the decline in beta-cell mass and to sustain the improvement in beta-cell function for a long period would be a milestone in diabetes care, with the potential to influence the natural history of the disease [15].

Prevention of Type 2 Diabetes Mellitus is Still Hard and Puzzling

The history of diabetes prevention is pretty long [1, 16, 17]. Until now it has been demonstrated that intensive lifestyle modification can significantly decrease the rate of onset for type 2 diabetes. Large international studies are sustaining this concept [2, 4].

Regarding the drug therapy for diabetes prevention, it is important to note that metformin has a convincing history and can be used for very high-risk-subjects. Other pharmacologic agents (α-glucosidase inhibitors, thiazolidinediones, orlistat *etc.*) were also tried in preventive trials for type 2 diabetes and some interesting results were achieved [4].

The International Diabetes Federation considers that a real challenge for diabetes prevention is to modify the obesogenic environment [2]. According to Hawley and Gibala (2012) [16], the primary goals should be the supervised exercise and diet programs, along with increased education and public awareness of the benefits.

INCRETIN THERAPY: AN ALTERNATIVE

Insulin resistance and beta-cell pancreatic dysfunction are considered the two main physiopathological mechanisms underlying hyperglycemia in type 2 diabetes. Abnormal secretion of incretin hormones, mainly glucagon-like peptide (GLP-1), and of amylin from pancreatic beta-cell may contribute to the onset of type 2 diabetes mellitus.

Excessive secretion of glucagon, reflecting a subsequent beta-cell dysfunction and paradoxically increasing after oral administration of glucose or carbohydrate, associated with alteration in gastric emptying are additional factors that determine postprandial hyperglycemic excursions.

All these factors originate from genetic and/or environmental factors [18-20].

As we already mentioned, conventional treatment in type 2 diabetes does not significantly influence the progressive decline in pancreatic beta-cell function. Moreover, this approach has the disadvantage of weight gain, increasing cardiovascular risk and worsening the beta-cell insulinosecretor defect.

Administration of exogenous insulin and insulin secretagogues is associated with an increased risk of hypoglycemia, therefore limiting patient compliance in achieving glycemic targets.

Age and diabetes-related deterioration in renal and hepatic function may affect the pharmacokinetic and pharmacodynamic characteristics of antidiabetic and other associated treatments, especially in elderly people, which lead to differences in drug metabolism and increased risk of hypoglicemia and drug interaction.

Difficulties in achieving and maintaining good metabolic control with existing therapies (two-thirds of patients still do not meet current glycemic targets), with significant side effects to these therapies have led to the development of new therapeutic options: incretin-mimetics.

Incretin Effect

In the past 40 years interest continued to grow in pathophysiological link between intestinal hormones (incretin) and insulin secretion, because of the demonstrated role of these hormones in glycemic homeostasis in people without diabetes, due to their action in stimulating glucose-dependent secretion of postprandial insulin. Incretin name is an acronym derived from "**In**testinal **Secret**ion of **In**sulin".

The incretinic effect was first illustrated in the 1960s, when significantly increased insulin secretion response after oral glucose administration, compared to insulin response obtained after intravenous isoglicemic administration were observed (Fig. **1**).

This increased secretion of insulin has been attributed to glucose load in the gut and has been described as "incretin effect".

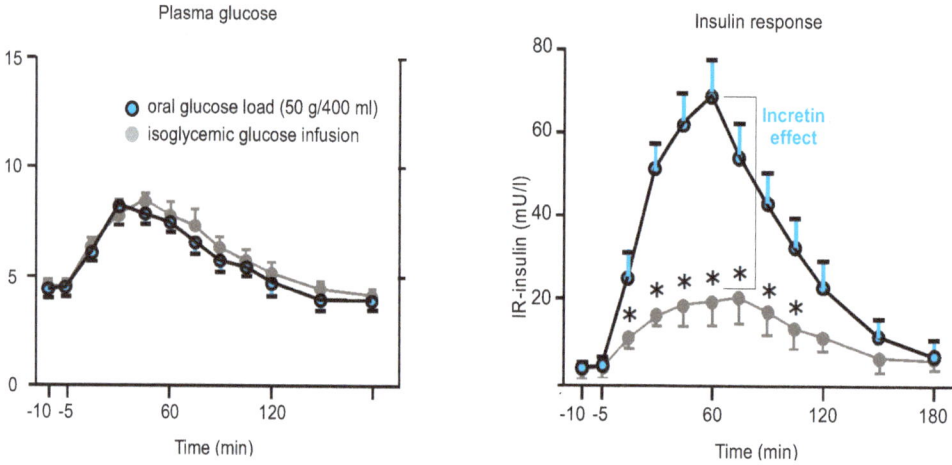

Figure 1: The incretin effect (Derived from [22]).

Subsequently, it was found that the incretin effect is due to the insulinotropic action of two intestinal hormones: glucagon like peptide (GLP-1) and glucose-dependent insulinotropic polypeptide (GIP), called incretin hormones [21-25].

Plasma levels of GLP-1 and GIP are minimal in the fasting state, but increase by 10-fold within minutes of eating [26-28].

In healthy subjects GLP-1 and GIP are released from L- and K-intestinal cells a few minutes after food intake and stimulate glucose-dependent insulin secretion, restoring postprandial glycemic control [26, 29] (Fig. **2**).

GLP-1 is produced by L-cells of the gastrointestinal tract distal (ileum and colon) as an active peptide, activation occurs as a result of posttranslational cleavage of six amino acids from the amino end [29, 30]. The active form of GLP-1 is the most common circulating form. The GLP-1 release is prompted by glucose only.

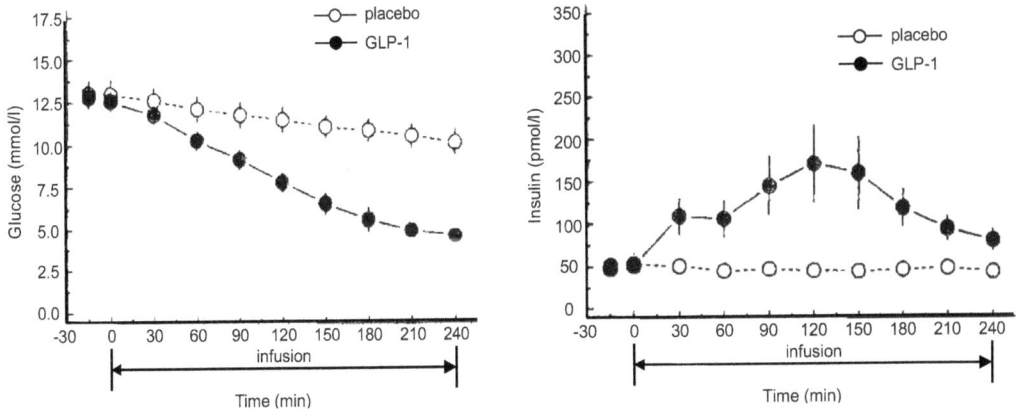

Figure 2: Effects of placebo or glucagon-like peptide-1 (GLP-1) infusion on glucose and insulin levels (Derived from [29]).

Approximately 70% of the insulin secretory response to oral glucose administration is due to incretin hormone action [25, 26]. This effect is secondary to GLP-1 interaction with specific receptors on pancreatic cell membrane. Binding to these receptors triggers an adenylate cyclase, with consequent accumulation of intracellular cAMP and activation of protein kinase A, increased intracellular calcium concentration, and consecutive stimulation of insulin secretion. The physiological cascade is strictly dependent on glucose and does not occur with low blood sugar values [31].

GIP is secreted by K-cells in the lining of the proximal intestine (duodenum and proximal jejunum) and demonstrated in both animal and human models, that is an important factor in postprandial glycemic homeostasis [21, 23, 25, 32-36]. Its release is stimulated by both fat and glucose ingestion [27].

GIP receptors are found in a large number of cells types, including A and B pancreatic cells, gastric cells, fat and brain cells, whereas GLP-1 receptors are located in the B and D pancreatic cells, gastric parietal cells, as well as lung and brain cells [26, 32, 37].

Physiologic Role of Incretin

GLP-1 has the following extrapancreatic effects, *via* its action on specific receptors outside the pancreas [26, 29, 30, 32, 38-42]:

- it inhibits (postprandial) glucose dependent glucagon secretion; note that GIP does not affect glucagon secretion. In patients with diabetes, glucagon levels may paradoxically rise after ingestion of a carbohydrate meal and enhance the production of glucose through gluconeogenesis. Consequently, not only are glucose level increased due to insulin deficiency, but also more glucose is being produced in these patients after a meal. Thus, there actually two mechanisms by which postprandial hyperglicemia occurs in patients with diabetes mellitus (Fig. **3**).

- it slows gastric emptying and contributes to lower postprandial glucose excursions;

- it increases satiety, decreases appetite and limits weight gain;

- it leads to cardioprotection, by improving post myocardial infarction left ventricular ejection fraction.

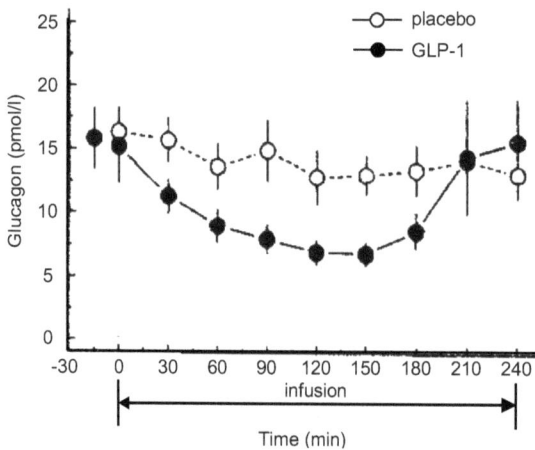

Figure 3: Effects of placebo or glucagon-like peptide-1 (GLP-1) infusion on glucagon level (Derived from [29]).

Several studies have shown that GLP-1 has trophic effects on primary cultures of beta-cell lines from animal models of type 2 diabetes. In pre-clinical studies it was demonstrated that GLP-1 promotes pancreatic beta-cell proliferation and prevents apoptosis [32, 38, 39, 43-45].

In addition, GLP-1 promotes cell differentiation. Therefore, insulin nonsecreting cells (*i.e.,* ductal exocrine cells or immature progenitors islanders) differentiate to cells able to synthesize and secrete insulin (*i.e.,* differentiated beta-cell phenotypes) [21, 46]. This effect is of particular importance, considering the fact that beta-cell mass and function are already significantly decreased at the time of diagnosis of T2DM. Therefore, after three years of treatment, approximately 50% of patients will require more than one drug to achieve glycemic control, while after nine years, 25% of patients can control the disease with only one drug [47]. Loss of glycemic control is due to the progressive loss of beta-cell function.

A summary of the clinical effects of the incretins is presented in Table **1**.

According to studies *in vitro* and *in vivo*, effects of GIP on pancreatic beta cell are the increased replication and survival [46].

In healthy human subjects, GIP and GLP-1 contribute equally to the incretin response when were secreted in physiological amounts [21].

Role of Incretin in Patients with T2DM

In people with type 2 diabetes, incretin effect is reduced or absent, most likely due to poor response to postprandial secretion of GLP-1 [23, 36]. Although the activity of GIP is also diminished, it is secreted in nearly normal amounts [20] (Fig. **4**). It is possible that GLP-1 deficiency plays a role in the progressive decline in function of pancreatic beta-cells in type 2 diabetes patients (Fig. **5**). Exogenous administration of GLP-1 in type 2 diabetes has a significant effect in lowering blood glucose, while GIP administration has very little or no effect on insulin secretion. Thus, GLP-1 is an attractive therapeutic alternative in the pharmacological treatment of type 2 diabetes.

The therapeutic value of GLP-1 in diabetes mellitus is limited by its short half-life. After secretion, GIP and GLP-1 are rapidly metabolized by dipeptidyl peptidase 4 (DPP-4), their half-lives are 5 and 2 minutes for GIP and GLP-1 respectively. Inactivation occurs at a rate of at least 50% in the intestinal capillaries, close to the release locus for GLP-1 and GIP, which limits the therapeutic use of these incretin hormones [3]. Peptidase is a ubiquitous enzyme, present in high concentrations in kidney cells, liver, intestine, and endothelial capillaries and in a soluble form in plasma; the enzyme is present in two forms: a surface, closely linked to the cell membrane, and other form with low molecular weight, which is present in the circulation. DPP-4 specific action is to split a number of chemokines and peptide hormones [28, 31, 35, 42, 45].

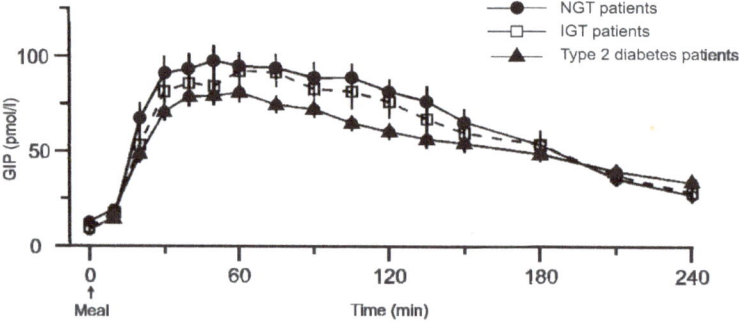

Figure 4: Plasma GIP concentration in type 2 diabetes mellitus (Derived from [20]).

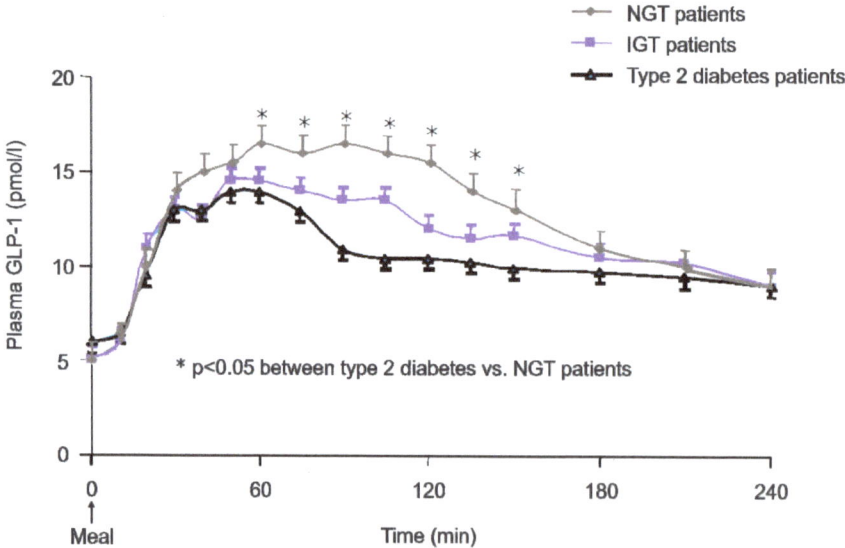

Figure 5: Plasma GLP-1 concentration in type 2 diabetes mellitus (Derived from [20]).

Development of Incretin-Related Therapies: Two Alternative Approaches

Continuous 24-hour infusion of pharmacological levels of GLP-1 would be required to obtain optimal, sustained glycemic control, and therefore, this is not a practical long-term solution [48]. To overcome this limitation, the potency of

GLP-1 may be modified in two ways: one way is the synthesis of peptides resistant to DPP-4 degradation and therefore have longer half-lives than GLP-1, so-called incretin-mimetics; another way is to use molecules that inhibit the action of DPP-4 degradation, named DPP-4 inhibitors, which increase plasma levels of endogenous GLP-1 (Fig. **6**) and GIP.

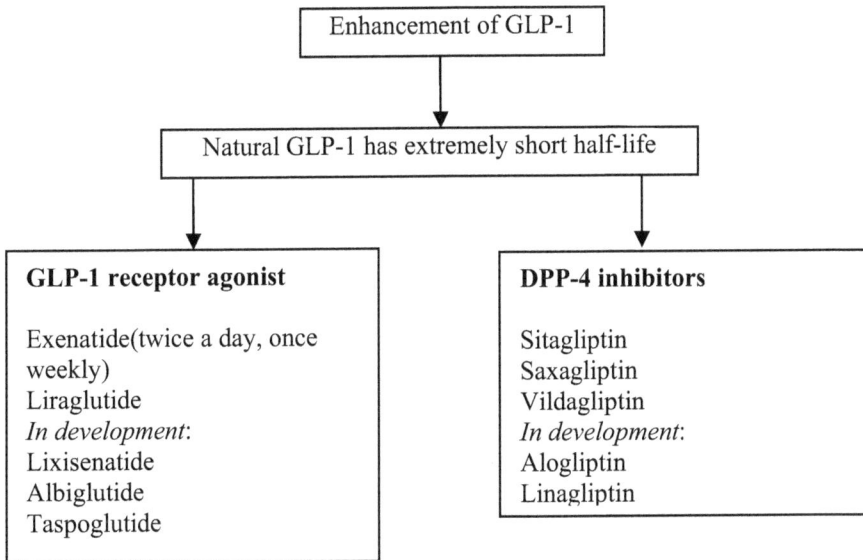

Enhancement of GLP-1

Natural GLP-1 has extremely short half-life

GLP-1 receptor agonist

Exenatide(twice a day, once weekly)
Liraglutide
In development:
Lixisenatide
Albiglutide
Taspoglutide

DPP-4 inhibitors

Sitagliptin
Saxagliptin
Vildagliptin
In development:
Alogliptin
Linagliptin

Figure 6: Incretin drug therapy.

Both classes of antidiabetic drugs demonstrate effectiveness with tolerability profiles, but do not produce identical effects on key parameters, such as degree of glycemic control and effect on body weight (Table **1**).

Losing weight is an important goal for obese individuals with T2DM as shown by Aylwin S and Al Zaman (2008) [49]: a weight loss of 5-10% was associated with a 0.5% decrease in HbA1c in patients with diabetes. Anderson JW and Konz EC (2001) [50] shown for every kilogram of weight loss, a reduction of 3,6 mg/dl in fasting plasma glucose levels was apparent in non-diabetic patients. Therefore, there are important benefits associated with GLP-1 receptor agonist therapies.

Both GLP-1 receptor agonist and DPP-4 inhibitors minimized the risk of hypoglycemia seen with some oral antidiabetic drugs and exogeneous insulin.

Both GLP-1 receptor agonist and DPP-4 inhibitors do not need dose titration relative to food intake or ambient blood glucose level.

Table 1: Overview of effects of GLP-1 agonists and DPP-4 inhibitors

General Characteristic/Action	GLP-1 Receptor Agonists	DPP-4 Inhibitors
Mode of administration	injected	orally
Known pharmacologic target	single GPCR target	GLP-1, another proteins (neuropeptides)
Incretin activity enhanced	GLP-1	GLP-1, GIP
Duration of activity	longer acting (daily to weeks)	shorter acting (hours)
Glucose–dependent insulin secretion	enhanced	enhanced
Glucose–dependent glucagon secretion	suppressed	suppressed
Hepatic glucose output	suppressed	suppressed
Gastric emptying	delays significantly	no effect
Food intake (Appetite)	suppressed	no effect
Satiety	induced	no effect
Weight loss	induced	no effect
Restore 1st-phase insulin	induced	induced
Postprandial hyperglycemia	reduced	reduced
Expected decrease from baseline HbA1c	0,8-1,89%	0,5-1,10%
Effect on beta-cell neogenesis/apoptosis	increased	increased
Antibody formation	variability	variability
Risk of hypoglycemia	low	low
Cardiovascular markers	improvement	minimal impact
Gastrointestinal side effects	yes	few, not clearly elucidated
Common adverse effects	nausea	headache, infection, dermatological effects

GLP-1= glucagon-like peptide-1, DPP-4= dipeptidyl-peptidase-4, GIP= glucose-dependent insulinotropic polypeptide, GPCR= G protein-coupled receptor, HbA1c= glycated hemoglobin.

DPP-4 inhibitors have the advantage of oral administration and excellent tolerability, but their efficacy is somewhat limited by their dependence on endogenous GLP-1 secretion. Therefore, DPP-4 inhibitors are most suitable in patients who preserve some beta-cell activity.

Diabetes is a cardiovascular risk factor and incretin-related therapies also promise therapeutic benefits with regard to the cardiovascular system. Potentially welcome effects have been demonstrated with GLP-1 and GLP-1 receptor agonist, including reductions in systolic blood pressure, plasminogen activator inhibitor-1 and B-type natriuretic peptide (BNP), improved vasodilatatory function, and protect against myocardial ischemia [41].

As a result of proven effectiveness in achieving glycemic control and safety management, incretins found their place in the list of recommended antihyperglycemic agents in pharmacotherapy of type 2 diabetes mellitus, as mentioned in the consensus American Diabetes Association/European Association for the Study of Diabetes (ADA/EASD) as well in American Association of Clinical Endocrinologists (AACE).

Additional data from randomized controlled trials are needed to establish the persistent of the glycemic response and the long-term impact on beta-cell function of incretin-based antidiabetic therapies alone and in combination with other therapies.

WHAT IS LIRAGLUTIDE?

This part of the chapter includes: general information about liraglutide; its structure and pharmacology; pertinent preclinical studies; early clinical studies; practical recommendations and side effects.

General Information

GLP-1 is a natural incretin hormone produced by L-cells in the distal small intestine and the colon as response to food intake [10, 15]. It determines several metabolic effects:

- on glucose control

  - glucose-dependent insulin release

  - increased insulin synthesis and secretion

  - increased beta-cell glucose sensitivity

- • glucose-dependent inhibition of glucagon secretion

- -   on food intake and body weight

- • reduced food intake

- • reduction in body weight due to increased satiety

- • delayed gastric emptying [15].

GLP-1 stimulates insulin secretion and suppresses glucagon secretion, but only when glucose levels are elevated.

These physiological effects of GLP-1 are mediated by the GLP-1 receptor (GLP-IR), which is found in pancreatic islets and other tissues of the body.GLP-1 cannot be a therapeutically viable agent for diabetes mellitus because of its rapid degradation by DPP-4 [51].

Liraglutide has 97% (36/37) amino acid sequence identity to native GLP-1: the lysine at position 34 is replaced with arginine. Additionally, a fatty acid moiety is linked at position 26 *via* a glutamoyl spacer.

These chemical modifications result in increased self-association with haptomer formation, albumin binding, and reduced susceptibility to DPP-4. The plasma half-life of liraglutide is approximately 13h, compared with 1.5min for native GLP-1 [10].

Such an important increase of the half-life means that therapeutic levels of liraglutide are sustained over 24 h. As a result of its kinetic properties, liraglutide can be given once daily without regard to mealtimes [4].

Victoza® is the commercial name for liraglutide launched by Novo Nordisk A/S. In Europe, it is indicated for treatment of adults with type 2 diabetes in combination with metformin plus sulfonylureas or metformin plus thiazolidinediones, when previous therapy does not bring about good metabolic control [51, 52]. In addition, the US Food and Drug Administration (FDA) has granted approval for its use as an adjunct to diet and exercise when metformin or sulfonylureas fail [51]. Recently, the addition of Detemir insulin (Levemir) to liraglutide treatment has also been approved [53].

Structure and Pharmacology

The amino acid structure of GLP-1 and liraglutide is presented in Fig. **7**.

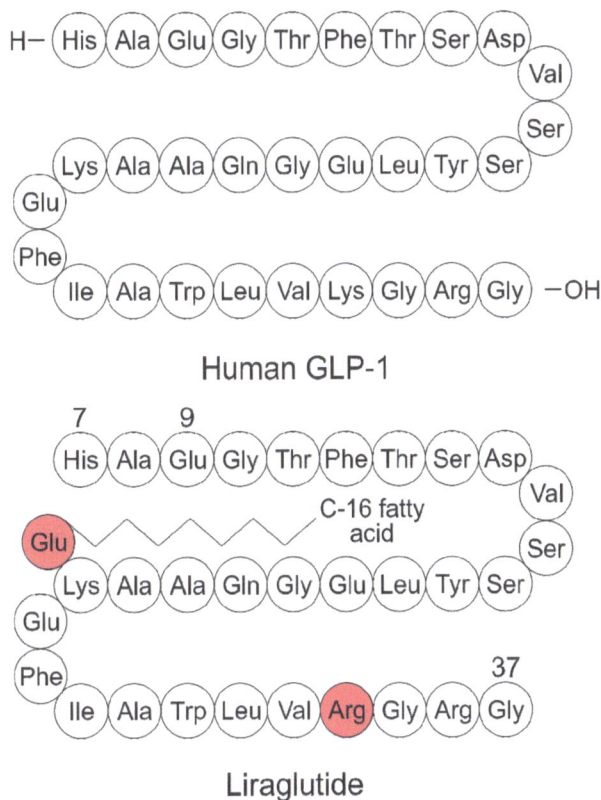

Figure 7: Structure of human glucagon-like peptide-1 (GLP-1) and liraglutide (Derived from [54]).

As we have already mentioned, their molecules differ in two aspects. First, a C16 fatty acid chain (palmitic acid) is attached to lysine at position 26 *via* a glutamic acid linker. Second, lysine is replaced by arginine at position 34. The fatty acid chain allows the binding of liraglutide to the albumin in the blood, prolonging the effect of liraglutide and increasing the resistance to degradation by the DPP-4 enzyme. This fatty acid chain also allows the self-association of liraglutide into heptamers at the injection site, delaying its absorption [54].

By comparison, exenatide, originally identified as an extract of Gila monster saliva, shares only 53% sequence identity with native GLP-1 [10].

Absorption of liraglutide is slow. It is completely metabolized within the body by DPP-4 and neutral endopeptidase (MEP) in a similar manner to GLP-1, but at a slower rate. The half-life of liraglutide was reported as 13 h, and this supports its once-daily administration. Liraglutide provides glucagon-like peptide-1 at high, pharmacological levels, and can restore the insulin secretion of pancreatic beta-cells to values similar to those of healthy subjects [10].

Preclinical Studies

The interesting therapeutic potential of liraglutide was revealed in animal models of diabetes mellitus and obesity. The experimental research demonstrated that utilization of liraglutide improves beta-cell function and glycemic control, reduces body weight and has a positive influence on the cardiovascular system [10, 54].

Beta-Cell Regulation. The action of liraglutide on beta-cell mass is difficult to evaluate directly in humans and the results of animal studies are very important for the understanding of this aspect [51]. Liraglutide significantly increases beta-cell mass in diabetic rats and significantly increases beta-cell mass and proliferation in diabetic mice [10]. On the other hand, liraglutide has a greater anti-apoptotic effect in rat islet cells than native GLP-1. Also, a reduced beta-cell apoptosis was noticed *in vitro*, using human pancreatic islet cells [55, 56].

Glycemic Control. Liraglutide presented strong, long-lasting dose and glucose-dependent antihyperglycemic effects in different animal models: *ob/ob* mice, *db/db* mice, younger Zucker diabetic fatty (ZDF) rats, older ZDF rats [54]. Investigations on the efficacy of liraglutide were also conducted in minipigs made diabetic with streptozotocin. Short -term studies confirmed that liraglutide has a glucose-dependent antihyperglycemic action, whereas long-term studies showed a reduction of the gastric emptying and an improvement of the glucose tolerance and insulin sensitivity [57].

Weight Loss. The effects of liraglutide on food intake and body weight were evaluated in normal rats and in a number of rat models. These investigations showed liraglutide reduced food intake, body weight, fat mass and glucose tolerance. The main mechanism of weight loss is a lowered energy intake [10, 51]. Similar studies were also conducted in minipigs. It was demonstrated that liraglutide reduced food intake and body weight in obese, hyperphagic minipigs [58].

Cardiovascular Function. Liraglutide has a positive influence on the cardiovascular system, increasing survival, reducing cardiac rupture and improving cardiac function in a mouse model of myocardial infarction [10, 59]. Liraglutide has anti-inflammatory properties in human vascular endothelial cells by increasing nitric oxide production and suppressing NF-kappa B activation [10].

Other Data. Liraglutide has been shown to delay the onset of diabetes in an obese rat model with susceptibility to this condition. It may also ameliorate the function of the nervous system and increase neuronogenesis. Liraglutide exposure was shown to cause C-cell hyperplasia and C-cell adenomas in rodents [10, 51].

Early Clinical Studies

Starting from encouraging results of the preclinical studies, the investigators decided to extend their research to humans. The pharmacokinetics and pharmacodynamics of liraglutide were evaluated in a total of 26 trials, including 19 trials in healthy subjects and 7 trials in patients with type 2 diabetes mellitus [10, 60-62]. Basically, it was confirmed that liraglutide has a good influence on the human beta-cells, and also on glycemic control and levels of insulin and glucagon [10].

Effects on Beta-Cell Function: The studies exploring the effects of liraglutide on beta-cell function demonstrate its ability to ameliorate and to maintain beta-cell function [15]. Early pharmacological studies in patients with type 2 diabetes mellitus showed that liraglutide increased beta-cell function in the fasting state by 30% compared with baseline [61]. The maximum beta-cell secretor capacity was significantly higher after liraglutide and the proinsulin: insulin ratio was reduced by 40-50%. In patients with type 2 diabetes, liraglutide increases the insulin secretion in a glucose-dependent manner, and improves the first and second phase of insulin response [10, 60, 61].

New data suggest that the actions of liraglutide on the beta-cell extend beyond its role as an insulin secretagogue [15, 56].

Main Effects on Circulating Glucose. As a consequence of the improvement of the insulin secretion in type 2 diabetes, liraglutide lowers fasting plasma glucose (FPG) concentrations with a small risk of hypoglycemia [15]. Additionally, liraglutide demonstrated the ability to reduce postprandial plasma glucose (PPG) excursions. This is an important aspect, because PPG may be an independent risk factor for cardiovascular disease, as well a significant contribution to raised HbA1c [61, 62]. Liraglutide has also been shown to reduce blood glucose levels during the night, even when administered in the morning [15].

Weight Loss. The improvements in glycemic control were associated with weight loss. Liraglutide delays the rate of gastric emptying, reduces energy intake and induces a moderate suppression of hunger [10].

Administration of liraglutide has positive effects on plasma glucagon secretion and does not impair the counter regulation glucagon response to hypoglycemia in individuals with type 2 diabetes mellitus [10].

Influence on Cardiovascular Risk Factors. Weight loss is accompanied by amelioration of other cardiovascular risk factors: PPG (as we already mentioned), blood pressure, triglycerides, cardiovascular risk biomarkers (plasminogen-activator inhibitor-1, B-type natriuretic peptide) [15].

Adverse Effects. The most common negative effects were gastrointestinal symptoms, predominantly a mild and transient nausea [15, 60].

In conclusion, these preliminary good results supported liraglutide as an adequate candidate for a comprehensive clinical development program [15].

Practical Recommendations

In June 2009, the EMA approved liraglutide (Victoza®) as a second- and third-line option for the therapy of type 2 diabetes in Europe [63]. Liraglutide was recommended in combination with:

- metformin or a sulfonylurea, in patients with poor glycemic control despite maximal tolerated dose of monotherapy with those drugs;

- metformin and a sulfonylurea or metformin and a thiazolidinedione, in patients with poor glycemic control despite dual therapy.

In January 2010, the FDA approved liraglutide for USA and indicated it as an adjunct to diet and physical exercise.

Liraglutide can be used:

- as monotherapy, although it is not recommend as first-line therapy;

- in combination with oral antidiabetic drugs.

In Japan liraglutide was approved in January 2010 and it was indicated:

- as monotheraphy;

- in combination with sulfonylureas.

As we can see, there are some differences regarding the prescribing regimen of liraglutide from a one region to another [10, 63]. The combination with insulin was only recently permitted [53].

The dosing regulation, as approved by the EMA, is based on the clinical development program. In clinical trials with liraglutide, patients were given a starting dose of 0.6 mg/day for a week, before increasing to 1.2 mg/day. In terms of clinical response and after at least one week, the dose can be increased to 1.8 mg/day (Fig. **8**). Daily doses higher than 1.8 mg are not recommended [61, 64].

Liraglutide is administrated once daily at any time, independent of meals, and can be injected subcutaneously in the abdomen, thigh, or upper arm. It should not be administered intravenously or intramuscularly [52].

Liraglutide is available in a prefilled pen device. The pen contains 3 ml of solution, delivering 30 doses of 0.6 mg, 15 doses of 1.2 mg, or 10 doses of 1.8 mg. These devices should be stored in a refrigerator (2-8°C) prior to the first use, although pen in use can be stored at room temperature for one month [63].

Liraglutide has no important drug-drug interactions.

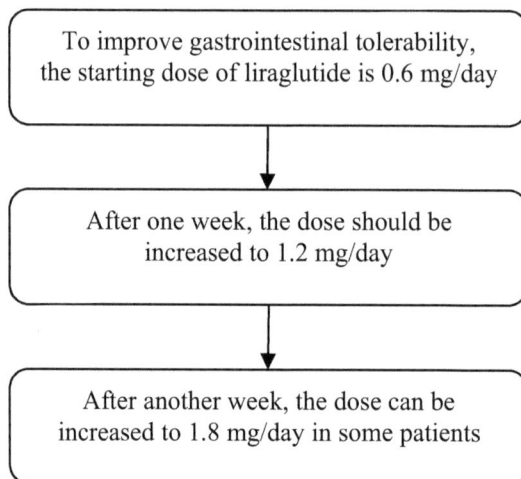

To improve gastrointestinal tolerability, the starting dose of liraglutide is 0.6 mg/day

↓

After one week, the dose should be increased to 1.2 mg/day

↓

After another week, the dose can be increased to 1.8 mg/day in some patients

Figure 8: Posology of liraglutide (Derived from [10]).

The use of liraglutide in special populations. Liraglutide should not be used during pregnancy and breastfeeding. Gender, race and ethnicity do not require special consideration when initiating liraglutide therapy. Age has no clinically relevant influence on the pharmacokinetics of this compound. Therapeutic experience with liraglutide in patients with hepatic impairment is too limited to recommend its utilization in such patients. Similarly, the limited experience in patients with renal insufficiency means liraglutide administration is not recommended in such situations [10, 52, 63].

Side Effects

As liraglutide is a peptide, it has the potential to be immunogenic. Clinical trials showed that 8.6% of patients developed antibodies to liraglutide, but this phenomenon was not associated with a reduction of the efficacy [63, 64].

The most frequently reported side effects are nausea and diarrhoea. Less common reactions include: headache, vomiting, dyspepsia, upper abdominal pain, gastritis, constipation, flatulence, abdominal distention, gastroesophageal reflux, dizziness, bronchitis, decreased appetite [52]. Major hypoglycemic episodes are rare. However, patients receiving liraglutide in combination with sulfonylurea may have an increased risk of hypoglycemia. Few cases of acute pancreatitis have also been reported in the long-term treatment with liraglutide [52].

Since liraglutide induces thyroid C-cell tumors in rodents and the importance of this finding for humans could not be determined, in the USA liraglutide is contraindicated in patients with a history of medullary thyroid carcinoma (HTC) or multiple endocrine neoplasm syndrome type 2 (MEN-2) [63].

CLINICAL EVALUATION OF LIRAGLUTIDE

Liraglutide is a relatively new addition to the range of treatments for type 2 diabetes, having been approved in Europe in mid-2009 and in the United States and Japan separately in early 2010.

The Novo Nordisk company sponsored a program including randomized, controlled, multicenter, multinational trials, named Liraglutide Effect and Action in Diabetes (LEAD), which examined the efficacy and safety of once-daily liraglutide treatment in monotherapy or in combination with commonly oral antidiabetic therapy.

Overall, more than 4400 people with type 2 diabetes from 40 countries were recruited, and 2739 patients were treated with liraglutide at doses of 0.6, 1.2, and 1.8 mg. The study population was similar global except for the duration of diabetes, which varied from shorter to longer across the progression of type 2 diabetes from monotherapy to combined oral agent therapy, respectively. Trial durations ranged from 26 to 52 weeks, and several trials have extension phases ranging from 1 to 2 years. The primary outcome in all trials was change in glycated hemoglobin (HbA1c) levels. Additional parameters included change in fasting plasma glucose (FPG) and postprandial plasma glucose (PPG), beta-cell function, weight and systolic/diastolic blood pressure (SBP).

The LEAD trials are the largest development program ever conducted by Novo Nordisk (Fig. **9**).

LEAD-1 and LEAD-2 were double-blind and double-dummy study, and examined the association of liraglutide to sulphonylurea or metformin monotherapy.

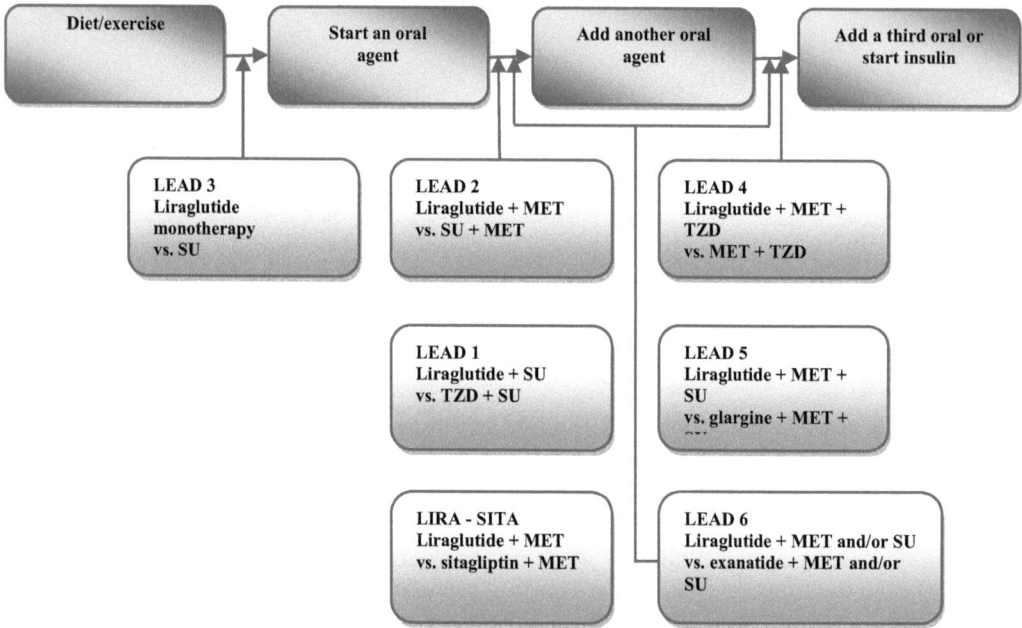

Figure 9: The LEAD program.

The LEAD-3 trial, which examined liraglutide monotherapy, was double-blind and double-dummy study, but did not include a placebo group.

LEAD-4 was double-blind and placebo-controlled study which examined the addition of liraglutide to metformin and TZDs, and it did not include an active comparator group.

In LEAD-5, which examined the addition of liraglutide to metformin and sulphonylurea (background therapies), the liraglutide and liraglutide – placebo arms were blinded, but the insulin glargin arm was open-label because of the need to titrate the insulin dose.

In LEAD-6 patients inadequately controlled on maximal doses of metformin, sulphonylurea or both, which received either liraglutide or exenatide administered open-label were included.

The active comparators in the LEAD trials were rosiglitazone (LEAD-1), glimepiride (LEAD-1 and LEAD-2), insulin glargin (LEAD-5) and exenatide(LEAD-6) (show Table **2**).

Table 2: Overview of LEAD and Lira-DPP-4i trials

Trial	Length	Background Therapy	Comparator Therapy
LEAD 1	26 weeks	Glimepiride 2-4 mg/day	Lira 0.6 mg
			Lira 1.2 mg
			Lira 1.8 mg
			Placebo
			Rosiglitazone 4mg/day
LEAD 2	26 weeks	Metformin 1 g BID	Lira 0.6 mg
			Lira 1.2 mg
			Lira 1.8 mg
			Placebo
			Glimepiride 2-4 mg/day
LEAD 3	52 weeks	None	Lira 1.2 mg
			Lira 1.8 mg
			Glimepiride 2-4 mg/day
LEAD 4	26 weeks	Metformin 1 g BID + Rosiglitazone 4mg/day	Lira 1.2 mg
			Lira 1.8 mg
			Placebo
LEAD 5	26 weeks	Metformin 1 g BID + Glimepiride 2-4 mg/day	Lira 1.8 mg
			Placebo
			Insulin glargine 24 UI/day
LEAD 6	26 weeks	Metformin 1 g BID and/or Glimepiride 2-4 mg/day	Lira 1.8 mg
			Exen 10 mg
Lira *vs.* DPP-4	26 weeks	Metformin 1 g BID	Lira 1.8 mg
			Lira 1.2 mg
			Sita 100 mg

Another trial, part of the phase 3b clinical program, was Lira-DPP-4i which compared the effect of liraglutide with that of sitagliptin, when added to metformin background therapy. The Lira-DPP-4i study is the first direct comparison of liraglutide with a DPP-4 inhibitor [65].

In this issue, we shall summarize the key results from the LEAD program, focusing primarily on glycemic control.

Parameters of Metabolic Control

HbA1c

Across the LEAD 1-5 trials, liraglutide treatment, both as monotherapy and in combination with oral glucose-lowering therapies, provided substantial reductions in HbA1c (the primary outcome) ranging from 0.84% to 1.5% compared with reductions of between 0.4% and 1.1% observed for active comparators [66-70]. The except was LEAD-2 trial, where the reduction in HbA1c were the same 1.0% for liraglutide 1.2 mg and 1.8 mg and glimepiride 4 mg; for patients of this trial who received liraglutide as an add-on to a pre-existing monotherapy [66-70]. HbA1c reductions were greater than in patients who substituted an existing oral therapy with liraglutide (1.3% *vs.* 0.8% for an 1.8 mg liraglutide dose).

Liraglutide treatment in monotherapy reduced HbA1c by 0.84% and 1.14% with two higher doses (1.2 and 1.8 mg) compared with baseline, *vs.* 0.51% with glimepiride in the overall study population LEAD-3 (Fig. **10**). The significantly reduction in HbA1c in LEAD-3 trial was demonstrated with 1.8 mg liraglutide compared with liraglutide 1.2 mg [66].

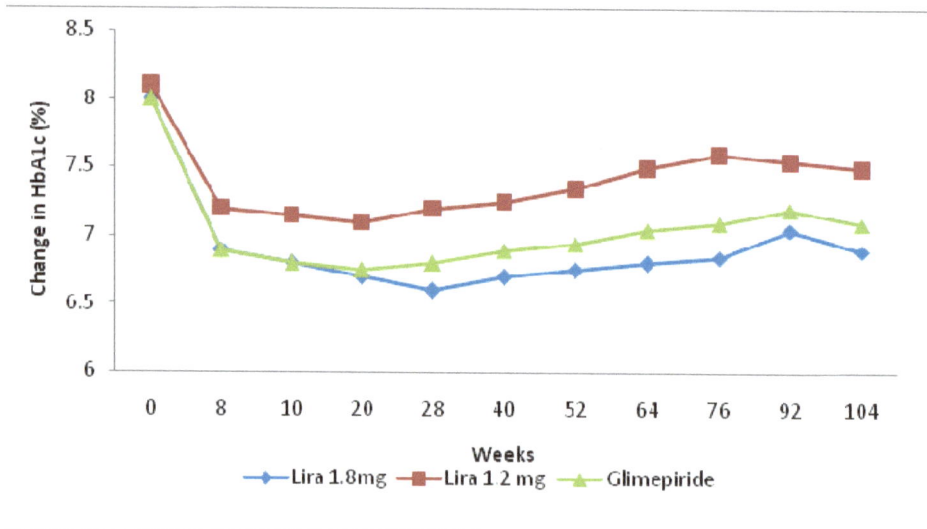

Figure 10: Change in HbA1c from baseline in LEAD-3 study (Derived from [10]).

The reductions in HbA1c in the LEAD 1-5 trials occurred within the first 8-12 weeks after initiation of liraglutide treatment and these reductions were sustained for the remainder of the trials. Thus, at the end of a one-year extension of LEAD-3, HbA1c was reduced at 104 weeks to a significantly greater extent by 1.2 and 1.8 mg liraglutide *vs.* glimepiride: -0.9% and -1.1% *vs.* -0.6% (p=0.037 and p=0.001) [71].

In the LEAD-6 head–to-head study of liraglutide 1.8 mg/d *vs.* exenatide 10μg twice daily, added to background therapy of metformin and/or sulphonylurea, liraglutide was more effective in decreasing HbA1c (-1.12% *vs.* -0.79%, p<0.0001), and both drugs brought about similar weight loss [72]. In a 14–week extension phase of LEAD-6 trial, the HbA1c decreased with 0.3% (p<0.0001) between week 26-week 40 in the patients who switched from exenatide to liraglutide [73] (Fig. **11**).

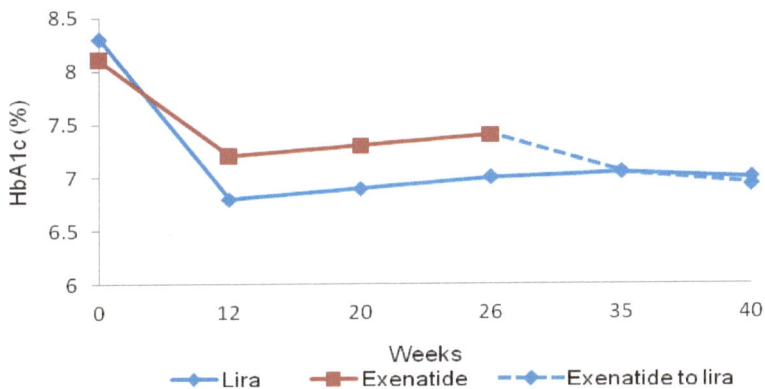

Figure 11: Change in HbA1c during LEAD-6 and LEAD-6 extension period (Derived from [73]).

In Lira-DPP-4i study liraglutide reduced HbA1c by -1.2 and -1.5% (p <0.0001) from baseline, *vs.* -0.9% (p<0.0001) with sitagliptin when added onto to existing metformin treatment [65].

Intermediate results from the MET-Lira-DET trial show a reductions in HbA1c with -1.3% in an 61% of patients treated with combination of liraglutide and metformin during a 12-week run-in period [74].

Importantly, it was shown that across the LEAD trials the reductions in HbA1c with liraglutide were independent of weight loss [75].

The majority of patients treated with liraglutide across the LEAD 1-5 trials reached the American Diabetes Association (ADA) and American Association of Clinical Endocrinologist (AACE) HbA1c targets ≤7% and ≤6.5%, respectively, compared with active comparators (except LEAD-2) [66-70, 72] (Fig. **12**).

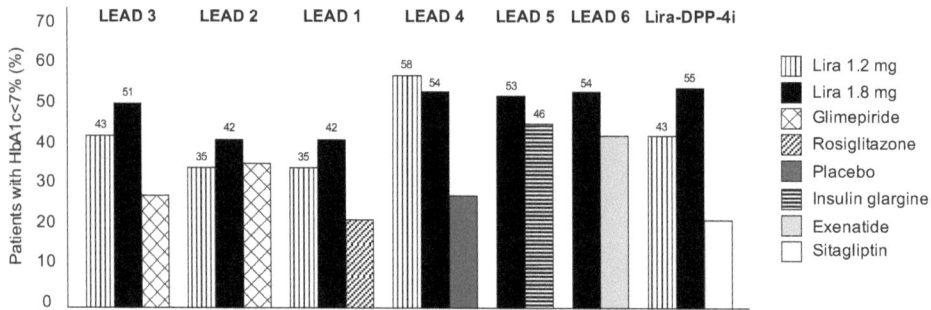

Figure 12: Proportion of patients achieving HbA1c <7% in LEAD program and in Lira-DPP-4i study (Derived from [10]).

The proportion of patients attaining the target of HbA1c <7% at 26 weeks was significantly higher for liraglutide at a dose of 1.8 mg (65%) *vs.* liraglutide 1.2 mg (56%), comparator treatments (30-53%) and placebo (17%)(p<0.05 for all). At the end of LEAD-3 trials (52 weeks duration) the proportion of the drug-naïve patients who achieved the HbA1c target <7% were 58.3% for liraglutide 1.2 mg, 62% for liraglutide 1.8 mg and 30.8% for glimepiride [66]. After a 1 year LEAD-3 extension period, 58% and 53% of the patients treated with liraglutide 1.8 mg and 1.2 mg respectively, achieved HbA1c <7%, *vs.* 37% with glimepiride [71] (Fig. **13**).

In the LEAD-6 and 14-week extension of the LEAD-6 trials, the proportion of patients achieving HbA1c <7% was significantly higher in the liraglutide group than in the exenatide group (54% *vs.* 43%, respectively; p=0.001).

In the Lira-DPP-4i trial a greater proportion of patients obtained ADA HbA1c targets <7% with liraglutide therapy (54.6% with 1.8 mg and 43.4% with 1.2 mg) *vs.* sitagliptin therapy (22.4%) (p<0.0001) [76]. After an supplementary 26-week extension period of this trial, 63.3% and 50.3% of patients treated with liraglutide 1.8 mg and 1.2 mg, respectively, reached HbA1c <7%, compared with 27.1% with sitagliptin treatment (p<0.0001) [76].

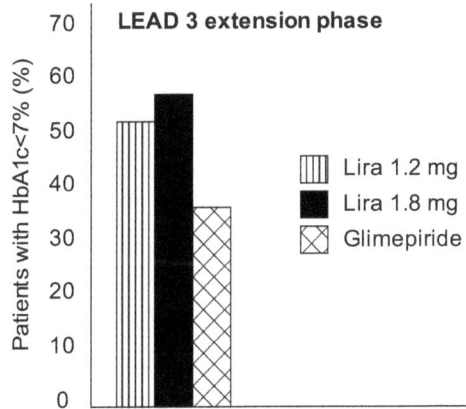

Figure 13: Proportion of patients achieving HbA1c <7% after 1 year LEAD-3 extension phase (Derived from [10]).

Overall, the endpoint of HbA1c <7% without weight gain or hypoglycemia was achieved in 40% and 32% of patients treated with liraglutide 1.8 mg and 1.2 mg, respectively (Table **3**). This result was statistical significantly greater than with any comparator therapy (6-25%, all p<0.01). The proportion of patients treated with liraglutide 1.8 mg which achieved HbA1c <7% and no weight gain was 49.6%, and 50.5% for HbA1c <7% and no hypoglycemia (Table **3**).

Table 3: Proportion of patients achieving HbA1c <7% treated with liraglutide or comparator

Drug/Dose	HbA$_{1c}$ <7.0 %	HbA$_{1c}$ <7.0 %, no Weight Gain, no Hypoglycemia	HbA$_{1c}$ <7.0 %, no Weight Gain	HbA$_{1c}$ <7.0 %, no Hypoglycemia
Liraglutide 1.8 mg	65%	40%	50%	51%
Liraglutide 1.2 mg	56%	32%	39%	43%
Sulphonylurea	48%	8%	15%	33%
Thiazolidinedione	34%	6%	9%	23%
Glargine	53%	15%	16%	48%
Exenatide	45%	25%	32%	34%
Sitagliptin	30%	11%	17%	21%
Placebo	18%	8%	11%	12%

Fasting Plasma Glucose (FPG)

Liraglutide treatment reduced FPG levels, which also contributes to improvements in overall glycemic control [65-70, 72].

Across the LEAD 1-5 trials, liraglutide in monotherapy and in combination with oral glucose-lowering agent reduced FPG from baseline by 13 to 43 mg/dl. In comparison, FPG levels decreased between 5.4 mg/dl (0.8 mmol/l) and 32.4 mg/dl (1.8 mmol/l) following treatment with active comparators [66-70] (Fig. **14**).

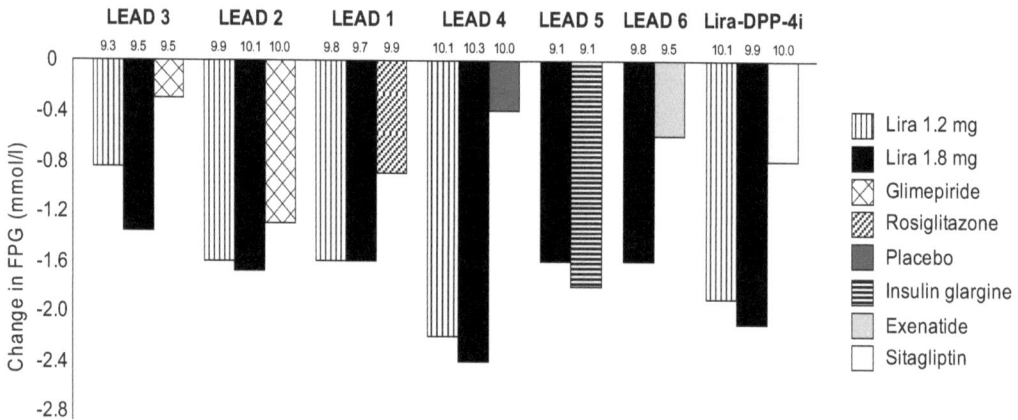

Figure 14: Change in fasting plasma glucose (FPG) from baseline in LEAD program and Lira-DPP-4i (Derived from [10]).

In the LEAD-6 trial, liraglutide reduced FPG from baseline by 29 mg/dl, compared with 10.8 mg/dl with exenatide treatment (p<0.0001) [72] (Fig. **14**). Patients who switched from exenatide to liraglutide in the 14-weeks LEAD-6 extension experienced a additional reduction in FPG of 16.2 mg/dl (0.9 mmol/l), by week 40 relative to week 26 [73].

FPG was decreased significantly more from baseline with liraglutide therapy than with sitagliptin, when added to preexisting metformin treatment in Lira-DPP-4i study (-1.9 mmol/l with 1.2 mg and -2.1 mmol/l with 1.8 mg liraglutide, *vs.* -0.8 mmol/l with sitagliptin) [65]. After 52 weeks, in liraglutide group reduction in FPG remained greater than in sitagliptin group (-1.7 mmol/l with 1.2 mg and -2 mmol/l with 1.8 mg, *vs.* 0.6 mmol/l with sitagliptin, both p<0.0001) [76] (Fig. **14**).

Postprandial Plasma Glucose (PPG)

In addition to lowering FPG, liraglutide has demonstrated the ability to reduce PPG excursions. This effect is a key finding with major clinical implications as PPG may be an independent risk factor for cardiovascular disease, as well as an important contributor to raised HbA1c.

Across LEAD 1-5 trials a mean PPG reduction over three meals was observed with liraglutide as monotherapy, and in combination with oral glucose-lowering drugs [66-70, 72] (Fig. **15**).

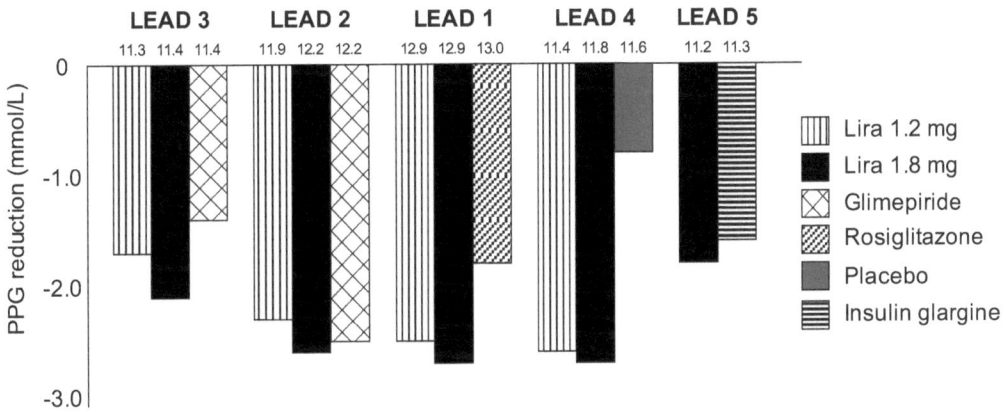

Figure 15: Mean postprandial glucose (PPG) reduction over three meals in LEAD 1-5 trials (Derived from [10]).

In LEAD-6 study, exenatide reduced PPG increment more than liraglutide after breakfast (3.9 *vs.* 3.2 mmol/l, p=0.012) and dinner (3.6 *vs.* 3.1 mmol/l, p=0.038) [72]. However, treatment differences after lunch were not significant. An explanation for these PPG differences was the different injection schedules: liraglutide was injected once daily, at the same time each day, while the exenatide was administered twice daily, before the morning and evening meals [72].

Beta-cell Function

One of the most encouraging features of incretin-based therapies is their potential to help preserve β-cell function. Apoptotic β-cell loss appears to play a central role in the development of insulin deficiency and the onset and progression of type 2 diabetes mellitus. The β-cell function is reduced by up to 50% by the time type 2 diabetes is diagnosed [77, 78].

Liraglutide treatments are associated with improvements in parameters of β-cell function such as the proinsulin: insulin ratio, proinsulin: C peptide ratio and homeostasis model of assessment (HOMA-B). Across all LEAD 1-6 trials an improvement of between 20-44% from baseline in HOMA-B was observed in the liraglutide groups [66-70, 72]. In LEAD-1 study, following treatments with liraglutide 1.2 and 1.8 mg, both in combination with sulphonylurea, substantial increases in HOMA-B of 28 and 35%, respectively, were observed [67] (Fig. **16**). These improvements were greater than those reported for rosiglitazone (+13%, p<0.05) [68].

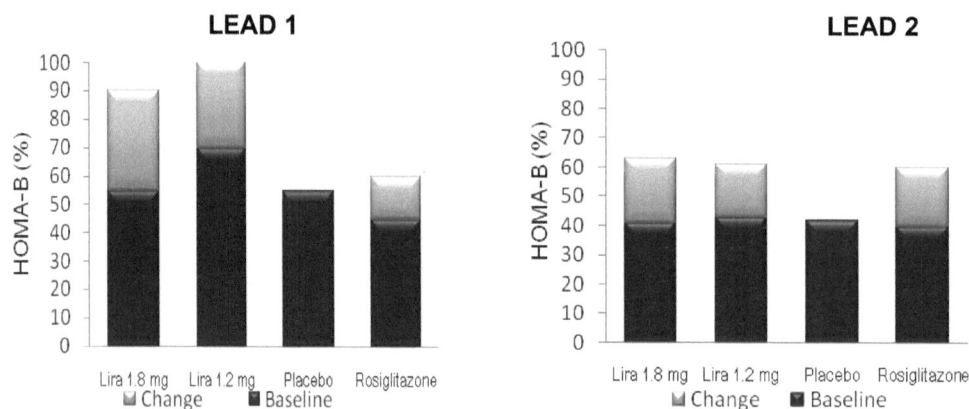

Figure 16: Change in HOMA-B function from baseline in LEAD-1 study and LEAD-2 study (Derived from [67, 68]).

More, were changes in the proinsulin: insulin ratio with ranged between -0.11 and 0.01 across the LEAD trials.

In the LEAD-6 trial, HOMA-B was reduced from the baseline to a greater extent with 1.8 mg liraglutide (32%) than with 10 µg exenatide (3%) after the addition of each agent to metformin and/or a sulfonylurea. No differences in the proinsulin: insulin ratio and fasting C-peptide were observed [72, 73, 79].

In Lira-DPP-4i study, compared with sitagliptin, 1.2 and 1.8 mg liraglutide therapy improvement HOMA-B levels and proinsulin: insulin ratio from the baseline [65, 76] (Fig. **17**).

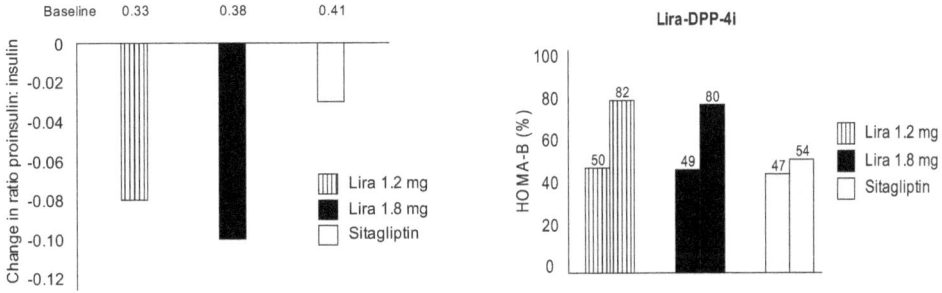

Figure 17: Change in proinsulin:insulin ratio and in HOMA-B function in Lira-DPP-4i study (Derived from [10]).

Body Weight Reduction

In contrast to the weight gain seen with the standard treatments for diabetes and weight neutrality of DPP-4 inhibitors, liraglutide has consistently shown a beneficial effect on body weight. The LEAD trials showed that liraglutide 1.8 mg/day reduced weight by 1.8-3.2 kg from baseline to 26 weeks and by 2.4 kg from baseline to 52 weeks [66-70, 72] (Table **4**).

Table 4: Change from baseline weight in LEAD program

Change from Baseline Weight (kg)	Lira 1.2 mg	Lira 1.8 mg	Comparator	Placebo
LEAD 3	-2.0	-2.4	1.1	NA
LEAD 2	-2.6	-2.8	1.0	-1.5
LEAD 1	0.3	-0.2	2.1	-0.1
LEAD 4	-1.0	-2.0	NA	0.6
LEAD 5	NA	-1.8	1.6	-0.4
LEAD 6	NA	-3.2	-2.9	NA

In the LEAD-1 trial, liraglutide 1.2 mg and 1.8 mg showed less weight benefit when associated with sulphonylurea (+0.3 kg and -0.2 kg, respectively), although both liraglutide doses were better than rosiglitazone therapy (+2.1 kg, p<0.0001) [66-70, 72].

In the LEAD-3 trial, liraglutide 1.2 and 1.8 mg used as monotherapy reduced body weight on average by -2.05 kg and -2.45 kg, respectively, compared with glimepiride group where there was a mean weight increase of 1.12 kg [66].

Weight loss for both doses of liraglutide and weight gain for glimepiride was maintained at the end of the LEAD-3 extension period (-2.1 kg and -2.7 kg *vs.* +1.1 kg) [71].

In LEAD-6 trial which compared the effects of liraglutide 1.8 mg with those of exenatide 10 µg, when added to background therapy, both drugs promoted similar reductions in body weight (-3.24 kg *vs.* -2.87 kg) [72]. In a 14-week extension to the LEAD-6 study, in patients who switched from exenatide to liraglutide, body weight reduced by an additional 0.9 kg, whereas those who continued with liraglutide lost an additional 0.4 kg [73].

Liraglutide treatment was associated with significantly greater weight loss from baseline compared with sitagliptin treatment in the Lira-DPP-4i study. The estimated mean weight reductions in favor of liraglutide over sitagliptin were -1.9 kg for the 1.2 mg dose and -2.42 kg for the 1.8 mg dose [65].

Dual energy X-ray absortiometry, used in a sub-study of the LEAD-2 and LEAD-3 trials, demonstrated that liraglutide treatment, either in monotherapy or in combination with metformin, generally reduced fat mass more than lean body mass (Fig. **18**).

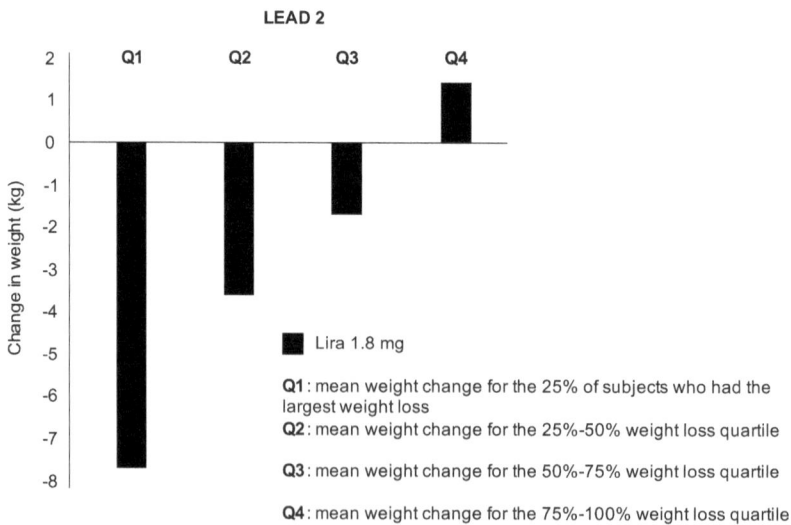

Figure 18: Mean weight loss for patients in each quartile (Q) of weight loss in LEAD-2 trial (Derived from [10]).

Computerized tomography (CT) was used to evaluate visceral and abdominal subcutaneous adipose tissue. In LEAD-2 trial, CT showed that, compared with combination glimepiride/metformin, liraglutide added to metformin therapy was associated with significantly greater reductions in visceral fat (16-17%) and subcutaneous fat (8-9%), although differences between liraglutide and placebo were not significant [80] (Fig. **19**).

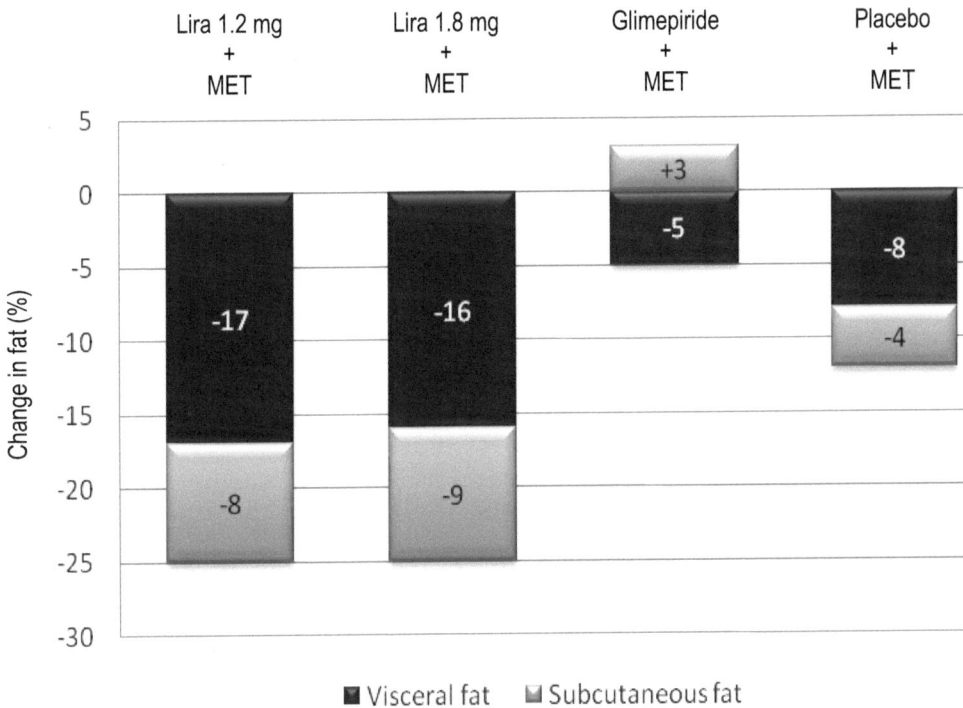

Figure 19: Change in visceral and subcutaneous adipose tissue from baseline in the LEAD-2 study (Derived from [80]).

An analysis of baseline body mass index (BMI) data in LEAD 1-5 trials showed that liraglutide resulted in weight loss in all BMI quartiles, and that the greatest weight loss was reported in patients who were the most overweight [81].

Cardiovascular Risk Biomarkers

The beneficial weight loss seen with liraglutide was associated with improvements in other cardiovascular risk factors, including blood pressure, lipid profile, brain natriuretic peptide (BNP), and high-sensitivity C-reactive protein.

Sistolic Blood Pressure (SBP)

Across the LEAD 1-5 trials, liraglutide therapy has consistently led to clinically relevant reductions of 2.1 to 6.7 mmHg in SBP [70, 72] (Fig. **20**). For example, in LEAD-4 trial, following treatment with liraglutide 1.2 or 1.8 mg added to metformin and rosiglitazone, important reductions in SBP of 6.7 and 5.6 mmHg, respectively, were observed. These reductions were significantly greater than reduction in SBP observed in placebo group (1.1 mmHg) [70].

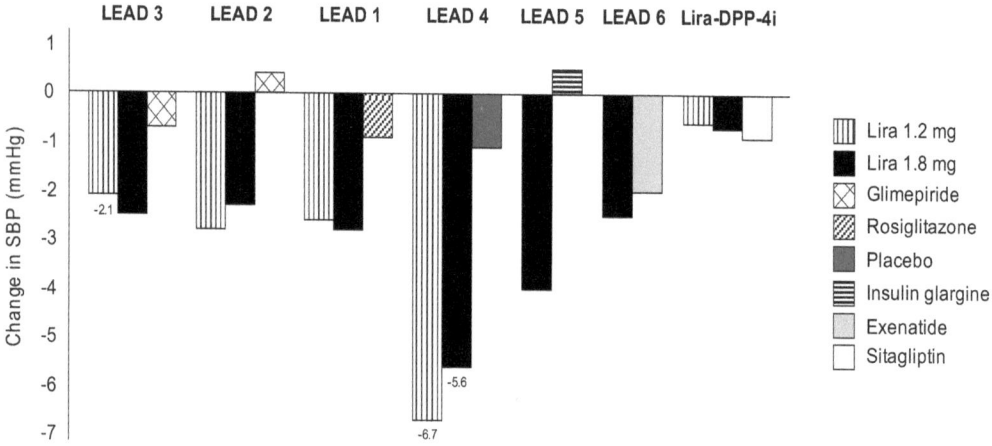

Figure 20: Change in systolic blood pressure (SBP) from baseline in LEAD program (Derived from [10]).

In LEAD-6 study, equivalent reductions in SBP were shown with liraglutide and exenatide (-2.5 and -2.0 mmHg, respectively). During LEAD-6 extension period, further significant reduction in SBP were observed in patients who switched from exenatide to liraglutide (-3.8 mmHg, p<0.0001) and in patients who continued with liraglutide (-2.2 mmHg, p=0.012).

It is important to note that reduction in SBP was before any major effect on body weight became evident [82].

Moreover, the reduction in SBP occurred immediately after initiation of liraglutide treatment, and was maintained for 26 weeks; liraglutide reduce SBP to a greater extent in patients with a high baseline SBP (Fig. **21**) and this effect was independent of concomitant antihypertensive treatment [82, 83].

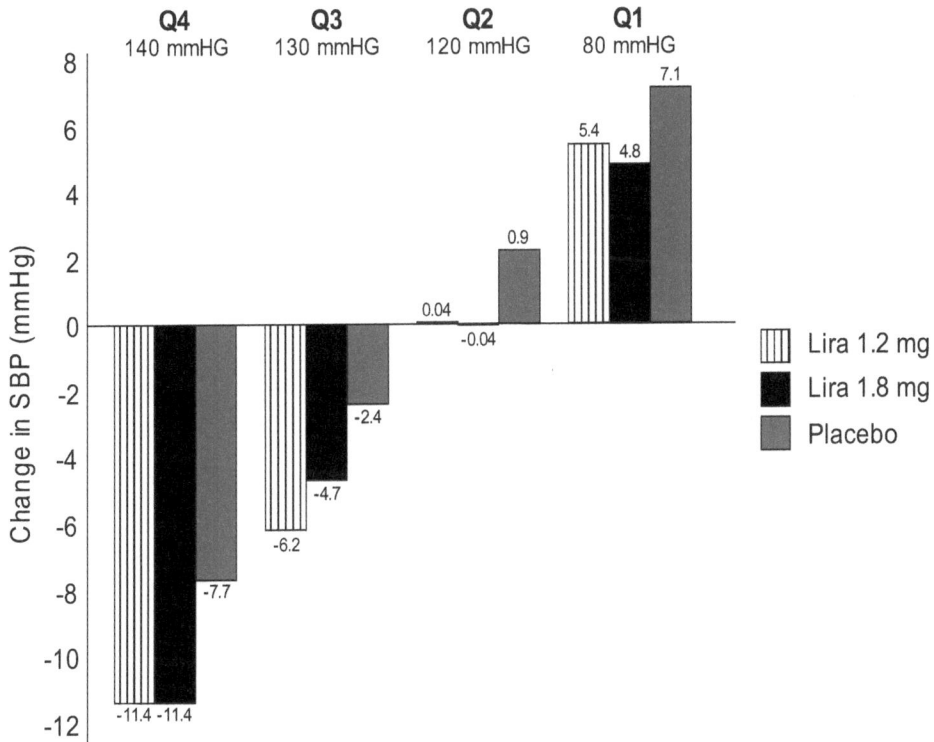

Figure 21: Effect of liraglutide on SBP by baseline SBP quartile (Q) (Derived from [10]).

Across the LEAD trials, treatment with liraglutide did not appear to have a significant effect on diastolic blood pressure (DBP).

Of note, in the LEAD trials, liraglutide treatment was associated with minor increases in pulse rate of 2-4 beats/minute compared with baseline.

Lipids and Other Markers of Cardiovascular Risk

In addition to the SBP reductions, liraglutide also appears to significantly improve levels of total cholesterol (TC), low-density lipoprotein cholesterol (LDL), free fatty acids and triglycerides compared with comparators [84] (Fig. **22**).

For example, liraglutide reduced levels of brain natriuretic peptide (BNP) by -11.9% and high-sensitivity C-reactive protein (hsCRP) by -23.1% compared with baseline, two markers of cardiovascular risk [84] (Fig. **23**).

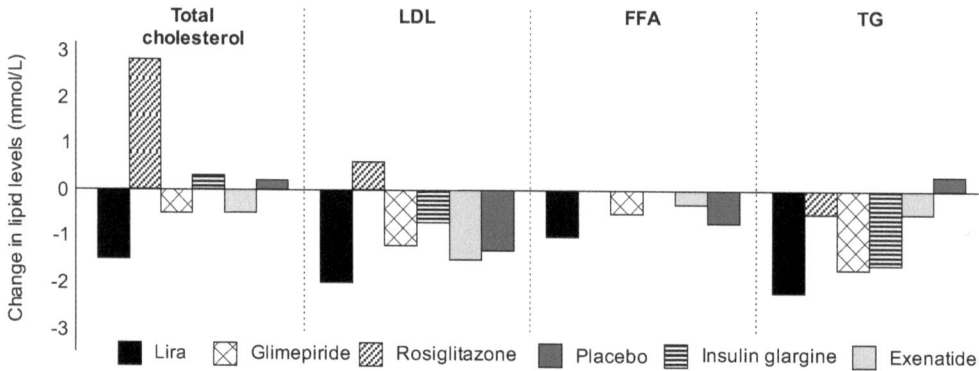

Figure 22: Effect of liraglutide and comparators on lipid levels from baseline (Derived from [10]).

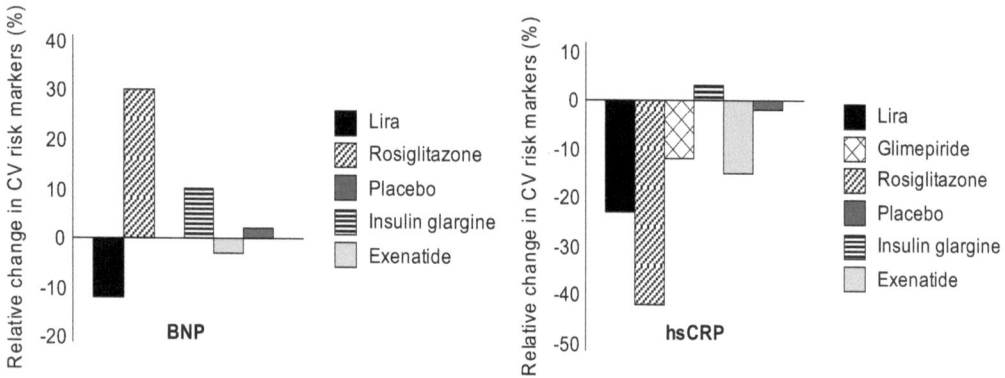

Figure 23: Effect of liraglutide and comparators on BNP and hsCRP from baseline (Derived from [10]).

Safety and Tolerability

Hypoglycemia

An overall low risk of hypoglycemia was observed for the patients treated with liraglutide across the LEAD 1-5 studies [66-70]. This contrasts with the pharmacological properties of other insulin secretagogues, and draw attention to this advantage of liraglutide as a therapeutic agent.

When liraglutide was used in combination with sulphonylurea an increased risk in hypoglycemia was observed (1.1 hypoglycemic episode/patient/year) [69]. On the other hand, liraglutide can be used safety in combination with metformin because a very low rate of hypoglycemia has been reported (0.03 - 0.09 hypoglycemic episode/patient/year) [68] (Fig. **24**).

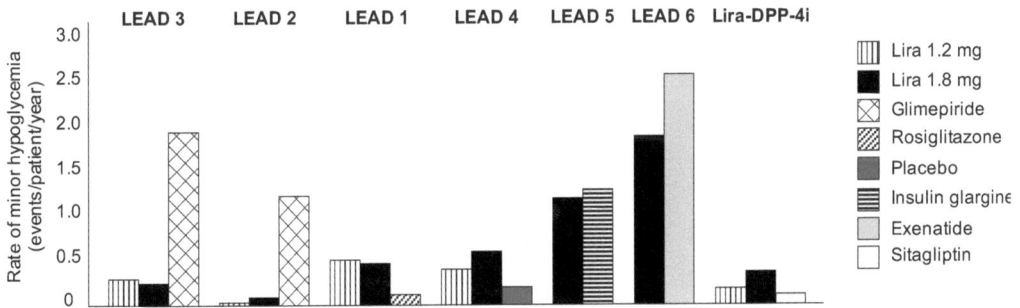

Figure 24: Rate of minor hypoglycemic events in LEAD program (Derived from [10]).

Major hypoglycemic episodes (defined as those requiring third-party assistance or medical intervention) were rare.

In LEAD-1 trial, one patient treated with liraglutide 1.8 mg in combination with glimepiride experienced a major hypoglycemic episode, whereas a total of six major hypoglycemic events were reported in five patients treated with liraglutide in LEAD-5 trial [67, 69]. All patients were receiving glimepiride treatment.

In LEAD-6 study, no major hypoglycemic events occurred in the liraglutide group, but two patients in the exenatide group experienced one hypoglycemic event each, both patients were receiving sulphonylurea [72]. In terms of minor hypoglycemic events, in LEAD-6 trial the liraglutide group experienced significantly lower rates of these episode compared with the exenatide group (1.93 *vs.* 2.6 episode/patient/year) [72]. A switch from exenatide to liraglutide in the extension of LEAD-6 trial was associated with reduced rate of minor hypoglycemia and one major hypoglycemic event [73].

Tolerability

Across the LEAD program, liraglutide was generally well tolerated. The most common adverse events associated with liraglutide treatment were gastrointestinal in nature (in particular, nausea and diarrhoea), but these adverse events were mild and transient and decreased over time [85]. Thus, the majority of episodes of nausea were transient and the proportion of patients reporting this adverse event decreased after two weeks of liraglutide treatment.

During the LEAD program, the frequency of nausea with liraglutide ranged from 10.5% when combined with a sulfonylurea to 40% with a thiazolidinedione [67, 70]. In contrast, the frequency of nausea and/or vomiting experienced with metformin is approximately 25% [86].

In LEAD-2 and LEAD-4 trials the proportion of patients who experienced nausea with liraglutide did not differ significantly from placebo after 9 weeks and 15 weeks, respectively [67, 70].

In LEAD-6 study, incidence of nausea, initially similar between groups, dropped more rapidly in liraglutide group; after 26 weeks 3% of patients in the liraglutide group experienced nausea *vs.* 9% in the exenatide group ($p<0.0001$) [72]. In the LEAD-6 extension period, nausea was reported in 3.2% of patients switching from exenatide to liraglutide and in 1.5% of those remaining on liraglutide [73].

In the Lira-DPP-4i study, liraglutide in doses of 1.8 and 1.2 mg was associated with a significant rate of nausea (27% and 21%, respectively) compared with sitagliptin (5%) [65].

Nausea led to withdrawal in 1-9% of patients receiving the 1.8 mg dose during the LEAD program.

Overall rates of withdrawal because of adverse events were low across the LEAD trials, with 7.8% of patients treated with liraglutide and 3.4% of patient treated with active comparators [85].

Immunogenicity

Liraglutide is a peptide, thus it is potentially immunogenic.

Low titers of anti-liraglutide antibodies were detected in 8.6% of patients tested during treatment with liraglutide in four LEAD trials, with no negative effect on HbA1c [67-69, 85, 87].

In LEAD-6 study, of those patients initially randomized to exenatide who entered the liraglutide extension study, 61% had anti-exenatide antibodies at week 26, ranging from 2.4 to 60.2% Bound/Total (B/T). After switching from exenatide to liraglutide at week 26, 50% and 17% of patients had persistent anti-exenatide antibodies at week 40 and week 78, respectively [73] (Fig. **25**).

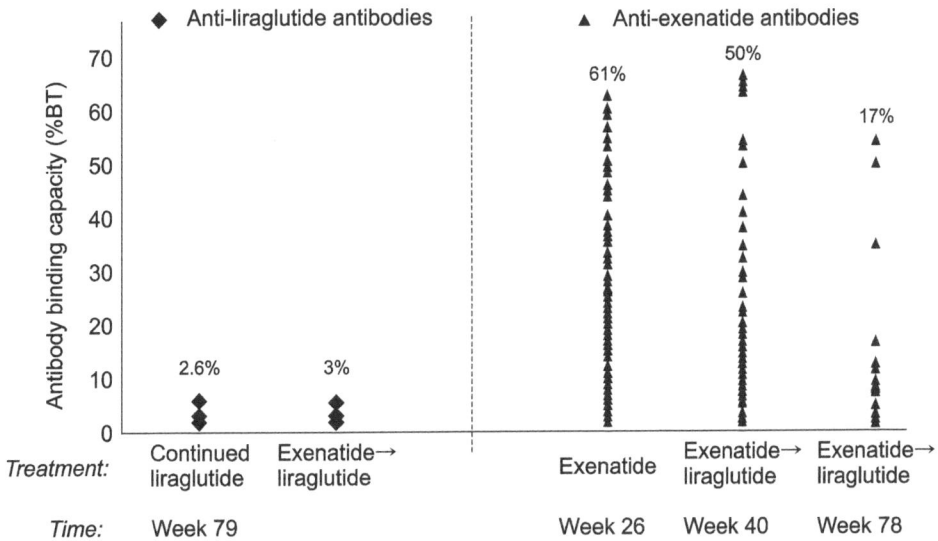

Figure 25: Anti-liraglutide and anti-exenatide antibody levels in LEAD-6 study (Derived from [73]).

Persistent anti-exenatide antibodies did not compromise the further glycemic response to liraglutide. After 78 weeks of total liraglutide treatment only 2.6% of patients recorded low titre anti-liraglutide antibodies. One year after switching to liraglutide therapy, 3% of patients possessed anti-liraglutide antibodies, whilst 17.5% of patients still possessed anti-exenatide antibodies [73].

In conclusion, liraglutide was less immunogenic than exenatide, and antibody titres were not associated with poor efficacy.

C-cell and Calcitonin

In animal studies, liraglutide activated thyroid C-cells *via* GLP-1 receptor, and caused C-cell hyperplasia with increased blood calcitonin levels [88]. Because liraglutide causes thyroid C-cell tumors in animal model and the relevance of this finding in humans could not be determined by clinical and non-clinical studies, liraglutide is contraindicated in patients with a personal or family history of medullary thyroid carcinoma, or in patients with multiple enocrine neoplasia syndrome type 2 (MEN-2).

During the LEAD trials, no evidence of C-cell activation, as assessed by calcitonin analysis, was found [89]. Across the LEAD trials, four cases of thyroid C-cell hyperplasia were observed for liraglutide group and one case in comparator group. In the liraglutide group none of patients developed C-cell carcinoma [66-70, 72, 90].

Adjusted mean serum calcitonin levels were not increased following liraglutide treatment compared with active comparators, and within normal ranges, across the clinical trials (Fig. **26**).

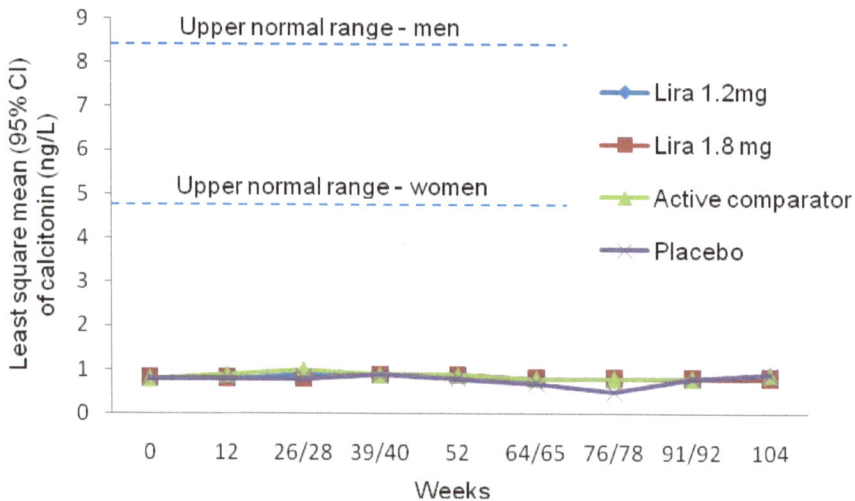

Figure 26: Least square mean calcitonin levels over two years among patients in LEAD-1 and LEAD-3 trials (Derived from [88]).

Pancreatitis

Because patients with type 2 diabetes have a 2.8 –fold higher risk than the general population of developing pancreatitis, there has been concern about an increased risk of pancreatic effects with incretin therapy [91].

Among patients in the LEAD trials, treated with liraglutide or comparator, the incidence of pancreatitis was consistent with the expected rate for type 2 diabetes patients. Rate of acute pancreatitis episodes were low (below 0.2 %) during phase 3 liraglutide trials, but it is recommended that incretin-based therapy be used with caution in patients with a history of pancreatitis or risk factors for pancreatitis [65-70, 72].

In the LEAD-6 study or extension period, no episode of acute pancreatitis was reported either in liraglutide or exenatide group [72]. Similarly, no incidences of pancreatitis was reported in the Lira-DPP-4i study [65].

Liraglutide in Special Population

Elderly Patients

In a meta-analysis of the six LEAD trials, with the upper age limit for enrollment of ≤ 80 years, age had no significant impact on glycemic response evaluated by HbA1c to liraglutide treatment [92].

The proportion of patients who experienced minor hypoglycemia with liraglutide 1.8 mg dose was low and similar for patients aged ≤ 65 years (13.4%) and those aged > 65 years (13.9%) [92].

Hepatic or Renal Impairment

Liraglutide is bound to albumin, it is not excreted in urine and studies have shown that its pharmacokinetic properties are unchanged in various degrees of renal impairment.

A meta-analysis of the LEAD program revealed that mild renal impairment (creatinine clearance ≈ 60-90 ml/min) has no effect on the safety and efficacy of liraglutide, and the treatment with liraglutide has no significant effect on renal function [93, 94].

Therapeutic experience with liraglutide in patients with hepatic impairment is currently limited, thus its use is not recommended in such patients.

Pediatric Population

Because the prevalence of type 2 diabetes in adolescents is increasing, thus there is a medical need for more treatment options. Until now, no large trials that have investigated the safety, tolerability and efficacy of liraglutide in the pediatric population with type 2 diabetes melitus. There are two recent small studies [95, 96] have studied the use of liraglutide in subjects with type 2 diabetes and aged between 10-17 years and showed that this analog of GLP-1 is safety, well tolerated and PK parameters was similar to adults.

Final Remarques

The phase 3 LEAD clinical program has shown that liraglutide treatment constitutes effective therapy for type 2 diabetes.

Liraglutide offers a significant improvement of glycemic control, produce sustained reduction in HbA1c, FPG and PPG levels and has a positive effect on β-cell function. The glycemic benefits of liraglutide are not associated with an increased risk of hypoglycemia compared with other conventional treatments.

Furthermore, liraglutide therapy results in important weight loss and decreases in SBP and additional relevant benefits on other cardiovascular risk biomarkers.

Liraglutide was generally well tolerated. Nausea, the most common adverse event, was mild and transient in nature.

The LEAD trials have also shown that liraglutide is not associated with an increased risk of pancreatitis.

Finally, no evidence of C-cell activation could be found in liraglutide group.

Thus, liraglutide is a promising new treatment option for type 2 diabetes.

CONCLUSIONS AND PERSPECTIVES

In the final part of the paper we shall discuss the advantages of liraglutide in metabolic control, news regarding hypoglycemia and weight loss, its influence on cardiovascular parameters, other new and interesting features and some presumptive ways of development for this drug.

Advantages of Liraglutide in the Metabolic Control of Diabetic Patients

As we already explained in this text, liraglutide is a GLP-1 receptor agonist, recently introduced in the treatment of type 2 diabetes mellitus [10]. It exerts its action trough: stimulation of insulin secretion in a glucose dependent manner, suppression of glucagon, reduction in appetite and food intake, deceleration of gastric emptying; additionally, it brings about weight loss [97].

Liraglutide has been shown to substantially reduce the levels of HbA1c and to increase the proportion of patients reaching HbA1c targets [10]. It has been also demonstrated that liraglutide reduces fasting plasma glucose (FPG) and decreases postprandial glucose (PPG). Considerable improvements of beta-cell function in type 2 diabetes patients were observed across LEAD trials 1-6 [10, 70, 72, 87, 97]. These trials include a number of endpoints to assure beta-cell function, such as HOMA-B and proinsulin: insulin ratio. The LEAD-6 study evaluated the addition of liraglutide or exenatide to metformin and/or sulfonylurea. Improvement in HOMA-B was significantly higher in the liraglutide 1.8 mg qd. group [72].

Hypoglycemia

Basically, liraglutide was associated with a hypoglycemia rate of 0.03-1.96 events per patient/per year in subjects with type 2 diabetes mellitus. There was a higher risk of hypoglycemia in combination with sulfonylureas: 0.47-1.93 events per patient per year in LEAD-1, LEAD-5, LEAD-6 [10]. No major hypoglycemic events were reported for liraglutide in three of LEAD trials. A switch from exenatide to liuraglutide in the LEAD-6 was associated with a decreased rate of minor hypoglycemia [73].

Weight Loss

A lot of studies confirmed that liraglutide reduces body weight in patients with type 2 diabetes mellitus. The most of the fat lost is visceral but weight loss with liraglutide also increases with increasing body mass index (BMI) [10]. A recent large analysis proved that therapy with GLP-1R agonists determines weight loss in overweight or obese patients with or without type 2 diabetes mellitus, in spite of improved metabolic regulation. The effect of GLP-1R agonists was associated with nausea, diarrhoea, and vomiting but not with hypoglycemia [98].

Effects on Cardiovascular Parameters

Liraglutide utilization produced significant reductions in systolic blood pressure (SBP) across the LEAD trials. It is important to note that this reduction in SBP occurred early after the introduction of liraglutide and before any major influence on body weight [51].

Additionally, total cholesterol, low density lipoprotein cholesterol (LDL), free fatty acids and triglycerides all decreased significantly from baseline with liraglutide [10]. The drug also significantly reduced brain natriuretic peptide (BNP) and high sensitivity C-reactive protein (hsCRP) [10, 84].

Liraglutide treatment did not appear to have a significant effect on diastolic blood pressure (DBP). On the other hand, a minor increase in heart rate was observed in most LEAD trials. The mechanism behind the effect of liraglutide on blood pressure and heart rate is yet to be explained [51].

Other Interesting Observations

New and just unexpected data referring to liraglutide were published in the last period.

The experimental research of McClean *et al.* [99], has shown that liraglutide can present neurodegenerative modifications found in Alzheimer's disease, suggesting that GLP-1 analogues represent a possible strategy for this condition.

- Hunter and Hölcher [100] confirmed that liraglutide and lixisenatide cross the blood brain barrier and enhance neurogenesis.

- A short-term treatment with liraglutide improved visceral fat obesity, appetite, food preference and the urge for fat intake in obese Japanese subjects with type 2 diabetes mellitus [101].

- Liraglutide has been reported to increase in tropical activity in cardiac myocytes and to protect against myocardial injury. Shiraki *et al.* [102] evaluated its effects against inflammation and oxidation stress on human endothelial cells.

- Ryan *et al.* [103] noted that American Diabetes Association recommended a GLP-1 agonist along with metformin as a second-tier therapy for type 2 diabetes. Although the American Association of Clinical Endocrinologists/American College of Endocrinologists' guidelines recommended it for first-line monotheraphy in subjects with HbA1c between 6.5% and 7,5% and with metformin if HbA1c is 7.6%-8.5%, liraglutide should be used in patients who cannot tolerate first-line drugs or if an additional compound is needed to reach target HbA1c.

- Vilsbøll and Garber [104] discuss the non-glycemic effects mediated *via* GLP-1 receptor agonists and the potential for using these in therapy, with focus on liraglutide. These aspects are currently being investigated in ongoing clinical trials.

Presumptive Developments

It is clear at this moment that liraglutide offers unique advantages for the treatment of type 2 diabetes mellitus, by enhancing metabolic control while also reducing body weight and avoiding hypoglycemia [105]. It seems to represent an exciting and very promising option for this type of diabetes [51]. However, the long-term clinical impact of its qualities remains to be established. We know, for instance, that cardiovascular safety of liraglutide will be prospectively evaluated in the international LEADER^{TM} trial, which is enrolling 9000 subjects with type 2 diabetes and a broad range of cardiovascular risk factors. Patients will be randomized 1:1 to liraglutide or placebo and will be followed for up to 5 years for macrovascular events [106].

It is logical to think that other different control studies will be taken into consideration.

ACKNOWLEDGEMENTS

We thank two groups of collaborators: 1. Dr. Lawrence Nwabudike, Ciprian Zetu and Dana Diaconu for technical assistance; 2. Dr. Mihai Briciu, Dr. Emiliana Mocanu, Dr. Madalina Avram (all from Novo Nordisk, Romania) for scientific information. Also we thank you to the authors of figures derived.

CONFLICT OF INTEREST

The authors confirm that this chapter contents have no conflict of interest.

REFERENCES

[1]   Cheţa DM. Sleeping with the enemy (Editorial). Romanian Journal of Diabetes, Nutrition and Metabolic Diseases 2012; 19(1):05-06.
[2]   IDF Diabetes Atlas, fifth edition. International Diabetes Federation, Brussels, 2011.
[3]   Edavalath M, Stephens JW. Liraglutide in the treatment of type 2 diabetes mellitus: clinical utility and patient perspectives. Patient Prefer Adherence 2010; 4:61-8.
[4]   American Diabetes Association. Standards of Medical Care in Diabetes -2012. Diabetes Care 2012; 35(Suppl 1):S11-S65.
[5]   Cheţa D, Rusu E, Constantin C. Farmacoterapia diabetului zaharat tip 2. In: Lichiardopol R, editor. Manual de Diabetologie. Bucureşti, Edit. Ilex 2011; p 95-121.
[6]   Aguilar RB. Evaluting treatment algoritms for the management of patients with type 2 diabetes mellitus: a perspective on the definition of treatment success. Clin Ther. 2011; 33(4):408-24.
[7]   Blevins T. Therapeutic options that provide glycemic control and weight loss for patients with type 2 diabetes. Postgrad Med. 2010, 122 (1):172-83.
[8]   Brunton S. Beyond glicemic control: treating the entire type 2 diabetes disorders: Postgrad Med. 2009, 121(5):68-81.
[9]   Dailey G. Early and intensive therapy for management of hyperglycemia and cardiovascular risk factors in patients with type 2 diabetes. Clin Ther. 2011; 33(6):665-78.
[10]   Liraglutide Scientific Synopsis. Novo Nordisk A/S Bagsvaerd, 2011.
[11]   Ovalle F. Cardiovascular implications of antihyperglycemic therapies for type 2 diabetes. Clin Ther. 2011; 33(4):393-407.
[12]   Pi-Sunyer FX. The impact of weight gain on motivation, compliance and metabolic control in patients with type 2 diabetes mellitus. Postgrad Med. 2009; 121(5):94-107.
[13]   Unger J. Liraglutide: can it make a difference in the treatment of type 2 diabetes? Int J Clin Pract. 2010; 64 (Suppl. 167):1-3.
[14]   DeFronzo RA. Banting Lecture. From triumvirate to the ominous octet: a new paradigm for the treatment of type 2 diabetes mellitus. Diabetes 2009; 58(4):773-95.
[15]   Schmidt WE. Early clinical studies with liraglutide. Int J Clin Pract. 2010; 64(Suppl. 167):12-20.
[16]   Hawley JA, Gibala MJ. What's new since Hippocrates? Preventing type 2 diabetes by physical exercise and diet. Diabetologia 2012; 55(3):535-9.

[17] Cheţa DM. Preventing Diabetes. Theory, Practice and New Approaches. Chichester. J Wiley & Sons Ltd, 1999.

[18] Burant CF. Medical Management of Type 2 Diabetes. 6th ed. Alexandria, VA: American Diabetes Association, Inc, 2008.

[19] Dunning BE, Foley JE, Ahren B. Alpha cell function in health and disease: influence of glucagon-like peptide-1. Diabetologia 2005; 48(9):1700-13.

[20] Toft-Nielsen MB, Damholt MB, Madsbad S, *et al.* Determinants of the impaired secretion of glucagon-like peptide-1 in type 2 diabetic patients. J Clin Endocrinol Metab. 2001; 86(8):3717-23.

[21] Holst JJ, Gromada J. Role of incretin hormones in the regulation of insulin secretion in diabetic and nondiabetic humans. Am J Physiol Endocrinol Metab. 2004; 287(2):E199-206.

[22] Nauck M, Stöckmann F, Ebert R, Creutzfeldt W. Reduced incretin effects in Type 2 (non-insulin-dependent) diabetes. Diabetologia 1986; 29(1):46-52.

[23] Nauck MA, Baller B, Meier JJ. Gastric inhibitory polypeptide and glucagon like peptide-1 in the pathogenesis of type 2 diabetes mellitus. Diabetes 2004; 53(Suppl. 3):S190-6.

[24] Nauck MA, Wollschläger D, Werner J, *et al.* Effects of subcutaneous glucagon-like peptide-1 (GLP-1 s7-36 amidet) in patients with NIDDM. Diabetologia 1996; 39(12):1546-53.

[25] Vilsbøll T, Holst JJ. Incretins, insulin secretion and type 2 diabetes mellitus. Diabetologia 2004; 47(3):357-66.

[26] Drucker DJ, Nauck MA. The incretin system: glucagon-like peptide-1 receptor agonists and dipeptidyl peptidase-4 inhibitors in type 2 diabetes. Lancet 2006; 368(9548):1696-705.

[27] Dupre J, Ross SA, Watson D, Brown JC. Stimulation of insulin secretion by gastric inhibitory polypeptide in man. J Clin Endocrinol Metab. 1973; 37(5):826-8.

[28] Kieffer TJ, McIntosh CH, Pederson RA. Degradation of glucose-dependent insulinotropic polypeptide and truncated glucagon-like peptide 1 *in vitro* and *in vivo* by dipeptidyl peptidase IV. Endocrinology 1995; 136(8):3585-96.

[29] Nauck MA, Kleine N, Orskov C, Holst JJ, Willms B, Creutzfeldt W. Normalization of fasting hyperglycaemia by exogenous glucagon-like peptide-1 (7-36 amide) in type 2 (non-insulin-dependent) diabetic patients. Diabetologia 1993; 36(8):741-4.

[30] Svec F. Incretin physiology and its role in Type 2 Diabetes Mellitus. J Am Osteopath Assoc. 2010; 110(7):eS20-eS24.

[31] Vilsbøll T, Knop FK. DPP-4 inhibitors - Current Evidence and future Directions. Br J Diabetes Vasc Dis. 2007; 7(2):69-74.

[32] Holst JJ. On the physiology of GIP and GLP-1. Horm Metab Res. 2004; 36(11-12):747-54.

[33] Lewis JT, Dayanandan B, Habener JF, Kieffer TJ. Glucose-dependent insulinotropic polypeptide confers early phase insulin release to oral glucose in rats: demonstration by a receptor antagonist. Endocrinology 2000; 141(10):3710-6.

[34] Miyawaki K, Yamada Y, Yano H, *et al.* Glucose intolerance caused by a defect in the entero-insular axis: a study in gastic inhibitory polypeptide receptor knockout mice. P Natl Acad Sci USA. 1999; 96(26):14843-7.

[35] Takeda J, Seino Y, Tanaka K, *et al.* Sequence of an intestinal cDNA encoding human gastric inhibitory polypeptide precursor. Proc Natl Acad Sci USA. 1987; 84(20):7005-8.

[36] Vilsbøll T, Krarup T, Madsbad S, Holst JJ. Defective amplification of the late phase insulin response to glucose by GIP in obese Type II diabetic patients. Diabetologia 2002; 45(8):1111-9.

[37] Mayo KE, Miller LJ, Bataille D, *et al*. International Union of Pharmacology XXXV. The glucagon receptor family. Pharmacol Rev. 2003; 55(1):167-94.

[38] Gallwitz B. Therapies for the treatment of type 2 diabetes mellitus based on incretin action. Minerva Endocrinol. 2006; 31(2):133-47.

[39] Holst JJ, Vilsbøll T, Deacon CF. The incretin system and its role in type 2 diabetes mellitus. Mol Cell Endocrinol. 2009; 297(1-2):127-36.

[40] Keith Campbell R. Clarifying the role of incretin-based therapies in the treatment of type 2 diabetes mellitus, Clin Ther. 2011; 33(5):511-27.

[41] Mafong DD, Henry RR. The role of incretins in cardiovascular control. Curr Hypertens Rep. 2009; 11(1):18-22.

[42] Rosenstock J, Zinman B. Dipeptidyl peptidase-4 inhibitors in the management of type 2 diabetes mellitus. Curr Opin Endocrinol Diabetes Obes. 2007; 14(2):98-107.

[43] Robertson C. Incretin-related therapies in type 2 diabetes: a practical overview. Diabetes Spectrum 2011; 24(1):26-35.

[44] Del Prato S, Marchetti P. Beta and alpha-cell dysfunction in type 2 diabetes. Horm Metab Res. 2004; 36(11-12):775-81.

[45] McIntosh CH. Dipeptidyl peptidase IV inhibitors and diabetes therapy. Front Biosci. 2008; 13:1753-73.

[46] Farilla L, Hui H, Bertolotto C, *et al*. Glucagon-like peptide-1 promotes islet cell growth and inhibits apoptosis in Zucker diabetic rats. Endocrinology 2002; 143(11):4397-40.

[47] Turner RC, Cull CA, Frighi V, Holman RR. Glycemic control with diet, sulfonylurea, metformin, or insulin in patients with type 2 diabetes mellitus: progressive requirement for multiple therapies (UKPDS 49). UK Prospective Diabetes Study (UKPDS) Group. JAMA 1999; 281(21):2005-12.

[48] Larsen J, Hylleberg B, Ng K, Damsbo P. Glucagon-like peptide 1 infusion must be maintained for 24 h/day to obtain acceptable glycemia in type 2 diabetes patients who are poorly controlled on sulphonylurea treatment. Diabetes Care 2001; 24(8):1416-21.

[49] Aylwin S, Al Zaman Y. Emerging concepts in the medical and surgical treatment of obesity. Front Horm Res. 2008; 36:229-59.

[50] Anderson JW, Konz EC. Obesity and disease management: effects of weight loss on comorbid conditions. Obes Res. 2001; 9 (Suppl. 4):S326-34.

[51] Davies MJ, Kela E, Khunti K. Liraglutide - overview of the preclinical and clinical data and its role in the treatment of type 2 diabetes. Diabetes Obes Metab. 2011; 13(3):207-220.

[52] Victoza® (liraglutide injection) Europe.Summary of Product Characteristics (SPC). Novo Nordisk A/S Bagsvaerd; 2011.

[53] Levemir. Summary of Product Characteristics (SPC). Novo Nordisk A/S Bagsvaerd, 2011.

[54] Knudsen LB. Liraglutide: the therapeutic promise from animal models. Int J Clin Pract 2010; 64(Suppl. 167):4-11.

[55] Bregenholt S, Møldrup A, Blume N, *et al*. The long-acting glucagon-like peptide-1 analogue, liraglutide, inhibits beta-cell apoptosis *in vitro*. Biochem Bioph Res Co. 2005; 330(2):577-84.

[56] Rütti S, Prozak R, Ellingsgaard H, *et al*. Liraglutide induces human β-cell proliferation, conteracts low-density lipoprotein anti-proliferative effects and protects from IL-1 β induced apoptosis. Diabetes 2009; 58(Suppl1):A410.

[57] Ribel U, Larsen MO, Rolin B, *et al*. NN 2211: a long-acting glucagon-like peptide-1 derivative with antidiabetic effects in glucose intolerant pigs. Eur J Pharmacol. 2002; 451(2):217-25.

[58]   Raun K, von Voss P, Knudsen LB. Liraglutide a once daily human-glucagon peptide-1 analog, minimizes food intake in severely obese minipigs. Obesity (Silver Spring) 2007; 15(7):1710-6.

[59]   Noyan-Ashraf MH, Momen BA, Ban K, *et al*. GLP-1 R agonist liraglutide activates cytoprotective pathways and improve outcomes after experimental myocardial infarction in mice. Diabetes 2009; 58(4):975-83.

[60]   Chang AM, Jakobsen G, Sturis J, *et al*. The GLP-1 derivative NN 2211 restores beta cell sensitivity to glucose in type 2 diabetic patients after a single dose. Diabetes 2003; 52(7):1786-91.

[61]   Degn K, Juhl C, Sturis J, *et al*. One week's treatment with long-acting glucagon-like peptide 1 derivative liraglutide (NN 2211) markedly improves 24-h glycemia and α-and β-cell function and reduces endogenons glucose release in patients with type 2 diabetes. Diabetes 2004; 53(5):1187-94.

[62]   Juhl CB, Hollingdal M, Sturis J, *et al*. Bedtime administration of NN 2211, a long acting GLP-1 derivative, substantially reduces fasting and postprandial glycemia in type 2 diabetes. Diabetes 2002; 51(2):424-9.

[63]   Peterson GE, Pollom RD. Liraglutide in clinical practice: dosing, safety and efficacy. Int J Clin Pract. 2010; 64(Suppl. 167):35-43.

[64]   Montanya E, Sesti G. A review of efficacy and safety data regarding the use of liraglutide, a once-daily human glucagon-like peptide 1 analogue, in the treatment of type 2 diabetes mellitus. Clin Ther. 2009; 31(11):2472-88.

[65]   Pratley RE, Nauck M, Bailey T, *et al*. for the 1860-LIRA-DPP-4 Study Group. Liraglutide *vs*. sitagliptin for patients with type 2 diabetes who did not have adequate glycaemic control with metformin: a 26-week, randomised, parallel-group, open label trial. Lancet 2010; 375(9724):1447-56.

[66]   Garber A, Henry R, Ratner R, *et al*. Liraglutide *vs*. glimepiride monotherapy for type 2 diabetes (LEAD-3 Mono); randomised, 52-week, phase III, double-blind, parallel-treatment trial. Lancet 2009; 373(9662):473-81.

[67]   Marre M, Shaw J, Brändle M, *et al*. Liraglutide, a once-daily human GLP-1 analogue, added to a sulphonylurea over 26 weeks produces greater improvements in glycemic and weight control compared with adding rosiglitazone or placebo in subjects with type 2 diabetes (LEAD-1 SU). Diabet Med. 2009; 26(3):268-78.

[68]   Nauck MA, Frid A, Hermansen K, *et al*. Efficacy and safety comparison of liraglutide, glimepiride, and placebo, all in combination with metformin in type 2 diabetes: the LEAD (Liraglutide Effect and Action in Diabetes)-2 study. Diabetes Care 2009; 32(1):84-90.

[69]   Russell-Jones D, Vaag A, Schmitz O, *et al*. Liraglutide vs insulin glargine and placebo in combination with metformin and sulfonylurea therapy in type 2 diabetes mellitus (LEAD-5 met+SU): a randomised controlled trial. Diabetologia 2009; 52(10):2046-55.

[70]   Zinman B, Gerich J, Buse JB, *et al*. Efficacy and safety of the human GLP-1 analogue liraglutide in combination with metformin and TZD in patients with type 2 diabetes mellitus (LEAD-4 Met+TZD) Diabetes Care 2009; 32(7):1224-30.

[71]   Garber A, Henry RR, Ratner R, Hale P, Chang CT, Bode B. Liraglutide, a once-daily human glucagon-like peptide 1 analogue, provides sustained improvements in glycaemic control and weight for 2 years as monotherapy compared with glimepiride in patients with type 2 diabetes. Diabetes Obes Metab. 2011; 13(4):348-56.

[72]   Buse J, Rosenstock J, Sesti G, *et al.* LEAD 6 Study Group. Liraglutide once a day *vs.* exenatide twice a day for type 2 diabetes: a 26-week randomised, parallel-group, multinational, open-label trial (LEAD-6). Lancet 2009; 374(9683):39-47.

[73]   Buse JB, Sesti G, Schmidt WE, *et al.* LEAD 6 Study Group. Switching to once-daily liraglutide from twice-daily exenatide further improves glycemic control in patients with type 2 diabetes using oral agents, Diabetes Care 2010; 33(6): 1300-3.

[74]   Bain SC, DeVries JH, Rodbard H, *et al.* A new treat-to-target paradigm for patients with Type 2 diabetes: liraglutide added to metformin followed by intensification with insulin detemir in subjects not reaching HbA1c target. Diabet Med. 2011; Abstract Number p122.

[75]   Schmidt WE, Gough S, Madsbad NS, *et al.* Liraglutide, a human GLP-1 analogue, lowers HbA1c independent of weight loss. Diabetologia 2009; 52(Suppl. 1):S289.

[76]   Pratley R, Nauck M, Bailey T, *et al.* One year of liraglutide treatment offers sustained and more effective glycaemic control and weight reduction compared with sitagliptin, both in combination with metformin, in patients with type 2 diabetes: a randomised, parallel-group, open-label trial. Int J Clin Pract. 2011; 65(4):397-407.

[77]   Lupi R, Del Prato S. Beta-cell apoptosis in type 2 diabetes: quantitative and functional consequences. Diabetes Metab. 2008; 34 (Suppl.2):s56-64.

[78]   U.K. prospective diabetes study 16. Overview of 6 years' therapy of type II diabetes: a progressive disease. UK Prospective Diabetes Study Group. Diabetes 1995; 44(11):1249-58.

[79]   Blonde L, Russel-Jones D. The safety and efficacy of liraglutide with or without oral antidiabetic drug therapy in type 2 diabetes: an overview of the LEAD 1-5 studies. Diabetes Obes Metab. 2009, 11 (Suppl 3):26-34.

[80]   Jendle J, Nauck MA, Matthews DR, *et al.* Weight loss with liraglutide, a once-daily human glucagon-like peptide-1 analogue for type 2 diabetes treatment as monotherapy or added to metformin, is primarily as a result of a reduction in fat tissue, Diabetes Obes Metab. 2009; 11(12):1163-72.

[81]   Russell-Jones D, Shaw JE, Brandle M, *et al.* The once-daily human GLP-1 analogue liraglutide reduces body weight in subjects with type 2 diabetes, irrespective of body mass index at baseline. Diabetes 2008; 57:A593-A594.

[82]   Fonseca V, Falahati A, Zychma M, Madsbad S, Plutzky J. Liraglutide, a once-daily human GLP-1 analogue, reduces systolic blood pressure within 2 weeks in patients with type 2 diabetes. Poster session presented at: IDF 20th World Congress Abstract Book (Poster D-0908) 2009, IDF, Montreal, Canada.

[83]   Fonseca V, Plutzky J, Montanya E, *et al.* Liraglutide, a once-daily human glp-1 analog, lowers systolic blood pressure (sbp) independently of concomitant antihypertensive treatment. Diabetes 2010; 59 (Suppl. 1) (Abstract 296-OR)

[84]   Plutzky J, Garber A, Toft AD, Poulter NR. Meta-analysis demonstrates that liraglutide, a one-daily human GLP-1 analogue, significantly reduces lipids and other markers of cardiovascular risk in T2D. Diabetologia 2009; 52(Suppl. 1):S299-300.

[85]   Liraglutide Summary of Product Characteristics 2010. Available from URL: http://www.aace.com/meetings/ams/2009/pdf/ABSTRACTS April 2010.

[86]   Bristol-Meyers Squibb Co. Glucophage (metformin) package insert. Princeton, NJ; Rev January 2009. Available from URL: http://packageinserts.bms.com/pi/pi_glucophage.pdf.

[87]   Buse JB, Garber A, Rosenstock J, *et al.* Liraglutide treatment is associated with a low frequency and magnitude of antibody formation with no apparent impact on glycemic

response or increased frequency of adverse events: results from the Liraglutide Effect and Action in Diabetes (LEAD) trials. J Clin Endocrinol Metab. 2011; 96(6):1965-702.

[88]   Knudsen BL, Madsen LW, Andersen S, *et al.* Glucagon-Like Peptide-1 Receptor Agonists Activate Rodent Thyroid C-Cells Causing Calcitonin Release and C-Cell Proliferation. Endocrinology 2010; 151(4):1473-86.

[89]   Liraglutide US Prescribing Information. Available from URL: http://www.novo-pi.com/victoza.pdf.

[90]   Matthews D, Marre M, Le-Thi T, *et al.* Liraglutide, a once-daily human GLP-1 analog, significantly improves β-cell function in subjects with type 2 diabetes. Diabetes 2008; 57(suppl. 1):A150(505-P).

[91]   Noel RA, Braun DK, Patterson RE, Bloomgren GL: Increased risk of acute pancreatitis and biliary disease observed in patients with type 2 diabetes: a retrospective cohort study. Diabetes Care 2009; 32(5):834-8.

[92]   Bode B, Falahati A, Brett J, Pratley R. Comparison of the efficacy and tolerability of the once-daily human GLP-1analogue, liraglutide, in elderly *vs.* younger patients with type 2 diabetes; a meta-analysis. Poster session presented at: the International Diabetes Federation, 20th World Diabetes Congress; 2009 October 18-22, Montreal, Canada.

[93]   Novo Nordisk A/S. Liraglutide (Victoza) Summary of Product Characteristics [EMA Approval: Document on the Internet]. EMC web site; 2009. http://emc.medicines.org.uk/medicine/21896/SPC/

[94]   Davidson J, Falahati A, Brett J. Mild renal impairement has no effect on the efficacy and safety of liraglutide (Abstract 235) AACE 2009. Available from URL: https://www.aace.com/files/2009-Abstracts.pdf

[95]   Petri KC, Jacobsen L, Ingwersen S, Klein D. Liraglutide in pediatric subjects with type 2 diabetes: a population pharmacokinetic (PK) analysis and comparison to adult subjects. Poster session presented at: EASD 48th Annual Meeting Abstract Book (Poster 805) 2012, EASD, Berlin, Germany.

[96]   Klein D., Battelino T, Chatterjee DJ, *et al.* Liraglutide trial in pediatric type 2 diabetes: safety, tolerability and pharmacokinetics/pharmacodynamics. Poster session presented at: ADA 72rd Scientific Sessions. Abstract Book (59-LB) 2012, ADA, Philadelphia, USA.

[97]   Bode B. Liraglutide: a review on the first once-daily GLP-1 receptor agonist. Am J Manag Care. 2011; 17 (2 Suppl): S59-70.

[98]   Vilsbøll T, Christensen M, Junker AT, Knop K, Glud LL. Effect on glucagon-like peptide-1 receptor agonists on weight loss: systematic review and meta-analyses of randomised controlled trials. Brit Med J. 2012; 344:d7771.

[99]   McClean PL, Parthsarathy V, Faivre E, Hölscher C. The diabetes drug, liraglutide prevents degenerative process in a mouse model of Alzheimer's disease. J Neurosci. 2011; 31(17):6587-94.

[100]   Hunter K, Hölscher C. Drugs developed to treat diabetes, liraglutide and lixisenatide, cross the blood brain barrier and enhance neurogenesis. BMC Neurosci. 2012; 13:33.

[101]   Inoue K, Maeda N, Kashine S, *et al.* Short-term effects of liraglutide on visceral adiposity, appetite, and food preference: a pilot study of obese Japanese patients with type 2 diabetes. Cardiovasc Diabetol. 2011; 10:109.

[102]   Shiraki A, Oyama J, Komoda H, *et al.* The glucagon-like peptide 1 analog liraglutide reduces TNF α-induced oxidative stress and inflammation in the endothelial cells. Atherosclerosis 2012; 221(2):375-82.

[103] Ryan GJ, Foster KT, Jobe LJ. Review of the therapeutic uses of liraglutide. Clin Ther. 2011; 33(7):793-811.

[104] Vilsbøll T, Garber AJ. Non glycaemic effects mediated *via* GLP-1 receptor and the potential for exploiting these for therapeutic benefit: focus on liraglutide. Diabetes Obes Metab. 2012; 14(Suppl.2):41-9.

[105] Zinman B, Schimdt WE, Moses A, Lund N, Gough S. Achieving a clinically relevant composite outcome of an HbA1c of <7% without weight gain or hypoglicemia in type 2 diabetes: a meta-analysis of the liraglutide clinical trial programme. Diabetes Obes Metab. 2012; 14(1):77-82.

[106] Marso SP, Lindsey JB, Stolker JM, *et al.* Cardiovacular safety of liraglutide assured in a patient-level pooled analysis of phase 2-3 liraglutide clinical development studies. Diab Vasc Dis Res. 2011; 8(3):237-40.

Send Orders for Reprints to reprints@benthamscience.net

CHAPTER 4

MicroRNA, A Diligent Conjurer in the Exposition of Diabetes

Puneetpal Singh[*]

Department of Human Genetics, Punjabi University, Patiala, Punjab, India

Abstract: MicroRNAs (miRNAs) are family of small, noncoding RNAs that regulate gene expression in sequence specific manner. Their non perfect pairing of 6-8 nucleotides with target mRNA subsequently forming miRNA Induced Silencing Complex (miRISC) generally results in translational repression, destabilization of mRNAs and gene silencing. miRNAs are transcribed as long primary transcripts (pri-miRNAs), which are processed in the nucleus to give one or more hairpin precursor sequences (pre-miRNAs). The hairpin precursor is exported to the cytoplasm where the mature miRNA is excised by the RNase III-like enzyme named Dicer. The human genome may encode over 1500 miRNAs (1527 listed in miRBase), which may target about 60 percent of mammalian genes and are abundant in many human cell types. Owing to their significant role in regulation of cellular events including the cellular proliferation and differentiation during tumorigenesis, organ development, cell fate determination and apoptotic pathways of various organisms, miRNAs have engrossed huge attention lately in relation to their role in several diseases including infectious, cardiovascular and neurological disorders alongwith cancer, diabetes and obesity. miRNAs facilitate profound functional consequences of signaling pathway involved in the development of diabetes, insulin resistance and obesity that offers novel ways of identification and validation of new targets of therapeutic intervention. Although, the understanding of their complex mechanism of gene regulation and expression is in infancy, analysis of differential miRNA expression holds a promise to add many clinical chapters in the treatise of 'prognosis, diagnosis and therapeutics' of diabetes.

Keywords: Adipose, beta cells, gene expression, gene repression, gene silencing, Insulin, insulin action, insulin production, insulin secretion, liver, microRNA, miRNA biogenesis, miRISC, miRNA targets, non-coding RNA, pre-miRNA, pri-miRNA, skeletal tissue, type2 diabetes, therapeutic strategies.

INTRODUCTION

Diabetes mellitus is a major public health issue with 285 million individuals affected worldwide in 2010 with the predicted increase upto 400 million in 2030 [1]. The chronic elevation of blood glucose due to the defects in either insulin

---

*Address correspondence to Puneetpal Singh:** Department of Human Genetics, Punjabi University, Patiala, Punjab, India; Tel: +91-9646448947; E-mail: puneetpalsingh@pbi.ac.in

production or insulin secretion is the primary causal factor for diabetes that triggers several micro and macro vascular complications and confers a huge risk for oxidative stress, neonatal complications, heart and kidney failure, peripheral vascular disease, lower limb amputations and blindness [2]. Owing to its severe influence on the development of other diseases many scientists have developed novel methods to predict its manifestation and occurrence of its debilitating sequels. Lately, it has been revealed that a family of small endogenous ribonucleic acids termed as micro RNAs (miRNAs) play direct role in insulin secretion, pancreatic islet development, beta cell differentiation, indirect glucose control alongwith lipid metabolism and are involved in secondary complications associated with diabetes [3-6].

MicroRNAs (miRNA) are a family of small (~22 nucleotides) noncoding, single stranded, highly conserved RNA molecule that regulate gene expression at the post transcriptional level inhibiting the translation of proteins from messenger RNA (mRNA) by promoting the degradation of mRNA. These endogenously produced RNAs play significant role in mammalian gene expression and regulate several key cellular functions [7]. It is estimated that there are over 1500 human miRNAs and that about 60% of the human protein coding genes can be targeted by these miRNAs making these wonder molecules as the key regulators of several proteins. A plethora of miRNAs prevalent in disease conditions are helpful in the development of miRNA based diagnosis and therapies. Lin-4 and let-7 were the first identified miRNAs in *Caenorhabditis elegans* as the regulators of the developmental timing in the nematode [8-10]. Primarily, these are transcribed as an imperfect hairpin structure in a larger transcript (primary miRNA or pri-miRNA), the mature single stranded miRNA being contained in the stem of two hairpins. Biogenesis of miRNAs

The processing of the miRNA takes place in two stages. First, the tip of the hairpin known as the precursor miRNA or pre-miRNA is liberated from the primary transcript. This stage takes place in the nucleus by two nuclear proteins DROSHA and DGCR8 (known as PASHA in other organisms) which forms complex known as microprocessor. DROSHA chops off the stem loop to give rise to the subsequent ~70 nt pre-miRNA leaving a characteristic overhang on the 3ᴵ end of the hairpin. Exportin-5 and Ran-GTP export these small duplex pre-miRNA species into the cytoplasm where the second stage of processing takes place. Here, another enzyme DICER cleaves the pre-miRNA of ~22nt or 2 helical turns from the site of previous cleavage, again leaving the 3ᴵ overhang removing the loop region and leaving the 22nt double stranded miRNA duplex. Only one strand of this duplex becomes the mature miRNA and the other strand is degraded. This duplex is unwound by an unknown helicase protein and mature miRNA is loaded into an argonaute family protein which alongwith dicer and several other proteins form the microRNA induced silencing complex (miRISC). The relative stability of two ends of the duplex decides that which strand is to be loaded into miRISC. Generally the 5' end at the less stable end of the duplex is loaded into miRISC (Fig. **1**).

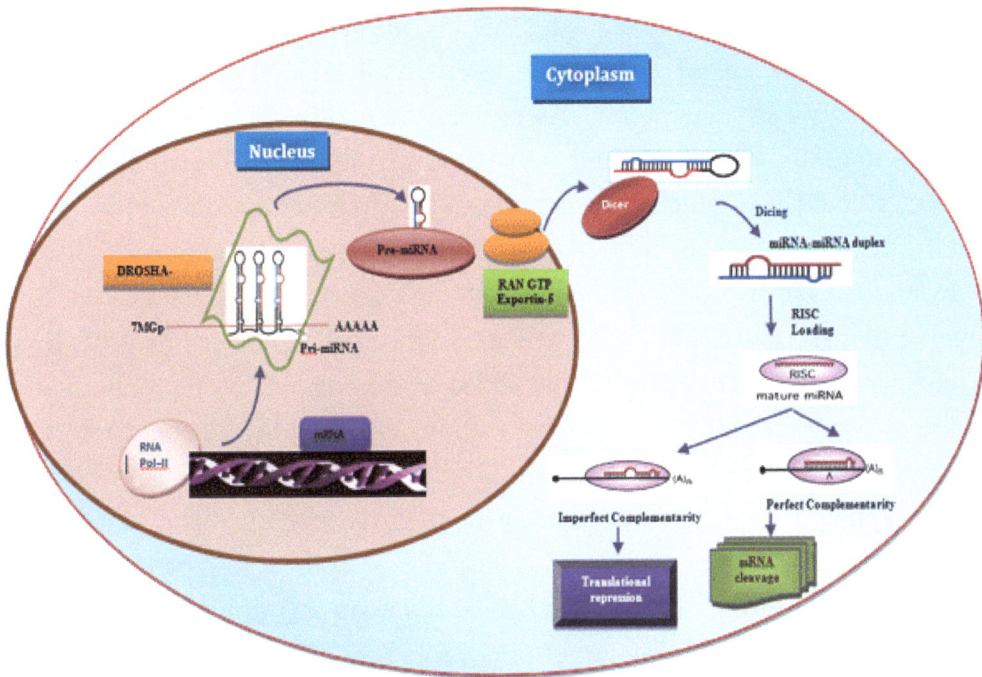

Figure 1: The miRNA biogenesis and mechanism of action. RNA polymerase transcribes the primary transcript of miRNA *i.e.* pri-miRNA from mRNA or other genomic locations. The hairpin in the pri-miRNA is cleaved in the nucleus by microprocessor complex, consisting of dicer and DGCR8 (DiGeorge syndrome critical region gene 8), a double stranded RNA binding protein to give a 60-70 nucleotide long stem loop structure known as pre-miRNA. These pre-miRNAs are then exported into the cytoplasm by RAN GTP dependent nuclear transport receptor, exportin- 5. Once in the cytoplasm, these pre-miRNAs are further processed into 21-23 nucleotide long miRNA-miRNA duplex by type III RNase, dicer and double stranded RNA protein; TRBP (TAR RNA binding protein) as a co-factor. One of the strand of this miRNA duplex along with dicer, TRBP and a member of the Argonaute protein family (AGO2) assembles and loads into the miRNA Induced Silencing Complex (miRISC). Mature miRNA then regulates gene expression by guiding the miRISC to their target complementary mRNA, represses translation of mRNA and may also promote the degradation of the mRNA.

How miRNA Induced Silencing Complex (miRISC) Operates?

miRISC containing miRNA attenuates the expression of a set of genes which is determined by the sequence of the miRNA. It is still not well understood and controversial that in which way the miRISC silences the expression of the genes. It is believed that mRNA is cleaved at specific position only when there is perfect complementarity of miRNA with some sequences of mRNA. However this mechanism is rare in animals since targets are not perfectly complementary to the sequences of the miRNA. On the other hand it appears that seed sequence (6 or 7 nt sequences at the 5′ end) of the miRNA are complementary to the target sequences in the 3′ UTR of the mRNA. Initially it was believed that most important mechanism by which miRNAs repress the expression of proteins was by inhibiting translation by disrupting translation initiation or elongation. Now, we have ample evidence which suggests that while translational inhibition is a significant effect, miRNAs imperfectly matched to the targets and can reduce mRNA levels. How this process happens is though not fully understood but, it may involve the destabilization of the mRNA by promoting the removal of the methylguanosine cap (m^7G cap) which is added to the mRNAs to promote their nuclear export and prevent nuclease degradation and also the removal of the polyadenine tail. miRNAs, their target mRNAs as well as Argonaute proteins alongwith enzymes involved in the removal of m^7G cap are all found localized together at places in the cell known as processing or P bodies. It is possible that destabilization of the mRNA by the removal of m^7G cap and polyadenine tail is due to the localization of miRNA target mRNA in P bodies where the required enzymes are found. Both these processes- translational repression and mRNA destabilization appears to be different for different miRNAs and different targets.

This chronic disease of severe ramifications is caused by defects in insulin production, insulin secretion, beta cell development and insulin action. In type I diabetes, there is self destruction of the insulin producing beta cells in the pancreas and Type 2 diabetes is caused by the defects in insulin action and insulin production. Since, miRNAs play significant role in glucose homeostasis, hence understanding their role and relevance in post transcriptional regulation or translation repression of target mRNA have been discussed.

The Role of miRNA in β-cell Development

It has been suggested by some recent studies that miRNAs are inevitable players for appropriate development of pancreatic islet of langerhans and hence for their special functions like insulin production, insulin secretion and development of T2D. It has been demonstrated in a mouse model lacking the miRNA processing enzyme Dicer in pancreas [11]. This dicer-null mice exhibited severe defects in all endocrine pancreatic lineages with insulin producing beta cells being the most effective. Consistent with these observations, targeted inhibition of miR-375 in zebra fish resulted in major defects in pancreatic islet development [12]. Only few studies have identified miRNAs that are found to be expressed, preferentially in islets from mice and human, however their roles in development of islets are not well understood [13-15]. Nonetheless, some studies have reported that microRNA (miR)-124a , miR-195, miR-15a, miR-15b and miR-16 play significant regulatory role in the pancreas development by inhibiting the expression of pancreatic transcription factors [16, 17]. miR-7, miR-9, miR-375 and miR-376 are islet specific miRNAs which were expressed at high levels during human pancreatic islet development [15] and out of these miR-7 is observed to be highly expressed with approximately 200-fold increase in mature human islets than acinar tissue [13].

The Role of miRNA in Insulin Secretion

It is observed that miR-375 is one of the most widely studied islet specific miRNA. miR-375 has been observed to regulate glucose stimulated insulin secretion in a negative manner, since over expression of miR-375 inhibits insulin secretion. Consequently, inhibition of endogenous miR-375 function using antagonist miRs (Anti-miR) enhances insulin secretion [4]. Anti-miR stands for Anti-microRNA construct, which is a short, single-stranded anti-microRNA that competitively binds to target its miRNA; ultimately inhibiting its function. It has been suggested that effects of miR-375 on myotrophin expression and insulin secretion is *via* the transcription factor, nuclear factor-kappa B (NF-κB). Activation of NF-κB is associated with improved insulin secretion [18], whereas inactivation of NF-κB showed decreased insulin secretion from beta cells [19]. The expression of this most abundant miRNA in islets cells is under the control of pancreatic and duodenal homeobox-1 (Pdx-1) and neurogenic differentiation-1 (neuroD1), two critical transcription factors for the development of pancreas. It has been observed that genetic deletions of miR-375 in homozygous mice profoundly slow down the capacity of endocrine pancreas to compensate for insulin resistance and resulted in severally diabetic state [20]. Besides miR-375, other miRNAs play inhibitory roles in insulin secretion. miR-9 targets the transcription factor 'Onecut 2', which inhibits the expression of 'Granuphilin', a negative regulator of insulin exocytosis [21]. Two more miRNAs miR-124a and miR-96 are observed to regulate insulin exocytosis, hence affecting islets function [22]. It is suggested that over expression of miR-124a in pancreatic beta cell line MIN6 (a pancreatic beta cell line) increases insulin secretion under basal conditions but decreases insulin secretion mediated by high concentrations of glucose [22]. The effect of miR-124a on insulin secretion is observed to be regulated owing to the increased levels of protein Synaptosomal-associated protein 25 (SNAP25), Ras-related protein-3A (RAB3A), Synapsin Ia and decreased levels of Ras-related protein-27A (RAB27A) and Noc2. miR-124a directly targets RAB27A and its effect on the expression of the other genes involved in insulin secretion observed to be indirectly mediated by changes in their mRNA levels.

Some set of miRNAs play significant role in glucose metabolism in both neuronal tissue and pancreatic beta cells. Previously, miR-9 was considered to be brain specific miRNA only but it has been identified that it is expressed in pancreatic beta cells also. miR-9 has been reported as a strong candidate and regulator of insulin exocytosis from the pancreas. The pancreatic exocytotic machinery for insulin involves the participation of several proteins that are under direct or indirect control of several factors. Overexpression of miR-9 reduces glucose stimulated insulin exocytosis *via* targeting the transfector onecut2. Onecut2 represses the expression of garnuphilin/slp4 by directly binding to its promoter [21]. Granuphilin, a negative regulator of insulin exocytosis is a Rab GTPase effector which is associated with the secretory granules. It suggests that miR-9 inhibits insulin secretion by repressing the expression of onecut2 transcription factor and thereby increasing the levels of granuphilin that downregulates insulin secretion.

Another miRNA; miR-96 [22] has been observed to mediate granuphilin levels. It down-regulates insulin secretion *via* up-regulation of granuphilin mRNA and protein levels. In addition, it also decreases Noc2 expression, a Rab effector protein involved in endocrine and exocrine exocytosis. Among some of newly identified miRNAs is miR-30d, which is up-regulated by glucose, induces gene regulation in pancreatic beta cells. Over expression of miR-30d is associated with increased expression of MafA (mast cell function associated antigen), a beta cell specific transcription factor and prevents reduction of both MafA and insulin receptor substrate 2 (IRS2) with TNF alpha exposure. miR-30d directly targets mitogen activated protein 4 Kinase 4 (Map4K4), a TNF alpha activated kinase. It has been observed that miR-30d expression is decreased in islets of diabetic homozygous mice in which Map4K4 expression levels are elevated, hence, supporting the role of miR-30d in insulin transcription and beta cell protection from pro-inflammatory cytokines [23]. Genome wide screens recently showed that miR-33a expression is differentially regulated in human macrophages during cholesterol depletion or enrichment [24] and it has been shown that modulation of miR-33a expression inversely correlates with ABCA1 (ATP binding cassette transporter A1) expression [25]. The 3'UTR region of human ABCA1 consist of three highly conserved binding sites for miR-33a. miR-33a overexpression in human and mice islets reduces ABCA1 expression, decrease glucose stimulated insulin secretion and increase cholesterol levels [26].

miR-130a, miR-200, miR-410 have also been identified as regulators of insulin secretion. Overexpression of miR-410 enhances the level of glucose stimulated insulin secretion in MIN6 cells [30].

The Role of miRNA in Insulin Production

miR-15a has been observed to be a mediator of beta cell function and insulin biosynthesis [27]. It has been observed that expression of miR-15a is up-regulated in presence of high glucose for one hour, whereas prolonged periods of high glucose exposure results in depressed expression of miR-15a concerned with insulin biosynthesis. Moreover, atopic expression of miR-15a promotes insulin biosynthesis in Min-6 cells, whereas its repression is sufficient to inhibit insulin biosynthesis. It is verified that miR-15a directly targets and inhibits Uncoupling Protein2 (UCP2) gene expression resulting in increase in oxygen consumption and reduced ATP generation. Hence, increased expression of miR-15a contributes to intracellular accumulation of insulin by repressing UCP2 expression. Recently, it has been exposed that Lin-28a/b and Let-7 are important modulators of glucose metabolism through interactions with Insulin-PI3K-mTOR (insulin-phospatidylinositol 3-kinase-mammalian target of rapamycin) pathway and T2D associated SNPs identified in GWAS [28]. When over expressed in mice, both Lin-28a and Lin-28b promote an insulin sensitized state that resists high fat diet induced diabetes, whereas, muscle specific loss of Lin-28a and overexpression of Let-7 results in insulin resistance and impaired glucose tolerance. Let-7 family of miRNAs regulates glucose metabolism in multiple organs. However, pancreas specific overexpression of Let-7 in mice results in impaired glucose tolerance and reduced glucose induced pancreatic insulin production. These mice over expressing Let-7 had decreased fat mass and body weight, as well as reduced body cells. Their findings demonstrated that Let-7 regulates multiple aspects of glucose metabolism and suggested antimiR induced Let-7 knockdown as a potential treatment of T2D [29].

Expression of miR-21, miR-34a and miR-146a is increased in islets of non obese diabetic mice during development of 'Prediabetic Insulitus' [31]. Recently in an Asian Indian T2D patients [32], it has been reported that miR-146a are regulators associated with insulin resistance, poor glycemic control and circulatory levels of TNF alpha and IL6. Similarly increase in miR-29 levels may cause insulin resistance [33]. Another miRNA; miR-133a precursor decreases polypyrimidine tract binding (PTB) production levels and insulin biosynthesis similar to high glucose. Its inhibitor prevents the high glucose induced decrease in PTB protein and insulin biosynthesis through high glucose induced increase in miR-133a. It may decrease UCP2 activity because it targets UCP2 in islets cells [34]. Since miR-133a is highly expressed in muscle tissue, insulin mediated downregulation of miR-133a levels in human skeletal muscles are attenuated in T2D [35]. It has been observed lately that miR-126 is actively involved in the development of insulin resistance induced by mitochondrial dysfunction [36]. It directly targets the 3′UTR of IRS-1 (Insulin receptor substrate-1) and it's over expression in hepatocytes causes a substantial reduction in IRS1 protein expression and consequent impairment in insulin signaling.

The Role of miRNAs in Insulin Action

Insulin secretion, insulin gene transcription, mRNA stability, mRNA translation and insulin processing are mediated by glucose homeostasis in pancreatic beta cells [37]. Blood glucose homeostasis also requires the insulin target cells to respond effectively to the hormone. The peripheral tissues, namely the liver, adipose and skeletal muscles are three main target tissues in the body that engage in insulin action and hence, glucose uptake from the blood. Prolonged exposure of pancreatic β-cell line in MIN6 to high glucose results in the expression changes of several miRNAs [37]. Consequently miR-124a, miR-107, miR-30d in MIN6 cells are upregulated, whereas miR-296, miR-484 and miR-690 are significantly down regulated by high glucose treatment. miR369-5P, miR-130a, miR-410, miR-27a, miR-337, miR-200a, miR-320, miR-532, miR-192 and miR-379 are identified to be down regulators in glucose non responsive MIN6 cells, compared to glucose responsive cells [30]. The expression of miRNAs is often tissue specific hence, miRNAs play an important role in repression of gene expression at specific stages in various biological processes.

Liver

As the metabolic hub, liver plays a significant role in glucose homeostasis whereby miRNA participate in hepatic pathophysiology in the development of diabetics. Liver contributes hugely towards maintaining the balance as it can both stimulate and inhibit the glucose output. This balance is disturbed in diabetics whereby some miRNAs play role. Moreover, the insulin receptors in the liver serve as the treatment target in diabetics by regulating gluconeogenesis, glucogenolysis and hepatic glucose production [38]. The comparison of global miRNA expression in two target tissues- liver and adipose in diabetic Goto-Gakizaki rats with those in normogylcemic brown Norway rats [39] reveals that out of 170 miRNAs detected in both the tissues, miR-125a shows the most significant up-regulation in the liver. miR-222, miR-27a, miR-195, miR-103 and miR-10b are observed to have expression patterns consistent with glycemic status, suggesting their implications in primary events of T2D pathogenesis. Liver not only regulates glucose homeostasis but also participates in lipid metabolism. Several miRNAs that have been associated with hepatic pathogenesis are also associated with obesity, another hallmark of diabetes. miR-122 is a liver specific miRNA that is most abundantly expressed in liver. The inhibition of miR-122 in mice results in decreased hepatic fatty acids and cholesterol synthesis along with reduction in plasma cholesterol. Hepatic inhibition of miR-122 in diet induced obese mice leads to decreased plasma cholesterol and significant improvement in hepatic steatosis. It has been suggested that miR-122 indirectly regulates phosphorylated AMP activated protein kinase (AMPK) signaling which may correct the imbalance in liver insulin resistance [40, 41]. The dysregulation of miR-122 and miR-24a in diabetic livers further strengthens their implication in hepatic abnormalities and hyperglycemia. miR-33a and miR-33b have been reported as regulators of cholesterol homeostasis through interaction with sterol regulatory element binding proteins [42]. It has also been confirmed by another study by investigating the role of these two miRNAs in regulating fatty acid metabolism and insulin signaling [43]. miR-33a/b inhibit the expression of insulin receptor substrate2 (IRS2) in hepatic cells, subsequently reducing AKT and ERK. Anti-miRs of endogenous miR-33a/b regulate pathways controlling three of the risk factors of metabolic syndrome namely; levels of HDL, triglycerides and insulin signaling. Therefore, they are considered as useful in therapeutic modalities of metabolic syndrome. Reduction in miR-143 by transfecting 2'-o-methoxythyl Phosphorothioate modified antisense RNA, oligonucleotides inhibited the expression of adipocyte specific genes, insulin sensitive glucose transporter 4; Glut4, hormone sensitive lipase; HSL, fatty acid binding protein; activating protein 2 (AP2) and peroxysome proliferators activated receptor; (EPAR2-P) and the accumulation of triglycerides in drosophila. miR-17 regulates adipocyte droplet size and triglyceride levels [44] and miR-278 appears to control insulin sensitivity in adipose tissue [45]. These miRNAs have been found in insects only and there are no homologous of these miRNAs in mammals.

Insulin receptor substrate1 (IRS1) and IRS2 are significant mediators of insulin signaling pathway. IRS3 is limited only to mice and IRS4 is limited to brain, thymus, kidney and beta cells. In adipocytes, both IRS1 and IRS2 are expressed and have synergistic effects. IRS1 has been identified as the target of miRNA, especially miR-145 [46] whereas IRS2 plays a central role in the development of

T2 diabetes and its associated complications [47]. Mitochondrial dysfunction is associated with the development of insulin resistance, along with down regulation of IRS1 in myocytes [48].

Adipose Tissue

Adipose tissue is also a crucial regulator of glucose metabolism. Abnormally high amount of adipose tissue is associated with pathogenesis of diabetes in insulin induced glycogenesis, whereby, blood glucose is converted into fatty acids for energy storage. Lately, several miRNAs have been identified to be preferentially expressed in adipose tissue and to regulate several biological processes in the tissue including adipogenesis, adipocyte differentiation, insulin action and fat metabolism [49]. miR-103 and miR-107 are involved in the regulation of cellular Acetyl CoA and lipid levels [50]. miR-143 regulates adipocyte differentiation and its inhibition leads to reduced adipogenesis [51]. Levels of miR-143, miR-103 and miR-107 are reduced in adipocytes after treatment with TNF-alpha, which suggests that cytokines contribute to reduced adipogenesis and obesity [52]. The research on the association of miRNAs with adipose tissue reflects that obesity leads to loss of miRNA function, which is required for adipogenesis and suggests a mechanism for obesity induced insulin resistance. miR-320 and miR-27b are also involved in insulin resistance. miR-320 expression in insulin resistant adipocytes is observed to be 50 folds higher than normal 3T3-L1 adipocytes [53]. It is found that anti-miR-320 treatment of insulin resistant adipocytes increases the insulin sensitivity by targeting P85, which contributes to all growth by regulating the AKT phosphorylation and Glut4 levels. Overexpression of miR-27b is associated with adipogenesis and inhibits the expression of Peroxisome proliferator-activated receptor gamma (PPAR-γ) and CCAAT/enhancer binding protein alpha (C/EBP-α) in the early stage of adipogenesis [54]. Another miRNA that has been identified to target PPAR family, especially PPAR alpha is miR-519d. More recently miR-130 overexpression is also found to impair adipogenesis and to repress (PPAR-α) [55]. In the search of miRNAs that showed differential expression during adipogenesis of adipose tissue derived stem cells [56], miR-642a-3P is identified as an adipose specific miRNA and miR-30 family as an important regulator of human adipogenesis. Overexpression of miR-30a induces adipogenesis whereby silencing its expression blocks adipogenesis. Enhanced expression is observed in miR-29 family during hyperglycemia and hyperinsulinemia [39]. Introduction of miR-29 in 3T3-L1 adipocytes reduces insulin stimulated glucose uptake by inhibiting insulin signaling *via* the AKT signaling pathway. Similarly, elevated levels of miR-320 are observed in insulin resistant 3T3-L1 adipocytes [33] and antago-miR treatment improves insulin sensitivity *via* regulation of the insulin PI3K signaling pathways [53]. Although, these therapeutic effects are observed in insulin resistant adipocytes and not the normal cells, these results suggest the significant role of miRNAs as important players in insulin signaling pathway in the adipose tissue and reasonably, one may believe that these miRNAs could serve as a promising therapeutic agents against diabetes.

Skeletal Tissue

Skeletal muscles are the major site of glucose disposal, which account for approximately 75% of insulin stimulated glucose uptake, hence serving as the primary site for glucose uptake postprandially. Several studies have shown that both diabetic animal models and patients have revealed some miRNAs to be dysregulated in the skeletal muscles. The impact of insulin on the global miRNA expression in the skeletal muscles is investigated by analyzing miRNA expression in muscle biopsies taken from healthy subjects before and after 3hours euglycemic-hyperinsulinemic clamps [35]. Insulin down-regulation is observed in 39 miRNAs including miR-1, miR-133a, miR-206 and miR-29a/c, which are also involved in muscle development and are highly expressed in insulin sensitive tissue. Down-regulation of miR-1 and miR-133a by insulin is regulated by transcription factors; sterol-regulatory-binding protein (SREBP)-1C and myocyte enhancing factor 2C (MEF2C). The effect of insulin on these two miRNAs is observed to be altered in the skeletal muscles of T2D patients which corroborate impairment of MEF2C and SREBP1C stimulation in the same conditions [57]. However, in non insulin stimulated conditions, no significant differences are observed in miR-1 and miR-133a expression in the skeletal muscles of the diabetic *vs.* healthy subjects. Similar reduction in miR-133a levels in T2D patients is observed that correlates with higher fasting glucose levels and clinical parameters [58]. However, no change is detected in miR-1, miR-133a and miR-

206 expression in response to 3 hour hyperinsulinemic-euglycemic clamps [59]. The role of the muscle specific miR-1 and miR-133a in glucose homeostasis is also investigated in cardiomyocytes. Both miR-1 and miR-133a are found to suppress Glut4 expression resulting in reduced insulin stimulated glucose uptake and this effect is observed to be mediated by inhibition of Kruppel like transcription factor (KLTF-15) and is the direct target of miR-1 and miR133a/b [60]. Other identified targets of mir-133a/b include human ether–a-go-go (hERG) and potassium voltage-gated channel, KQT-like subfamily, member 1 (KCNQ1) which are involved in the formation of K+ current channel in the heart and is significantly associated with the long QT syndrome [61]. It has been observed that apoptosis in cardimyocytes exposed to high glucose is associated with elevated miR-1 levels [62]. Some studies are carried out in GK insulin resistant and diabetic rats. miRNA profiling in the skeletal muscles of GK rats is compared with non-diabetic control rats [33, 64, 65]. Almost 25 changes are reported and only two could detect the up-regulation of miR-130a. Overall, neither miR-1 nor miR-133a expression is altered in GK rats. It has been observed that reduced expression of miR-24 in diabetic rat muscles is inversely correlated to P38 mitogen-activated protein kinase (P38MAPK) expression [65]. P38MAPK promotes the activation of myocyte enhancer factor2, a muscle specific transcription factor that regulates the transcription of insulin responsive Glut4. Down-regulation of miR-24 with simultaneous increase of P38MAPK activity in GK rat muscles suggests an adaptation mechanism for skeletal muscles in response to high glucose concentration. It has been suggested that down-regulation of miR-126 shows a similar adaptive response [66].

These studies have investigated the role and relevance of miRNAs in diabetes and in different insulin target tissues. Such studies offer a comparison of dysregulated miRNAs across different tissues and hence, allow the identification of those miRNAs that are unique or similar in different tissues (Table **1**). This may prove to be useful in future therapeutic applications by delivering the specific miRNA, the one or a few targeted tissues effectively. However, future studies focusing on miRNA vis-à-vis hyperglycemia, miRNA dysregulation and insulin resistance will clear the overall implications of these miRNAs in the context of diabetes.

Table 1: List of significant miRNAs and their role in diabetes and diabetic complications

MicroRNA	Function of the MicroRNA	References
Let-7	Developmental timing and adipogenesis	[78]
Let-7f	Promotes endothelial progenitor cell-mediated angiogenesis	[65, 79]
Lin-28a/b	Promote an insulin-sensitized state that resists high-fat-diet induced diabetes	[28]
miR-1	Cardiac development, protection against cardiac hypertrophy, impaired insulin response in skeletal muscles of T2D patients, regulates the expression of GLUT4 by targeting KLF15 and is involved in metabolic control in cardiac myocytes	[35, 58-63, 80-84]
miR-7	Expresses in pancreatic and adult fetal endocrine cells	[13-15]
miR-9	Expresses in pancreatic development, impairs insulin secretion in β cells, gets up-regulated in cardiomyocytes of STZ induced diabetic mice	[15, 21, 85]
miR-10b	Expresses in skeletal muscles in hyperglycemic rats	[64]
miR-15a	Pancreatic regeneration	[15, 27]
miR-15b	Role in pancreatic regulation possibly by targeting Ngn3	[15]
miR-16	Pancreatic regeneration in fetus	[15, 86]
miR-17	Regulates adipocyte droplet size and triglyceride levels	[44, 87]
miR-17-5P	Role in omental and subcutaneous adiposity, down regulation leads to monocytopoesis	[88]
miR-19a	Diabetes in re-fed (obese) mice	[89]
miR-19b	Regulates differentiation and function of beta cells	[90, 91]
miR-21	Self-renewal, suppression, interstitial fibrosis, cardiac hypertrophy, diabetic nephropathy	[31, 92-96]
miR-23a/b	Cardiac hypertrophy	[97]
miR-24	Associates with diabetes through down-regulation of p38 MAPK	[65]
miR-24a	Dysregulation in diabetic liver, hepatic abnormalities and hyperglycemia.	[42]
miR-25	Diabetic nephropathy	[98]

Table 1: contd....

miR-27a/b	Role in omental and subcutaneous adiposity, impairs human adipocyte differentiation by suppression of PPAR-γ	[39, 54, 64, 88]
miR-29	Diabetic nephropathy	[33, 99, 100]
miR-29a/b/c	Glucose transport, over expression induces insulin resistance in 3t3 adipocytes, induced by high glucose and high insulin overexpression leads to insulin resistance, diabetic nephropathy	[33, 39, 64, 101]
miR-30	Regulates human adipogenesis	[102]
miR-30a/c	Down-regulates in T2D individuals	[51, 56, 103, 104]
miR-30d	Up-regulates in pancreas by high glucose up-regulating insulin gene transcription	[37, 56]
miR-30e	Role in omental and subcutaneous adiposity	[88]
miR-33a/b	Inhibit the expression of insulin receptor substrate2 (IRS2) in hepatic cells, reducing AKT and ERK. Their Anti-miRs regulate pathways controlling three of the risk factors; levels of HDL, triglycerides and insulin signaling.	[26, 43]
miR-34a/b	Contribute to fatty acid induced beta cell dysfunction, pro-inflammatory cytokines induce its over expression in human islets and MIN6cells	[31, 41, 105, 106]
miR-93	Gets down-regulated by high glucose through down-regulation of its host gene (MCN7), diabetic nephropathy	[107]
miR-96	Increases mRNA and protein levels of granulophilin (a negative regulator of insulin exocytosis)	[22]
miR-99a	Role in omental and subcutaneous adiposity	[88]
miR-103	Overexpression accelerates adipogenesis, gets reduced in response to TNFα, gets down-regulated in obesity	[37, 41, 50-52, 64, 108]
miR-107	High glucose down regulates its expression in insulima cells, gets up-regulated in βcells in presence of high glucose, over expression accelerates adipogenesis, reduces in response to TNFα	[41, 50-52, 108]
miR-122	Suppression in liver results in reduced fatty acid accumulation, gets down-regulated in liver of STZ mice	[40-42]
miR-124a	Regulation of insulin secretion mechanisms in fetus, regulation of transcription factor foxA2 in insulima cells, pancreatic islet development	[16, 22, 106, 109]
miR-125a/b	Epigenetic regulation of inflammatory genes in vascular smooth muscle cells	[39, 51, 52, 64, 110]

Table 1: contd....

miR-126	Expresses in pancreatic development, gets up-regulated in skeletal muscles of GK rats and in liver of obese mice	[11, 41, 65, 36]
miR-130	Suppresses adipogenesis by inhibiting peroxisome proliferator-activated receptor gamma expression	[55]
miR-130a	Down regulators in glucose non responsive MIN6 cells compared to glucose responsive cells	[65, 30]
miR-132	Significantly increases in retinal endothelial cells of diabetic rats, role in omental and subcutaneous adiposity	[88, 111]
miR-133	Role in long QT syndrome, cardiac hypertrophy	[61, 81, 82, 58]
miR-133a/b	Cardiomyocyte hypertrophy in diabetes, reduces in human skeletal muscles in T2D, high fasting glucose associates with lowered expression of miR-133a, regulates the expression of GLUT4 by targeting KLF15 and is involved in metabolic control in cardiac myocytes	[35, 58-60, 89]
miR-134	Brain specific spatiotemporal control of mRNA translation, ectodermal lineage differentiation, role in omental and subcutaneous adiposity	[88]
miR-140	Role in omental and subcutaneous adiposity	[88]
miR-142-3P	Significantly increases in retinal endothelial cells of diabetic rats	[111]
miR-143	Adipocyte differentiation and gets down-regulated in obesity	[51, 52, 112]
miR-144	Up-regulates in all insulin target tissues of T2D rats, a direct regulator of IRS1 which is the key component of insulin signaling cascade and down regulated in T2D.	[89, 113]
miR-145	Cell proliferation, role in omental and subcutaneous adiposity	[46, 88]
miR-145a	Up-regulates during adipocyte differentiation	[40]
miR-146	Potential target for diabetic retinopathy	[22, 89, 111]
miR-146a	Increased expression in islets from db/db obese mice, contributes to fatty acid- induced beta cell dysfunction, pro-inflammatory cytokines induce its expression in human islets and MIN6 cells, transcriptional circuitry regulating extracellular matrix protein production in diabetic mice	[22, 31, 32]
miR-147	Role in omental and subcutaneous adiposity	[88]
miR-150	Beta cell differentiation	[33]

Table 1: contd....

miR-155	Immune function, role in omental and subcutaneous adiposity	[88, 111]
miR-181a	Role in omental and subcutaneous adiposity	[88]
miR-181d	Most effective miRNA at reducing intracellular lipid content of hepatocytes	[114]
miR-192	Kidney and diabetic neuropathy development	[30, 75, 115-119]
miR-195	Role in pancreatic regulation possibly by targeting Ngn3	[17, 64]
miR-197	Role in omental and subcutaneous adiposity	[88]
miR-200	Regulator of insulin secretionby expression profiling of MIN6 cells	[30]
miR-200a	Prevention of progressive kidney diseases	[41, 75, 105, 116, 119, 120]
miR-200b	Diabetic retinopathy, pro-inflammatory role in vascular smooth muscle cells in diabetic mice	[105, 117, 119-121]
miR-200c	Diabetic nephropathy	[105, 117, 119]
miR-206	Contributes to high glucose mediated apoptosis in cardiomyocytes	[58, 83]
miR-208	Stress dependent cardiac remodeling, biomarker of cardiac injury	[122, 123]
miR-210	Role in omental and subcutaneous adiposity	[88]
miR-216a	Diabetic nephropathy	[75, 116]
miR-217	Kidney disorders, gets up-regulated by transforming growth factor β, activates Akt signaling through targeting of PTEN	[75, 116]
miR-218	Expresses in mouse early fetal pancreas, controls liver and pancreatic development regulator Onecut2 in liver embryonic cells	[124]
miR-222	Up-regulates in response to high glucose in adipose tissue of diabetic rats	[39, 64]
miR-223	Increased levels required for granulopoesis, up-regulates in insulin resistant heart increasing glucose uptake through increase of Glut4	[125]
miR-278	Regulates insulin sensitivity in adipose tissue	[45]
miR-296	Expresses in β cell islets, up-regulates by glucose	[109]
miR-320	Impaired angiogenesis in diabetes	[53, 126]

Table 1: contd....

miR-337	Down-regulator in glucose non responsive MIN6 cells	[30]
miR-369-5p	Down-regulator in glucose non responsive MIN6 cells	[30]
miR-373	Cardiomyocyte hypertrophy	[127]
miR-375	Pancreatic islet development, Regulates insulin secretion by inhibiting myotrophin, increases beta cell death by lipo-apoptosis, regulates cell proliferation, up-regulates in beta cells of T2D, regulates P13 pathway by regulation of PDK1 in insulima cells	[4, 11, 12, 15, 18, 19, 105, 109, 128-133]
miR-377	Fibronectin production in diabetic nephropathy	[87, 107]
miR-379	Down regulates in glucose non responsive MIN6 cells, compared to glucose responsive cells	[30, 33]
miR-410	Regulator of insulin secretionby expression profiling of MIN6 cells	[30]
miR-484	High glucose down regulates its expression in insulima cells	[37]
miR-503	Insulin secretion control, acts on progenitor cells, function similar as miR-375	[11, 134]
miR-519d	Targets PPAR family, especially alpha, regulation of adiposity and obesity	[135, 136]
miR-532	Down-regulator in glucose non responsive MIN6 cells	[30]
miR-657	Regulates insulin like growth factor regulator and variants in its regulation site confers a risk for diabetes	[137]
miR-690	Expresses in β cell islets, up-regulates by glucose	[37]

Futuristic miRNA Therapeutic Strategies

The function of individual miRNA can regulate the expression of multiple target genes, hence, modulating the expression of single miRNA may influence an entire gene network which may pose threat for the development of certain complex diseases. Those miRNAs which are highly expressed in diabetic conditions need to be inhibited to restore normal regulation of target genes while for those miRNA whose expression is lost or attenuated need to be replenished by supplementation with miRNA of interest. Some important theories have been developed to identify the miRNA target recognition and these have been applied to computationally predict targets of miRNA regulation. Experimental target identification have been applied by transcriptome analysis and biochemical approaches of mRNA target identification by immunoprecipitation of Ago proteins, either tagged or endogenous to analyse the associated mRNA as candidate mRNA targets. As the Ago tagging and immunoprecipitation can miss degraded mRNA, hence, these methods cannot be used to identify targets that are exclusively regulated at the level of translation.

Diabetes is associated with alterations of several miRNA levels either in insulin secreting cells or insulin target tissues. The techniques that allow *in vivo* delivery of miRNA mimics or anti-miRs especially in beta cells and insulin target tissue could permit correcting the level of key mRNA under diebetic condition. Different chemically modified oligonucleotides have been used *in vitro* to modulate mRNA expression (Kolfschoten IG, 2009). Out of them, miRNA-mimic oligonuceotides increase the expression of mRNA of interest. Locked nucleic acid (LNA) anti-miRs, antagomiR and morpholinos are very promising inhibitors of miRNA function and quite effective *in vivo* (Wu B, 2009, Elmen J, 2008, Krutzfeldt J, 2005).

Alternative strategies to chemically modify oligonucleotides have been developed to mitigate the miRNA function *in vivo*. We have abundant tools at our disposal for forced expression of miRNAs in specific tissues or to block the function of individual miRNAs *in vivo* using mRNA sponge [76] or target mimicry techniques [77]. The most important approach in this regard is the use of "miRNA sponges". These are artificial miRNA decoys that bind and hold native miRNA to compete with natural targets and then create a loss of function of the miRNA of interest. As it is known to us that one miRNA can potentially target several genes hence, miRNA based therapeutics have become challenging task sofar. Moreover, individual miRNA may express differently when other miRNAs are also participating which may hold the key for therapeutic intervention and clinical management of diabetes.

EPILOGUE

miRNAs are emerging as wonder molecules that regulate gene expression. Diabetes is associated with alterations in the level of several miRNAs in insulin secreting cells and insulin target tissues. There is spate of data regarding human miRNAs and their functional consequences in various tissues, but the functional role and relevance of these miRNAs will have to be validated since mRNA can be potentially targeted by numerous miRNAs. miRNAs can also stimulate the translation of miRNAs [67] or bind to promoter region of genes and activate transcription [68]. Moreover, miRNAs can regulate the abundance of other miRNAs [69].

It is now well understood that miRNA profile of insulin resistance tissues changes years before the onset or diagnosis of T2D. The plasma signature of 5 miRNAs *i.e.,* miR-15a, miR-29b, miR-126, miR-223 and miR-28-3p can accurately diagnose the likelihood of developing diabetes [70]. However, it is clarified that although these miRNAs may or may not serve as disease indicators of early onset of diabetes but are significant players for its progression, hence, good target for early intervention [71]. It is also possible that these miRNAs act far away from their original site of biogenesis but participate as extracellular signaling molecule [72]. It has been corroborated recently that miR-150 transmits through the blood stream and is uptaken by target endothelial cells [73]. miRNAs are now serving as therapeutic targets also. miR-192 can be a molecular target for preventing diabetic nephropathy as its action seems to enhance other renal miRNAs.

ACKNOWLEDGMENTS

I am very thankful to Dr. Monica Singh, Research Scientist, Department of Human Genetics, Punjabi University, Patiala for her scholarly and exhaustive help in compiling this chapter.

CONFLICT OF INTEREST

The authors confirm that this chapter contents have no conflict of interest.

REFERENCES

[1]   World Health Organization. Diabetes fact sheet no. 312. Available at http://www.who.int/mediacentre/factsheet/fs312/en/english.html accessed July, 2012.
[2]   McCarthy MI. Genomics, type2 diabetes and obesity. N Eng J Med 2010; 363: 2339-50
[3]   Fernandez-Valverde SL, Raft RJ, Mattick JS. MicroRNAs in β-cell biology, insulin resistance, diabetes and its compcations. Diabetes 2011; 60:1825-31
[4]   Poy MN, Eliasson L, Krutzfeldt J, Kuwajima S, Ma X, Macdonald PE, Pfeffer S, Tuschl T, Rajewsky N, Rorsman P, Stoffel M. A pancreatic islet-specific microRNA regulates insulin secretion. Nature 2004; 432: 226-30.

[5]   Small EM, Frost RJ, Olson EN. MicroRNA add new dimension to cardiovascular disease. Circulation 2010; 121: 1020-32.

[6]   Sayed D, Abdellatif M. MicroRNA in development and disease Physiol Rev 2011; 91: 827-87.

[7]   Bartel DP. MicroRNAs: genomics, biogenesis mechanism and function. Cell 2004; 116: 281-97.

[8]   Lee RC, Feinbaum RL, Ambros V. The *C. elegans* heterochronic gene lin-4 encodes small RNA with antisense complementarity to lin-14. Cell 1993; 75: 843-54.

[9]   Reinhart BJ, Slack FJ, Basson M, Pasquinelli AE, Bettinger JC, Rougvie AE, Horvitz HR, Ruvkun G. The 21-nucleotide let-7 RNA regulates developmental timing in *Caenorhabditis elegans*. Nature 2000; 403: 901-906.

[10]   Wightman B, Ha I, Ruvkun G. Posttranscriptional regulation of the heterochronic gene lin-14 by lin-4 mediates temporal pattern formation in *C. elegans*. Cell 1993; 75: 855-62.

[11]   Lynn FC, Skewes-Cox P, Kosaka Y, McManus MT, Harfe BD, German MS. MicroRNA expression is required for pancreatic islet cell genesis in the mouse. Diabetes 2007; 56(12): 2938-45.

[12]   Kloosterman WP, Lagendijk AK, Ketting RF, Moulton JD, Plasterk RH. Targeted inhibition of miRNA maturation with morpholinos reveals a role for miR-375 in pancreatic islet development. PLoS Biol 2007; 5(8): e203.

[13]   Bravo-Egana V, Rosero S, Molano RD, Pileggi A, Ricordi C, Domínguez-Bendala J, *et al.* Quantitative differential expression analysis reveals miR-7 as major islet microRNA. Biochem Biophys Res Commun 2008; 366(4): 922-6.

[14]   Correa-Medina M, Bravo-Egana V, Rosero S, Ricordi C, Edlund H, Diez J, *et al.* MicroRNA miR-7 is preferentially expressed in endocrine cells of the developing and adult human pancreas. Gene Expr Patterns 2009; 9(4): 193-9.

[15]   Joglekar MV, Joglekar VM, Hardikar AA. Expression of islet-specific microRNAs during human pancreatic development. Gene Expr Patterns 2009; 9(2): 109-13.

[16]   Baroukh N, Ravier MA, Loder MK, Hill EV, Bounacer A, Scharfmann R, *et al.* MicroRNA-124a regulates Foxa2 expression and intracellular signaling in pancreatic beta-cell lines. J Biol Chem 2007; 282(27): 19575-88.

[17]   Joglekar MV, Parekh VS, Mehta S, Bhonde RR, Hardikar AA. MicroRNA profiling of developing and regenerating pancreas reveal post-transcriptional regulation of neurogenin3. Dev Biol 2007; 311(2); 603-12.

[18]   Norlin S, Ahlgren U, Edlund H. Nuclear factor-{kappa}B activity in {beta}-cells is required for glucose-stimulated insulin secretion. Diabetes 2005; 54(1): 125-32.

[19]   Gauthier BR, Wollheim CB. MicroRNAs: 'ribo-regulators' of glucose homeostasis. Nat Med 2006; 12(1): 36-8.

[20]   Poy MN, Hausser J, Trajkovski M, Braun M, Collins S, Rorsman P, *et al.* miR-375 maintains normal pancreatic alpha- and beta-cell mass. Proc Natl Acad Sci U S A. 2009; 106(14): 5813-8.

[21]   Plaisance V, Abderrahmani A, Perret-Menoud V, Jacquemin P, Lemaigre F, Regazzi, R. MicroRNA-9 controls the expression of Granuphilin/Slp4 and the secretory response of insulin-producing cells. J Biol Chem 2006; 281(37): 26932-42.

[22]   Lovis P, Gattesco S, Regazzi R. Regulation of the expression of components of the exocytotic machinery of insulin-secreting cells by microRNAs. Biol Chem 2008; 389(3): 305-12.

[23]   Zhao X, Mohan R, Tang X. microRNA-30d induces insulin transcription factor MafA and insulin production by targeting mitogen-activated protein 4 kinase 4 (Map4k4) in pancreatic beta cells. J Biol Chem 2012.

[24]   Rayner KJ, Suárez Y, Dávalos A, Parathath S, Fitzgerald ML, Tamehiro N, *et al.* MiR-33 contributes to the regulation of cholesterol homeostasis. Science 2010; 328(5985): 1570-3.

[25]   Gerin I, Clerbaux LA, Haumont O, Lanthier N, Das AK, Burant CF, *et al.* Expression of miR-33 from an SREBP2 intron inhibits cholesterol export and fatty acid oxidation. J Biol Chem 2010; 285(44): 33652-61.

[26]   Wijesekara N, Zhang LH, Kang MH, Abraham T, Bhattacharjee A, Warnock GL, *et al.* miR-33a modulates ABCA1 expression, cholesterol accumulation, and insulin secretion in pancreatic islets. Diabetes 2012; 61(3): 653-8.

[27]   Sun LL, Jiang BG, Li WT, Zou JJ, Shi YQ, Liu ZM. MicroRNA-15a positively regulates insulin synthesis by inhibiting uncoupling protein-2 expression. Diabetes Res Clin Pract 2011; 91(1): 94-100.

[28]   Zhu H, Shyh-Chang N, Segrè AV, Shinoda G, Shah SP, Einhorn WS, *et al.* The Lin28/let-7 axis regulates glucose metabolism. Cell 2011; 147(1): 81-94.

[29]   Frost RJ, Olson EN. Control of glucose homeostasis and insulin sensitivity by the Let-7 family of microRNAs. Proc Natl Acad Sci USA 2011; 108(52): 21075-80.

[30]   Hennessy E, Clynes M, Jeppesen PB, O'Driscoll L. Identification of microRNAs with a role in glucose stimulated insulin secretion by expression profiling of MIN6 cells. Biochem Biophys Res Commun 2010; 396(2): 457-62.

[31]   Roggli E, Britan A, Gattesco S, Lin-Marq N, Abderrahmani A, Meda P, *et al.* Involvement of microRNAs in the cytotoxic effects exerted by proinflammatory cytokines on pancreatic beta-cells. Diabetes 2010; 59(4): 978-86.

[32]   Balasubramanyam M, Aravind S, Gokulakrishnan K, Prabu P, Sathishkumar C, Ranjani H, *et al.* Impaired miR-146a expression links subclinical inflammation and insulin resistance in Type 2 diabetes. Mol Cell Biochem 2011; 351(1-2): 197-205.

[33]   He A, Zhu L, Gupta N, Chang Y, Fang F. Overexpression of micro ribonucleic acid 29, highly up-regulated in diabetic rats, leads to insulin resistance in 3T3-L1 adipocytes. Mol Endocrinol 2007; 21(11): 2785-94.

[34]   Fred RG, Bang-Berthelsen CH, Mandrup-Poulsen T, Grunnet LG, Welsh N. High glucose suppresses human islet insulin biosynthesis by inducing miR-133a leading to decreased polypyrimidine tract binding protein-expression. PLoS One 2010; 5(5): e10843.

[35]   Granjon A, Gustin MP, Rieusset J, Lefai E, Meugnier E, Güller I, *et al.* The microRNA signature in response to insulin reveals its implication in the transcriptional action of insulin in human skeletal muscle and the role of a sterol regulatory element-binding protein-1c/myocyte enhancer factor 2C pathway. Diabetes 2009; 58(11): 2555-64.

[36]   Ryu HS, Park SY, Ma D, Zhang J, Lee W. The induction of microRNA targeting IRS-1 is involved in the development of insulin resistance under conditions of mitochondrial dysfunction in hepatocytes. PLoS One 2011; 6(3): e17343.

[37]   Tang X, Muniappan L, Tang G, Ozcan S. Identification of glucose-regulated miRNAs from pancreatic {beta} cells reveals a role for miR-30d in insulin transcription. RNA 2009; 15(2): 287-93.

[38]   Sekine S, Ogawa R, Mcmanus MT, Kanai Y, Hebrok M. Dicer is required for proper liver zonation. J Pathol 2009; 219(3): 365-72.

[39]   Herrera BM, Lockstone HE, Taylor JM, Wills QF, Kaisaki PJ, Barrett A, *et al*. MicroRNA-125a is over-expressed in insulin target tissues in a spontaneous rat model of Type 2 Diabetes. BMC Med Genomics 2009; 2:54.

[40]   Esau C, Davis S, Murray SF, Yu XX, Pandey SK, *et al*. miR-122 regulation of lipid metabolism revealed by *in vivo* antisense targeting. Cell Metab 2006; 3(2): 87-98.

[41]   Li S, Chen X, Zhang H, Liang X, Xiang Y, Yu C, Zen K, Li Y, Zhang CY. Differential expression of microRNAs in mouse liver under aberrant energy metabolic status. J Lipid Res 2009; 50(9): 1756-65.

[42]   Najafi-Shoushtari SH, Kristo F, Li Y, Shioda T, Cohen DE, Gerszten RE, *et al*. MicroRNA-33 and the SREBP host genes cooperate to control cholesterol homeostasis. Science 2010; 328(5985): 1566-9.

[43]   Dávalos A, Goedeke L, Smibert P, Ramírez CM, Warrier NP, Andreo U, *et al*. miR-33a/b contribute to the regulation of fatty acid metabolism and insulin signaling. Proc Natl Acad Sci USA 2011; 108(22): 9232-7.

[44]   Xu P, Vernooy SY, Guo M, Hay BA. The Drosophila microRNA Mir-14 suppresses cell death and is required for normal fat metabolism. Curr Biol 2003; 13(9): 790-5.

[45]   Teleman AA, Maitra S, Cohen SM. Drosophila lacking microRNA miR-278 are defective in energy homeostasis. Genes Dev 2006; 20(4): 417-22.

[46]   Shi B, Sepp-Lorenzino L, Prisco M, Linsley P, deAngelis T, Baserga R. MicroRNA 145 targets the insulin receptor substrate-1 and inhibits the growth of colon cancer cells. J Biol Chem 2007; 282(45): 32582-90.

[47]   White MF. Regulating insulin signaling and beta-cell function through IRS proteins. Can J Physiol Pharmacol 2006; 84(7): 725-37.

[48]   Kim JA, Wei Y, Sowers JR. Role of mitochondrial dysfunction in insulin resistance. Circ Res 2008; 102(4); 401-14.

[49]   Alexander R, Lodish H, Sun L. MicroRNAs in adipogenesis and as therapeutic targets for obesity. Expert Opin Ther Targets 2011; 15(5): 623-36.

[50]   Wilfred BR, Wang WX, Nelson PT. Energizing miRNA research: a review of the role of miRNAs in lipid metabolism, with a prediction that miR-103/107 regulates human metabolic pathways. Mol Genet Metab 2007; 91(3): 209-17.

[51]   Xie H, Sun L, Lodish HF. Targeting microRNAs in obesity. Expert Opin Ther Targets 2009; 13(10): 1227-38.

[52]   Xie H, Lim B, Lodish HF. MicroRNAs induced during adipogenesis that accelerate fat cell development are downregulated in obesity. Diabetes 2009; 58(5): 1050-7.

[53]   Ling HY, Ou HS, Feng SD, Zhang XY, Tuo QH, Chen LX, *et al*. Changes in microRNA profile and effects of miR-320 in insulin-resistant 3T3-L1 adipocytes. Clin Exp Pharmacol Physiol 2009; 36: e32-e39.

[54]   Karbiener M, Fischer C, Nowitsch S, Opriessnig P, Papak C, Ailhaud G, *et al*. microRNA miR-27b impairs human adipocyte differentiation and targets PPARgamma. Biochem Biophys Res Commun 2009; 390(2): 247-51.

[55]   Lee EK, Lee MJ, Abdelmohsen K, Kim W, Kim MM, Srikantan S, Martindale JL, Hutchison ER, Kim HH, Marasa BS, Selimyan R, Egan JM, Smith SR, Fried SK, Gorospe M. miR-130 suppresses adipogenesis by inhibiting peroxisome proliferator-activated receptor gamma expression. Mol Cell Biol 2011; 31(4): 626-38.

[56]   Zaragosi LE, Wdziekonski B, Brigand KL, Villageois P, Mari B, Waldmann R, *et al*. Small RNA sequencing reveals miR-642a-3p as a novel adipocyte-specific microRNA and miR-30 as a key regulator of human adipogenesis. Genome Biol 2011; 12(7): R64.

[57]   Ducluzeau PH, Perretti N, Laville M, Andreelli F, Vega N, Riou JP, *et al.* Regulation by insulin of gene expression in human skeletal muscle and adipose tissue. Evidence for specific defects in type 2 diabetes. Diabetes 2001; 50(5): 1134-42.

[58]   Gallagher IJ, Scheele C, Keller P, Nielsen AR, Remenyi J, Fischer CP, *et al.* Integration of microRNA changes *in vivo* identifies novel molecular features of muscle insulin resistance in type 2 diabetes. Genome Med 2010; 2(2): 9.

[59]   Nielsen S, Scheele C, Yfanti C, Akerström T, Nielsen AR, Pedersen BK, *et al.* Muscle specific microRNAs are regulated by endurance exercise in human skeletalmuscle. J Physiol 2010; 588: 4029-37.

[60]   Horie T, Ono K, Nishi H, Iwanaga Y, Nagao K, Kinoshita M, *et al.* MicroRNA-133 regulates the expression of GLUT4 by targeting KLF15 and is involved in metabolic control in cardiac myocytes. Biochem Biophys Res Commun 2009; 389(2): 315-20.

[61]   Xiao J, Luo X, Lin H, Zhang Y, Lu Y, Wang N, *et al.* MicroRNA miR-133 represses HERG K+ channel expression contributing to QT prolongation in diabetic hearts. J Biol Chem. 2007; 282(17): 12363-367.

[62]   Yu XY, Song YH, Geng YJ, Lin QX, Shan ZX, Lin SG, *et al.* Glucose induces apoptosis of cardiomyocytes *via* microRNA-1 and IGF-1. Biochem Biophys Res Commun 2008; 376(3): 548-52.

[63]   Elia L, Contu R, Quintavalle M, Varrone F, Chimenti C, Russo MA, *et al.* Reciprocal regulation of microRNA-1 and insulin-like growth factor-1 signal transduction cascade in cardiac and skeletal muscle in physiological and pathological conditions. Circulation 2009; 120(23): 2377-85.

[64]   Herrera BM, Lockstone HE, Taylor JM, Ria M, Barrett A, Collins S, *et al.* Global microRNA expression profiles in insulin target tissues in a spontaneous rat model of type 2 diabetes. Diabetologia 2010; 53(6):1099-109.

[65]   Huang B, Qin W, Zhao B, Shi Y, Yao C, Li J, *et al.* MicroRNA expression profiling in diabetic GK rat model. Acta Biochim Biophys Sin (Shanghai) 2009; 41(6): 472-7.

[66]   Guo C, Sah JF, Beard L, Willson JK, Markowitz SD, Guda K. The noncoding RNA,miR-126, suppresses the growth of neoplastic cells by targetingphosphatidylinositol 3-kinase signaling and is frequently lost in colon cancers. Genes Chromosomes Cancer 2008; 47(11): 939-46.

[67]   Vasudevan S, Tong Y, Steitz JA. Switching from repression to activation: microRNAs can up-regulate translation. Science 2007; 318(5858): 1931-4.

[68]   Place RF, Li LC, Pookot D, Noonan EJ, Dahiya R. MicroRNA-373 induces expression of genes with complementary promoter sequences. Proc Natl Acad Sci USA 2008; 105(5): 1608-13.

[69]   Pulkkinen K, Malm T, Turunen M, Koistinaho J, Ylä-Herttuala S. Hypoxia induces microRNA miR-210 *in vitro* and *in vivo* ephrin-A3 and neuronal pentraxin 1 are potentially regulated by miR-210. FEBS Lett 2008; 582(16): 2397-401.

[70]   Zampetaki A, Kiechl S, Drozdov I, Willeit P, Mayr U, Prokopi M, *et al.* Plasma microRNAprofiling reveals loss of endothelial miR-126 and other microRNAs in type 2diabetes. Circ Res 2010; 107(6): 810-7.

[71]   Regazzi R. Diabetes mellitus reveals its micro-signature. Circ Res 2010; 107(6): 686-8.

[72]   Dinger ME, Mercer TR, Mattick JS. RNAs as extracellular signaling molecules. J Mol Endocrinol 2008; 40(4): 151-9.

[73]   Zhang Y, Liu D, Chen X, Li J, Li L, Bian Z, *et al.* Secreted monocytic miR-150 enhances targeted endothelial cell migration. Mol Cell 2010; 39(1): 133-44.

[74]   Elmén J, Lindow M, Schütz S, Lawrence M, Petri A, Obad S, *et al.* LNA-mediated microRNA silencing in non-human primates. Nature 2008; 452(7189): 896-9.

[75]   Kato M, Putta S, Wang M, Yuan H, Lanting L, Nair I, *et al.* TGF-beta activates Akt kinase through a microRNA-dependent amplifying circuit targeting PTEN. Nat Cell Biol 2009; 11(7): 881-9.

[76]   Ebert MS, Neilson JR, Sharp PA. MicroRNA sponges: competitive inhibitors of small RNAs in mammalian cells. Nat Methods 2007; 4(9): 721-6.

[77]   Franco-Zorrilla JM, Valli A, Todesco M, Mateos I, Puga MI, Rubio-Somoza I, *et al.* Target mimicry provides a new mechanism for regulation of microRNA activity. Nat Genet 2007; 39(8): 1033-7.

[78]   Sun T, Fu M, Bookout AL, Kliewer SA, Mangelsdorf DJ. MicroRNA let-7 regulates 3T3-L1 adipogenesis. Mol Endocrinol. 2009 Jun;23(6):925-31.

[79]   Staszel T, Zapała B, Polus A, Sadakierska-Chudy A, Kieć-Wilk B, Stępień E, *et al.* Role of microRNAs in endothelial cell pathophysiology. Pol Arch Med Wewn 2011; 121(10):361-6.

[80]   Yang B, Lin H, Xiao J, Lu Y, Luo X, Li B, *et al.* The muscle-specific microRNA miR-1 regulates cardiac arrhythmogenic potential by targeting GJA1 and KCNJ2. Nat Med 2007; 13(4):486-91.

[81]   Carè A, Catalucci D, Felicetti F, Bonci D, Addario A, Gallo P, *et al.* MicroRNA-133 controls cardiac hypertrophy. Nat Med 2007; 13(5):613-8.

[82]   Liu N, Williams AH, Kim Y, McAnally J, Bezprozvannaya S, Sutherland LB, *et al.* An intragenic MEF2-dependent enhancer directs muscle-specific expression of microRNAs 1 and 133. Proc Natl Acad Sci USA 2007; 104(52): 20844-9.

[83]   Shan ZX, Lin QX, Deng CY, Zhu JN, Mai LP, Liu JL, *et al.* miR-1/miR-206 regulate Hsp60 expression contributing to glucose-mediated apoptosis in cardiomyocytes. FEBS Lett 2010; 584(16): 3592-600.

[84]   Katare R, Caporali A, Zentilin L, Avolio E, Sala-Newby G, Oikawa A, *et al.* Intravenous gene therapy with PIM-1 *via* a cardiotropic viral vector halts the progression of diabetic cardiomyopathy through promotion of prosurvival signaling. Circ Res 2011; 108(10):1238-51.

[85]   Ramachandran D, Roy U, Garg S, Ghosh S, Pathak S, Kolthur-Seetharam U. Sirt1 and mir-9 expression is regulated during glucose-stimulated insulin secretion in pancreatic β-islets. FEBS J 2011; 278(7):1167-74.

[86]   Shanmugam N, Reddy MA, Natarajan R. Distinct roles of heterogeneous nuclear ribonuclear protein K and microRNA-16 in cyclooxygenase-2 RNA stability induced by S100b, a ligand of the receptor for advanced glycation end products. J Biol Chem. 2008; 283(52):36221-33.

[87]   Wang Q, Li YC, Wang J, Kong J, Qi Y, Quigg RJ, *et al.* miR-17-92 cluster accelerates adipocyte differentiation by negatively regulating tumor-suppressor Rb2/p130. Proc Natl Acad Sci USA 2008; 105(8):2889-94.

[88]   Klöting N, Berthold S, Kovacs P, Schön MR, Fasshauer M, Ruschke K, *et al.* MicroRNA expression in human omental and subcutaneous adipose tissue. PLoS One 2009; 4(3):e4699.

[89]   Nakanishi N, Nakagawa Y, Tokushige N, Aoki N, Matsuzaka T, Ishii K, *et al*. The up-regulation of microRNA-335 is associated with lipid metabolism in liver and white adipose tissue of genetically obese mice. Biochem Biophys Res Commun 2009; 385(4):492-6.

[90]   He A, Liu X, Liu L, Chang Y, Fang F. How many signals impinge on GLUT4 activation by insulin? Cell Signal 2007; 19(1):1-7.

[91]   Zhang ZW, Zhang LQ, Ding L, Wang F, Sun YJ, An Y, *et al*. CB. MicroRNA-19b downregulates insulin 1 through targeting transcription factor NeuroD1. FEBS Lett 2011; 585(16):2592-8.

[92]   Thum T, Catalucci D, Bauersachs J. MicroRNAs: novel regulators in cardiac development and disease. Cardiovasc Res 2008; 79(4):562-70.

[93]   Roy S, Khanna S, Hussain SR, Biswas S, Azad A, Rink C, *et al*. MicroRNA expression in response to murine myocardial infarction: miR-21 regulates fibroblast metalloprotease-2 *via* phosphatase and tensin homologue. Cardiovasc Res 2009; 82(1):21-9.

[94]   Zhang Z, Peng H, Chen J, Chen X, Han F, Xu X, *et al*. MicroRNA-21 protects from mesangial cell proliferation induced by diabetic nephropathy in db/db mice. FEBS Lett 2009; 583(12):2009-14.

[95]   Vinciguerra M, Sgroi A, Veyrat-Durebex C, Rubbia-Brandt L, Buhler LH, Foti M. Unsaturated fatty acids inhibit the expression of tumor suppressor phosphatase and tensin homolog (PTEN) *via* microRNA-21 up-regulation in hepatocytes. Hepatology 2009; 49(4):1176-84.

[96]   Dey N, Das F, Mariappan MM, Mandal CC, Ghosh-Choudhury N, Kasinath BS, *et al*. MicroRNA-21 orchestrates high glucose-induced signals to TOR complex 1, resulting in renal cell pathology in diabetes. J Biol Chem 2011; 286(29):25586-603.

[97]   Lin Z, Murtaza I, Wang K, Jiao J, Gao J, Li PF. miR-23a functions downstream of NFATc3 to regulate cardiac hypertrophy. Proc Natl Acad Sci U S A 2009; 106(29):12103-8.

[98]   Fu Y, Zhang Y, Wang Z, Wang L, Wei X, Zhang B, *et al*. Regulation of NADPH oxidase activity is associated with miRNA-25-mediated NOX4 expression in experimental diabetic nephropathy. Am J Nephrol 2010; 32(6):581-9.

[99]   Pullen TJ, da Silva Xavier G, Kelsey G, Rutter GA. miR-29a and miR-29b contribute to pancreatic beta-cell-specific silencing of monocarboxylate transporter 1 (Mct1). Mol Cell Biol 2011; 31(15):3182-94.

[100]   Wang B, Komers R, Carew R, Winbanks CE, Xu B, Herman-Edelstein M, *et al*. Suppression of microRNA-29 expression by TGF-β1 promotes collagen expression and renal fibrosis. J Am Soc Nephrol 2012; 23(2):252-65.

[101]   Silva VA, Polesskaya A, Sousa TA, Corrêa VM, André ND, Reis RI, *et al*. Expression and cellular localization of microRNA-29b and RAX, an activator of the RNA-dependent protein kinase (PKR), in the retina of streptozotocin-induced diabetic rats. Mol Vis 2011;17:2228-40.

[102]   Duisters RF, Tijsen AJ, Schroen B, Leenders JJ, Lentink V, van der Made I, *et al*. miR-133 and miR-30 regulate connective tissue growth factor: implications for a role of microRNAs in myocardial matrix remodeling. Circ Res 2009; 104(2):170-8.

[103]   Ortega FJ, Moreno-Navarrete JM, Pardo G, Sabater M, Hummel M, Ferrer A, *et al*. MiRNA expression profile of human subcutaneous adipose and during adipocyte differentiation. PLoS One 2010; 5(2):e9022.

[104]   Karbiener M, Neuhold C, Opriessnig P, Prokesch A, Bogner-Strauss JG, Scheideler M. MicroRNA-30c promotes human adipocyte differentiation and co-represses PAI-1 and ALK2. RNA Biol 2011; 8(5).

[105]   Zhao E, Keller MP, Rabaglia ME, Oler AT, Stapleton DS, Schueler KL, *et al.* Obesity and genetics regulate microRNAs in islets, liver, and adipose of diabetic mice. Mamm Genome 2009; 20(8):476-85.

[106]   Lovis P, Roggli E, Laybutt DR, Gattesco S, Yang JY, Widmann C, *et al.* Alterations in microRNA expression contribute to fatty acid-induced pancreatic beta-cell dysfunction. Diabetes 2008; 57(10):2728-36.

[107]   Long J, Wang Y, Wang W, Chang BH, Danesh FR. Identification of microRNA-93 as a novel regulator of vascular endothelial growth factor in hyperglycemic conditions. J Biol Chem 2010; 285(30):23457-65.

[108]   Trajkovski M, Hausser J, Soutschek J, Bhat B, Akin A, Zavolan M, *et al.* MicroRNAs 103 and 107 regulate insulin sensitivity. Nature 2011; 474(7353):649-53.

[109]   El Ouaamari A, Baroukh N, Martens GA, Lebrun P, Pipeleers D, van Obberghen E. miR-375 targets 3'-phosphoinositide-dependent protein kinase-1 and regulates glucose-induced biological responses in pancreatic beta-cells. Diabetes 2008; 57(10):2708-17.

[110]   Villeneuve LM, Kato M, Reddy MA, Wang M, Lanting L, Natarajan R. Enhanced levels of microRNA-125b in vascular smooth muscle cells of diabetic db/db mice lead to increased inflammatory gene expression by targeting the histone methyltransferase Suv39h1. Diabetes 2010; 59(11):2904-15.

[111]   Kovacs B, Lumayag S, Cowan C, Xu S. MicroRNAs in early diabetic retinopathy in streptozotocin-induced diabetic rats. Invest Ophthalmol Vis Sci 2011; 52(7):4402-9.

[112]   Jordan SD, Krüger M, Willmes DM, Redemann N, Wunderlich FT, Brönneke HS, Merkwirth C, *et al.* Obesity-induced overexpression of miRNA-143 inhibits insulin-stimulated AKT activation and impairs glucose metabolism. Nat Cell Biol 2011; 13(4):434-46.

[113]   Karolina DS, Armugam A, Tavintharan S, Wong MT, Lim SC, Sum CF, *et al.* MicroRNA 144 impairs insulin signaling by inhibiting the expression of insulin receptor substrate 1 in type 2 diabetes mellitus. PLoS One 2011; 6(8):e22839.

[114]   Whittaker R, Loy PA, Sisman E, Suyama E, Aza-Blanc P, Ingermanson RS, *et al.* Identification of MicroRNAs that control lipid droplet formation and growth in hepatocytes *via* high-content screening. J Biomol Screen 2010 ; 15(7):798-805.

[115]   Krupa A, Jenkins R, Luo DD, Lewis A, Phillips A, Fraser D. Loss of MicroRNA-192 promotes fibrogenesis in diabetic nephropathy. J Am Soc Nephrol 2010; 21(3):438-47.

[116]   Kato M, Wang L, Putta S, Wang M, Yuan H, Sun G, *et al.* Post-transcriptional up-regulation of Tsc-22 by Ybx1, a target of miR-216a, mediates TGF-{beta}-induced collagen expression in kidney cells. J Biol Chem 2010; 285(44):34004-15.

[117]   Kato M, Arce L, Wang M, Putta S, Lanting L, Natarajan R. A microRNA circuit mediates transforming growth factor-β1 autoregulation in renal glomerular mesangial cells. Kidney Int 2011; 80(4):358-68.

[118]   Putta S, Lanting L, Sun G, Lawson G, Kato M, Natarajan R. Inhibiting microRNA-192 ameliorates renal fibrosis in diabetic nephropathy. J Am Soc Nephrol 2012; 23(3):458-69.

[119]   Long J, Wang Y, Wang W, Chang BH, Danesh FR. MicroRNA-29c is a signature microRNA under high glucose conditions that targets Sprouty homolog 1, and its *in vivo* knockdown prevents progression of diabetic nephropathy. J Biol Chem 2011; 286(13):11837-48.

[120]   Wang B, Koh P, Winbanks C, Coughlan MT, McClelland A, Watson A, *et al.* miR-200a prevents renal fibrogenesis through repression of TGF-β2 expression. Diabetes 2011; 60(1):280-7.

[121] Reddy MA, Jin W, Villeneuve L, Wang M, Lanting L, Todorov I, *et al*. Pro-inflammatory role of microrna-200 in vascular smooth muscle cells from diabetic mice. Arterioscler Thromb Vasc Biol 2012; 32(3):721-9.

[122] Zile MR, Mehurg SM, Arroyo JE, Stroud RE, DeSantis SM, Spinale FG. Relationship between the temporal profile of plasma microRNA and left ventricular remodeling in patients after myocardial infarction. Circ Cardiovasc Genet 2011; 4(6):614-9.

[123] Ji X, Takahashi R, Hiura Y, Hirokawa G, Fukushima Y, Iwai N. Plasma miR-208 as a biomarker of myocardial injury. Clin Chem 2009; 55(11):1944-9.

[124] Simion A, Laudadio I, Prévot PP, Raynaud P, Lemaigre FP, Jacquemin P. MiR-495 and miR-218 regulate the expression of the Onecut transcription factors HNF-6 and OC-2. Biochem Biophys Res Commun 2010; 391(1):293-8.

[125] Lu H, Buchan RJ, Cook SA. MicroRNA-223 regulates Glut4 expression and cardiomyocyte glucose metabolism. Cardiovasc Res 2010; 86(3):410-20.

[126] Wang XH, Qian RZ, Zhang W, Chen SF, Jin HM, Hu RM. MicroRNA-320 expression in myocardial microvascular endothelial cells and its relationship with insulin-like growth factor-1 in type 2 diabetic rats. Clin Exp Pharmacol Physiol 2009; 36(2):181-8.

[127] Shen E, Diao X, Wang X, Chen R, Hu B. MicroRNAs involved in the mitogen-activated protein kinase cascades pathway during glucose-induced cardiomyocyte hypertrophy. Am J Pathol 2011; 179(2):639-50.

[128] Tang X, Tang G, Ozcan S. Role of microRNAs in diabetes. Biochim Biophys Acta 2008; 1779(11):697-701.

[129] Poy MN, Hausser J, Trajkovski M, Braun M, Collins S, Rorsman P, *et al*. miR-375 maintains normal pancreatic alpha- and beta-cell mass. Proc Natl Acad Sci U S A 2009; 106(14):5813-8.

[130] Avnit-Sagi T, Kantorovich L, Kredo-Russo S, Hornstein E, Walker MD. The promoter of the pri-miR-375 gene directs expression selectively to the endocrine pancreas. PLoS One 2009; 4(4):e5033.

[131] Xia HQ, Pan Y, Peng J, Lu GX. Over-expression of miR375 reduces glucose-induced insulin secretion in Nit-1 cells. Mol Biol Rep 2011; 38(5):3061-5.

[132] Ling HY, Wen GB, Feng SD, Tuo QH, Ou HS, Yao CH.*et al*. MicroRNA-375 promotes 3T3-L1 adipocyte differentiation through modulation of extracellular signal-regulated kinase signalling. Clin Exp Pharmacol Physiol 2011; 38(4):239-46.

[133] Li Y, Xu X, Liang Y, Liu S, Xiao H, Li F, *et al*. miR-375 enhances palmitate-induced lipoapoptosis in insulin-secreting NIT-1 cells by repressing myotrophin (V1) protein expression. Int J Clin Exp Pathol 2010; 3(3):254-64.

[134] Caporali A, Emanueli C. MicroRNA-503 and the Extended MicroRNA-16 Family in Angiogenesis. Trends Cardiovasc Med 2011; 21(6):162-6.

[135] Martinelli R, Nardelli C, Pilone V, Buonomo T, Liguori R, Castanò I, *et al*. miR-519d overexpression is associated with human obesity. Obesity 2010; 18(11):2170-6.

[136] McGregor RA, Choi MS. microRNAs in the regulation of adipogenesis and obesity. Curr Mol Med 2011; 11(4):304-16.

[137] Lv K, Guo Y, Zhang Y, Wang K, Jia Y, Sun S. Allele-specific targeting of hsa-miR-657 to human IGF2R creates a potential mechanism underlying the association of ACAA-insertion/deletion polymorphism with type 2 diabetes. Biochem Biophys Res Commun 2008; 374(1):101-5.

Send Orders for Reprints to reprints@benthamscience.net

Frontiers in Clinical Drug Research: Diabetes and Obesity, Vol. 1, 2014, 149-210

CHAPTER 5

Drugs and the Peroxisome Proliferator Activated Receptors

Kevin Stuart[1], Andrew Hartland[2], Mithun Bhartia[3] and Sudarshan Ramachandran[4,*]

[1]Specialist Registrar in Chemical Pathology/Metabolic Medicine, Heart of England Foundation Trust, UK; [2]Consultant Chemical Pathologist, Walsall Manor Hospital, UK; [3]Specialist Registrar in Diabetes and Endocrinology, Sandwell Hospital, West Birmingham NHS Trust, UK and [4]Consultant Chemical Pathologist, Heart of England Foundation Trust, UK

Abstract: Fibrates and thiazolidinediones, modulators of PPARα and γ nuclear receptors respectively, are classes of drugs in current use. Their use in clinical medicine has varied depending on the evidence at the time. There have been agents within each group that have been withdrawn due to concern of adverse events. In the case of fibrates there possibly is a current resurgence based on early evidence of microvascular benefit in patients with T2DM. The use of pioglitazone, the only thiazolidinedione marketed, appears to be on the wane due to possible associations with cancer having already led to its withdrawal in certain countries. The dual PPARα/γ agonists are not in current use with muraglitazar and tessaglitazar withdrawn for differing reasons. Furthermore there is the possibility of PPARß/δ activators being developed.

It is clear that heterogeneity even within a class of drug exists and this may be dependent not just on the pharmacokinetics and pharmacodynamics of the drug, but also the complexity of the PPAR receptor with associated effects. We hope that this brief chapter, including a description of gene expression, nuclear receptors followed by details of PPAR and their ligands, will leave the reader with an appreciation of the complexity that PPAR agonism leads to and the heterogeneity of effects that has added to some of the confusion existing.

Keywords: Cholesterol, co-activators, co-repressor, fatty acids, fibrates, gene expression, glitazars, HDL-C, lipid metabolism, metabolic syndrome, nuclear receptors, paradoxical HDL-C change, PPARα, PPARß/δ, PPARγ, randomised controlled trials/meta-analysis, statins, thiazolidinediones, triglycerides, Type 2 diabetes.

*Address correspondence to Sudarshan Ramachandran: Department of Clinical Biochemistry, Good Hope Hospital, Heart of England NHS Foundation Trust, Rectory Road, Sutton Coldfield, West Midlands, B75 7RR, United Kingdom; Tel: +44 121 424 7246; Fax: +44 121 311 1800; E-mail: sud.ramachandran@heartofengland.nhs.uk

INTRODUCTION

PPAR is a family of nuclear receptors regulating many aspects of metabolism. Fibrates and thiazolidinediones are therapeutic agents considered to exert many of their actions *via* PPAR agonism. It was 50 years ago that the first PPARα agonist (clofibrate) was synthesised by Jeff Thorp and co-workers in the ICI laboratories in England in 1962 [1]. Since then use of both groups of drugs has varied considerably depending upon either benefits as demonstrated by clinical trials or reports of adverse effects. Currently fibrates appear to have gained a new lease of life following encouraging data regarding prevention of microvascular complications in patients with T2DM while the thiazolidinediones are waning due to published adverse effects.

The purpose of this chapter is to gain an understanding of the role of PPAR in the field of medicine. We will describe how PPAR agonists have affected human metabolism and the effect they have had on disease management. This will include both paths that have lead to significant progress as well as blind alleys. The phenomenon referred to as the paradoxical HDL-C decrease will be described, this outlining how effects opposite to those predicted are often encountered in clinical medicine. We will end by outlining further studies that are required to consolidate the role of PPAR agonists and speculate regarding new developments in this field.

To gain an understanding of this topic it is essential to grasp the process by which genotype is created into a phenotype; gene expression. Although most of the readers will possess greater knowledge than required, it is useful to remind us of some basic facts.

GENE EXPRESSION

This is a process by which the genetic code, carried within the gene by a stretch of DNA, expresses itself by forming the basis of a functional product which is usually a protein. There are non-protein coding genes which lead to the synthesis of transfer, ribosomal and small nuclear RNA. Regulation of the process takes place at various points including transcription, RNA splicing, translation and post-translational modification.

The process begins with transcription, a process whereby complimentary RNA, a copy of the 5' → 3' coding strand of the DNA, is created as shown in Fig. **1a** (the non-coding 3' → 5' DNA strand which is also complimentary to the 5' → 3' coding strand serving as the blueprint, with uracil occupying the position of thymine in the DNA). This is facilitated by the RNA polymerase enzyme of which there are many forms structurally and mechanistically related to each other.

The following, also outlined in Fig. **1b,** is a summary of the steps that lead to successful transcription and synthesis of a specific mRNA.

a. The breaking of hydrogen bonds between the coding and non-coding DNA strands.

b. The addition of RNA nucleotides that are complimentary and paired to the non-coding 3' → 5' DNA strand.

c. Formation of the RNA backbone.

d. Breaking the hydrogen bonds between the DNA strand and the newly formed mRNA, thus freeing the mRNA.

e. Modification of the mRNA such as addition of a 5' cap and the 3' poly A tail followed by migration from the nucleus to the cytoplasm.

Translation is the process by which this mRNA is decoded by the ribosome to form an amino acid sequence, a polypeptide or a larger protein. The individual amino acids are carried by tRNA, an adapter molecule (first proposed by Francis Crick in 1958) [2] which binds to the complimentary codons of the mRNA; this process being facilitated by ribosomes in the cytoplasm. Thus, ribosomes facilitate the base pairing of mRNA and tRNA leading to formation and elongation of the polypeptide chain (Fig. **1c**). The 3 termination codons UAA, UAG and UGA end elongation as no complimentary tRNA exists. At this point proteins referred to as release factors bind and result in dissociation of the complex.

Regulators of Gene Expression

Cell specialisation and functional heterogeneity are features of eukaryotes and each cell achieves this by activating a subset of its genes; differential gene

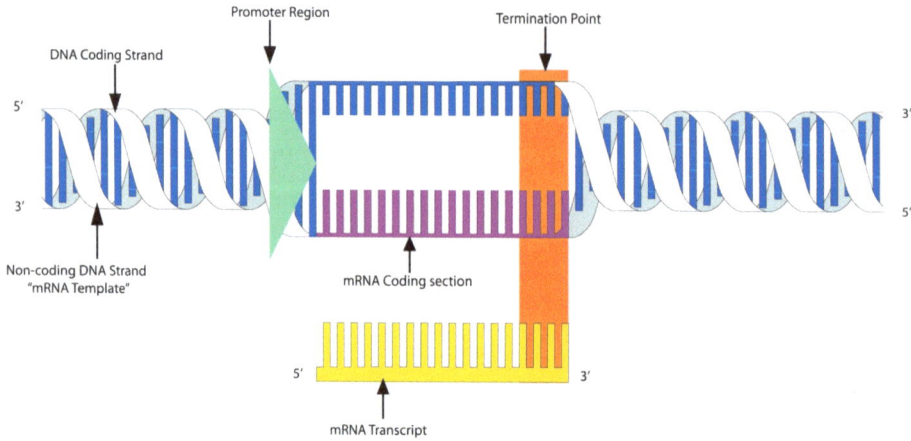

Figure 1a: A transcription unit containing the promoter, RNA coding section and the termination point.

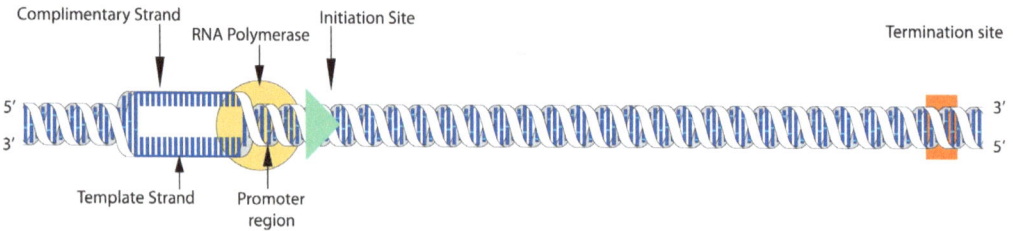

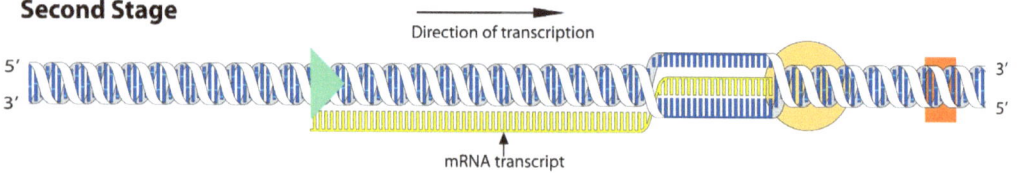

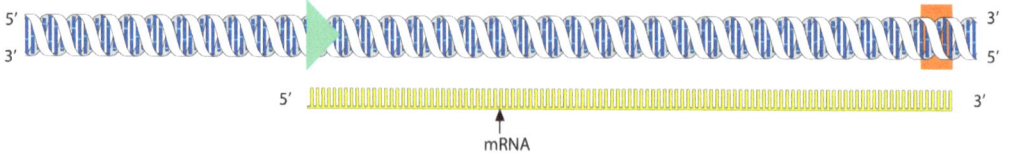

Figure 1b: The process of transcription of DNA to RNA. This figure illustrates the DNA partly unwound by the RNA polymerase and serving as the template for RNA synthesis. The 3 stages of transcription (first stage- initiation, second stage- elongation and third stage- termination) are shown separately.

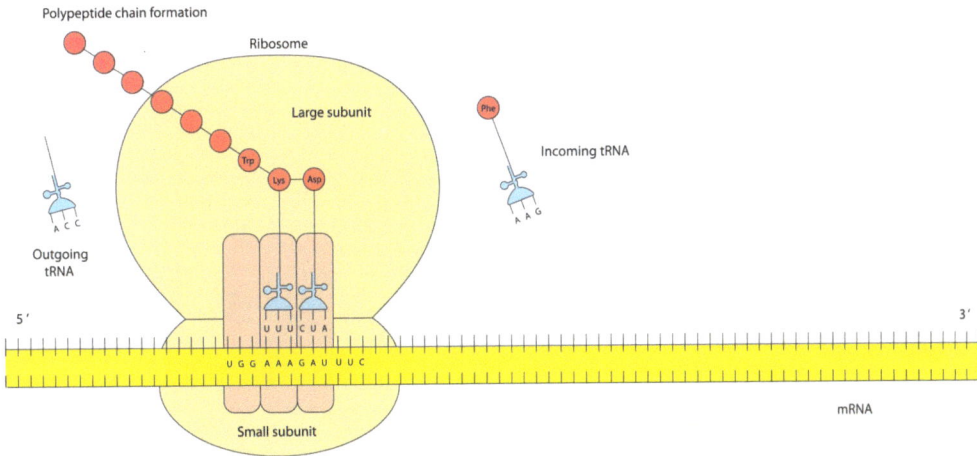

Figure 1c: mRNA translation- A ribosome decodes mRNA and pairs amino acid carrying tRNA to their matching sequences forming a growing polypeptide chain.

expression is a way to attain this complexity. Furthermore, this flexibility offers a way to adapt and respond to the environment. Gene expression can be regulated at various stages as outlined in Fig. **2** and points of control include epigenetic effects, transcription, post-translational modification, RNA transport, mRNA degradation and translation. As we are dealing with nuclear receptors in this chapter, only regulation of transcription will be dealt with at any length.

Eukaryote genes contain promoters (a DNA sequence where RNA polymerase binds prior to transcription) and one or more enhancers. Transcription factors (activators or repressors) bind to their specific enhancers. Within cells (or tissues) activity of transcription factors is regulated by environmental signals and ligands. While the promoter is responsible for initiating the start site and a relatively low degree of transcription, the enhancers (one or more DNA sequences that influence the transcription of a nearby gene) upon binding to transcription activator factors are responsible for enhancing the transcription rate. Transcription activator factors have functional domains for DNA binding and for transcription activation. They are classified by the structure of the DNA binding domains; examples include zinc finger proteins, helix turn proteins, helix loop helix proteins, leucine zipper proteins and nuclear receptors. Repressors on the other hand bind to non-coding sections of the DNA and impede the function of RNA polymerase leading to prevention of gene expression. Other proteins such as co-activators and co-repressors function in conjunction with transcription factors, but although altering gene expression they are not classified as transcription factors as they do not possess DNA binding sites.

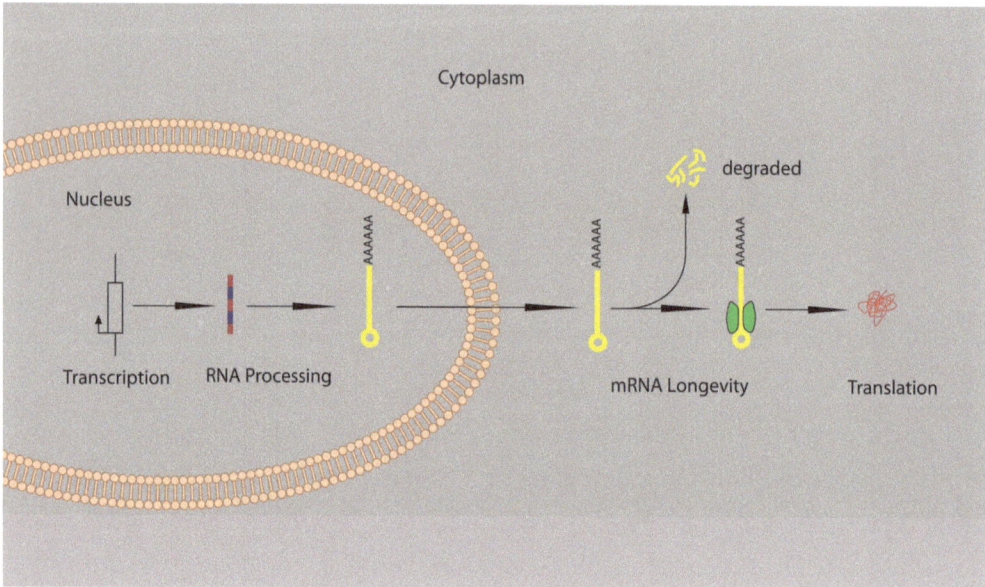

Figure 2: This figure illustrates the points at which gene expression can be regulated.

NUCLEAR RECEPTORS

The first of the nuclear receptor super-family to be identified was the oestrogen receptor by Elwood V Jensen at the University of Chicago in 1958 which he termed estrophilin [3]. Prior to that it was considered that oestradiol was an enzyme co-factor being converted to oestrone, thus facilitating the conversion of $NAD(P)^+$ to $NAD(P)H + H^+$, a redox reaction. Jensen, a man noted for following an alternative path as exemplified by his ascent of the Matterhorn, demonstrated that oestrogen did not alter in structure leading to him correctly surmising a receptor mediated action. Subsequently in 1968 he showed that the initial binding was in the cytoplasm followed by transfer of the complex into the cell nucleus. This seminal discovery of a nuclear receptor was initially presented by Jensen in Vienna to three other speakers and two attendees; a humble beginning to the field. The first nuclear receptor cloned was the human glucocorticoid receptor by Ronald M Evans [4]. Since then there has been an exponential increase in the study of the nuclear receptor super-family. Currently there are known to be 48 nuclear receptor genes in the human genome, but with alternative splicing and promoter usage there are 75 nuclear receptor proteins.

Nuclear receptors are characterised by their modular structure (Fig. **3**). They contain a hyper-variable (sequence and length) amino acid N-terminal domain (also referred to as the modulator or A/B domain) which can range from 6% (vitamin D receptor) to 50% (androgen receptors) of the total protein. It is ligand independent. The variability of this domain may be responsible for many of the species and cell type specific effects. This is attached to a highly conserved DNA binding domain (C domain) which binds to specific DNA sequences referred to as HRE. It is composed of 2 zinc fingers; the first one contains the proximal P-box region which is responsible for the high affinity recognition of the HRE, while the second contains the D-box region that mediates receptor dimerisation. Nuclear receptors bind to DNA as either hetero (*e.g.,* thyroid receptor, PPAR, vitamin D receptor bind together with RXRs) or homo (*e.g.,* glucocorticoid, oestrogen and androgen receptors) dimers and rarely as monomers. The binding affinity of this region is affected by the three dimensional structure of the heterodimer. This domain in turn is attached to the hinge region (D domain), a flexible domain which may influence intracellular movement and distribution. This domain may allow conformational changes to the protein structure following ligand binding and may also play a part in recognition of the 5' end of the HRE. The ligand binding domain (E domain) is less conserved than the DNA binding domain, but structural similarity is observed in this region between the various nuclear receptors. Ligand, co-activator/repressor binding sites are found within this domain with residues within the hydrophobic binding pocket conferring specificity. The C-terminal domain (F domain) is, like the N-terminal domain, highly variable in sequence.

Nuclear receptors are intra-cellular proteins specific to metazoans (nematodes, insects and vertebrates) which, following binding with a ligand and with the assistance of co-activators/repressors, regulate specific gene expression A conformational change is seen following ligand binding resulting in these protein complexes becoming active transcription factors leading to either up or down regulation of gene expression (Fig. **3b**). It is considered that the mechanism of action can take two forms. The steroid hormone receptors which are cytosolic are found in an inactive complex with proteins such as heat shock proteins. Following conformational change upon hormone binding, the monomeric receptor is released and has the capacity of binding to DNA as homo dimers. In contrast thyroid, vitamin D and PPAR receptors are found in the nucleus and do not bind to proteins such as heat shock proteins. They may be bound to co-repressors and in some cases may bind to DNA and act like transcription repressors. Conformational change following ligand binding allows dimerization with RXR (Fig. **4**). Co-activator binding can only follow the conformational change after ligand binding in both types of receptors.

Nuclear receptors have been classified in many ways, according to sequence homology, mechanism of action and function. We will try and describe these briefly in order to provide a background understanding prior to detailing the characteristics of PPAR.

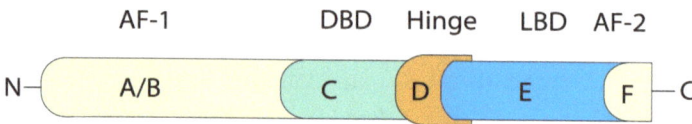

Figure 3a: The modular structure of the nuclear receptor family.

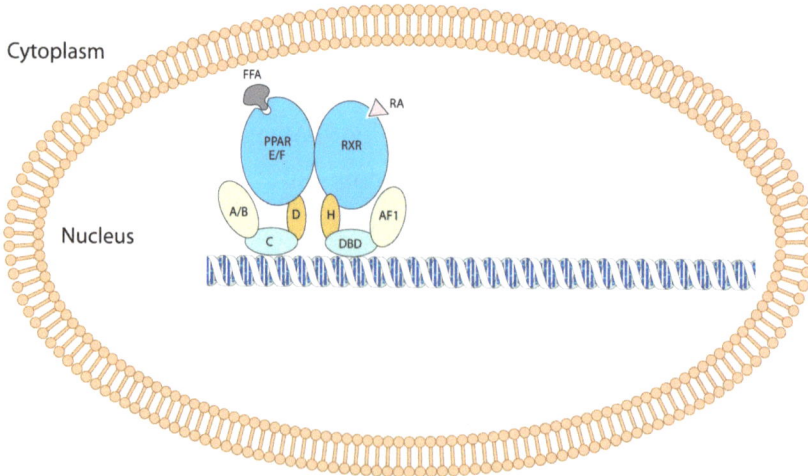

Figure 3b: The typical nuclear receptor structure is related to the modular DNA structure shown in Fig. **3a** by colour coding. PPAR on the left shows domains A-F; and RXR on the right shows the function of the domains with H being Hinge.

Activation of Nuclear receptors

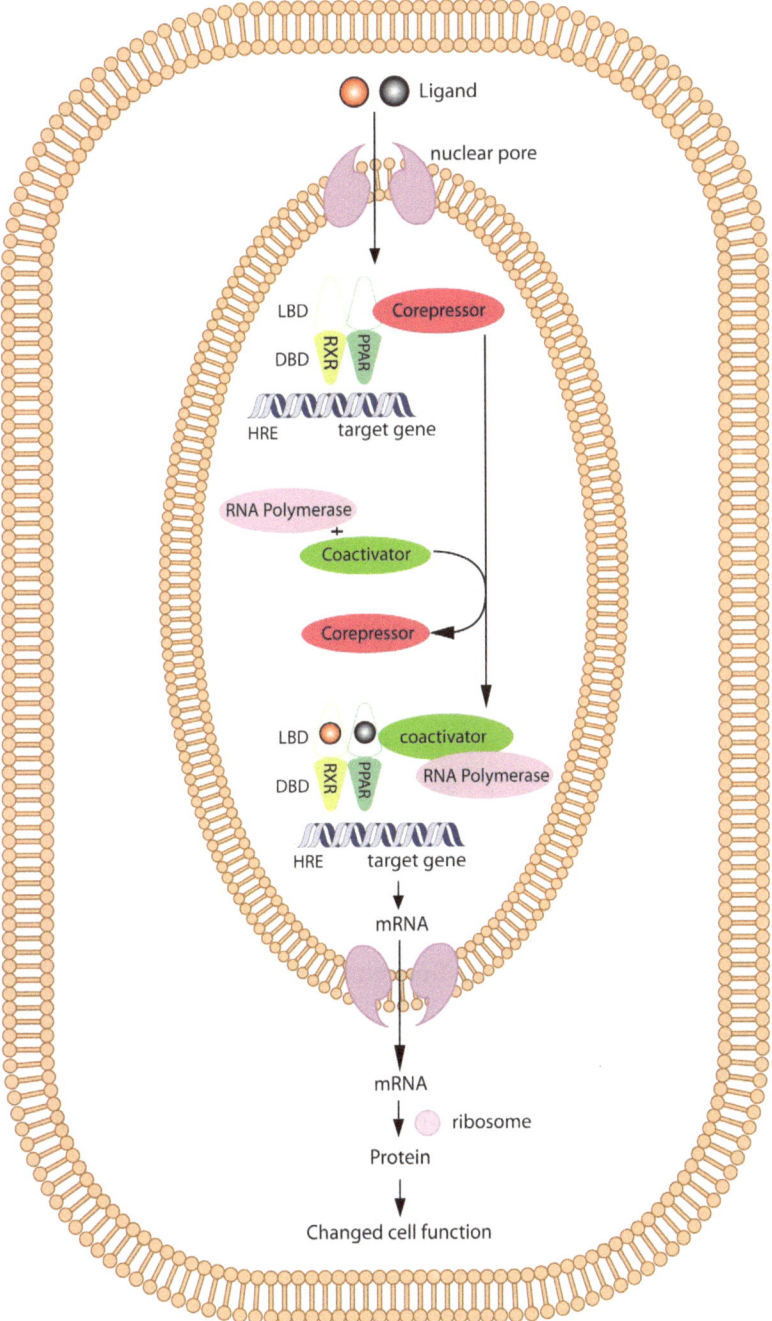

Figure 4: Ligand binding causes a conformational change leading to corepressor release and coactivator binding. The active transcription complex results in expression of specific target genes.

Nuclear receptors have been classified into 4 types by their mechanism of action and distribution in the unbound state. They are as follows:

a. Receptors located in the cytosol (type 1). They bind to HRE consisting of 2 half sites separated by a variable length of DNA with the second half site having an inverted DNA sequence (of the first sequence).

b. Receptors located within the nucleus (type 2). As mentioned earlier these bind to the HRE as hetero dimmers. In the absence of ligand binding they are often complexed with co-repressors and kept inactive (Fig. **4**).

c. Receptors similar to type 1 receptors, but bind to repeat HRE instead of inverted repeat HRE (type 3).

d. Receptors similar to type 2 receptors, but with only a single DNA receptor binding domain binding to a single half site HRE.

Nuclear receptors have also been functionally classified as classical (hormone receptors) or non-classical. Many of the latter were originally referred to as orphan receptors, but many have been adopted due to subsequent identification of the ligand. Many of these ligands have been found to be intermediary metabolites and their binding to the nuclear receptors provide feedback of the metabolic status at a cellular level. The PPAR family of nuclear receptors belongs to this group.

Classification Based on Phylogenetic Nomenclature

The Nuclear Receptors Nomenclature Committee in 1999 recommended a new phylogenetic based cataloguing system with 6 sub-families [5]. This system classifies the nuclear receptors based on homology of the DNA and Ligand binding domains. The format proposed is NRxyz; where x is the subfamily, y the group and z the gene. This phylogeny based system also parallels function. Unusual receptors containing only one of the two conserved domains (C or E) were grouped separately into sub-family 0. This system is shown in Table **1**. This method of classification has led to a precision in nomenclature. PPARα, PPARγ, and PPARß/δ have been named as NR1C1, NR1C3 and NR1C2 respectively.

Ligands and Activators

Nuclear receptor ligands are as varied as the nuclear receptors and are lipophilic as they have to cross the plasma membrane and in many cases the nuclear membrane as well. To be physiologically relevant the ligand concentration needs to approach the dissociation constant of the ligand-receptor complex. The phenomenon of reverse endocrinology is that in which physiological ligands are identified following the discovery of an orphan nuclear receptor. The metabolic pathways affecting PPAR followed the identification of the receptor. The detail of nuclear receptor activation by ligands (agonists, partial agonists and antagonists) is beyond this brief summary and would require a separate chapter.

Nuclear Receptor Co-Regulatory Factors

These are factors that lead to the function of nuclear receptors alternating between activation and repression depending on the need of the cell/tissue. Co-activators and co-repressors are included in this collective term. The first co-activator identified was the ERAP160 which was demonstrated to bind to the oestrogen receptor activating transcription. Co-repressors although similar in concept to co-activators, usually bind to the nuclear receptor in the absence of a ligand and have an opposite effect. Binding of co-activators often leads to weakening of the bond between histone and DNA, thus enhancing gene transcription. In contrast binding to co-repressors can have the opposite effect with strengthened histone DNA association.

Table 1: This table lists by groups the human nuclear receptors classified into subfamilies based on sequence similarity (Nuclear Receptor Nomenclature Committee Classification)

Nuclear Receptor Nomenclature Committee Classification
Subfamily 1: Thyroid Hormone Receptor-like
Group A: Thyroid hormone receptor (Thyroid hormone)
Group B: Retinoic acid receptor (Vitamin A and related compounds)
Group C: Peroxisome proliferator-activated receptor
Group D: Rev-erb
Group F: Retinoid-related orphan receptor
Group H: Liver X receptor-like
Group I: Vitamin D receptor-like
Subfamily 2: Retinoid X Receptor-like
Group A: Hepatocyte nuclear factor-4 (HNF4)

Table 1: contd....

Group B: Retinoid X receptor (RXR)
Group C: Testicular receptor
Group E: TLX/PNR
Group F: COUP/EAR
Subfamily 3: Oestrogen Receptor-like (Steroid hormone receptor)
Group A: Oestrogen receptor (Sex hormone receptors)
Group B: Oestrogen related receptor
Group C: 3-Ketosteroid receptors
Subfamily 4: Nerve Growth Factor IB-like
Group A: NGFIB/NURR1/NOR1
Subfamily 5: Steroidogenic Factor-like
Group A: SF1/LRH1
Subfamily 6: Germ Cell Nuclear Factor-like
Group A: GCN1
Subfamily 0: Miscellaneous
Group B: DAX/SHP

PPAR FAMILY

We have seen that cells require systems to deal with internal and external stimuli and altering gene expression is a way of meeting these challenges. It is essential for survival that cellular energy homeostasis is maintained. Thus, a cellular sensing system for fatty acid/metabolite concentrations leading to regulation of metabolic pathways is critical. PPARs are nuclear receptors (Table **1**: NR1C) that provide some of this function. Mouse PPARα was first cloned in 1990 by Issemann and Green [6]. Since then Dreyer *et al.* in 1992 cloned 3 types of PPAR (PPARα, PPARγ and PPARß/δ) from Xenopus in 1992 [7]. In addition to rodents and xenopus, they have subsequently been identified in many other species such as humans, amphibians, teleosts and cyclostome [8]. However, we will also see that these nuclear receptors have other functions.

PPARs were so named as they were considered to be activated by peroxisome proliferators which are hypolipidaemic chemicals. Since then it has been demonstrated that PPARs respond to substances such as endogenous fatty acids. Nearly 30 years before the discovery of PPAR, Rhodin in 1954 [9] first described the presence of a cytoplasmic organelle in the mouse proximal tubule which he termed a microbody. Subsequently it was found to contain H_2O_2 generating flavin oxidases and H_2O_2 degrading catalases; De Duve [10] who demonstrated this proposed the name peroxisome to focus on the H_2O_2 generation and degradation which was a feature of this organelle. Peroxisomes exist in nearly all eukaryotic cells. Proliferation of peroxisomes is observed following stimulation by a diverse group of chemicals including the fibrate group of lipid lowering drugs, industrial solvents, phthalates and certain herbicides.

The basic mechanisms of PPAR action have been similar, regardless of the species studied. They fall in line with that of nuclear receptors that we have previously described. PPAR tissue expression varies; although they can be co-expressed in some cell types with varying concentrations. Further, they have different ligands, target genes and function. PPARα is mainly expressed in hepatocytes, cardiac myocytes, proximal renal tubules and enterocytes; all these requiring fatty acid oxidation pathways. Expression of PPARß/δ is widespread and often at greater concentrations than the other subtypes. In contrast PPARγ is found in 3 forms (γ1, γ2 and γ3; this is due to transcription of the protein at different points of the gene and they also undergo different splicing) and is mainly located in adipocytes and the immune system. The γ3, γ1 and γ2 proteins are identical, but may undergo differing regulation. Following binding of a ligand PPARγ2 is approximately 10 times more active as a transcription factor than PPARγ1. There has been speculation that this feature may have use in differing ligand concentrations leading to differing effects; PPARγ2 when ligand concentrations are low as in early adipocyte differentiation and PPARγ1 activity predominant when ligand levels are high. The above subtypes are presented in Fig. **5**.

The domains of PPAR are similar to the basic nuclear receptor model seen in Figs. **3a, 3b**- A/B. C, E and F- and Fig. **5**. The hyper-variable A/B domains function has not been clearly defined. It contains the AF1 site which can undergo phorphorylation and as with other nuclear receptors its function may be modified by epigenetic factors. The C domain (DBD) contains the 2 zinc fingers; the P-box interacts with the ACCTCA half site on the DNA while the D-box may be involved in dimerisation and PPRE recognition. The hinge region (D domain) in addition to allowing for flexible linkage between the surrounding domains also contains the terminal extension of the C domain that may be involved in recognising the 5' end of the PPAR response element. The ligand binding E domain contains 13 α helices and a 4 strand ß sheet as initially shown by Escher and Wahli [8]. This E domain (which contains the activated function 2 site) is involved in functions other than ligand binding; these include hormone dependent activation and interactions with RXR and cofactors.

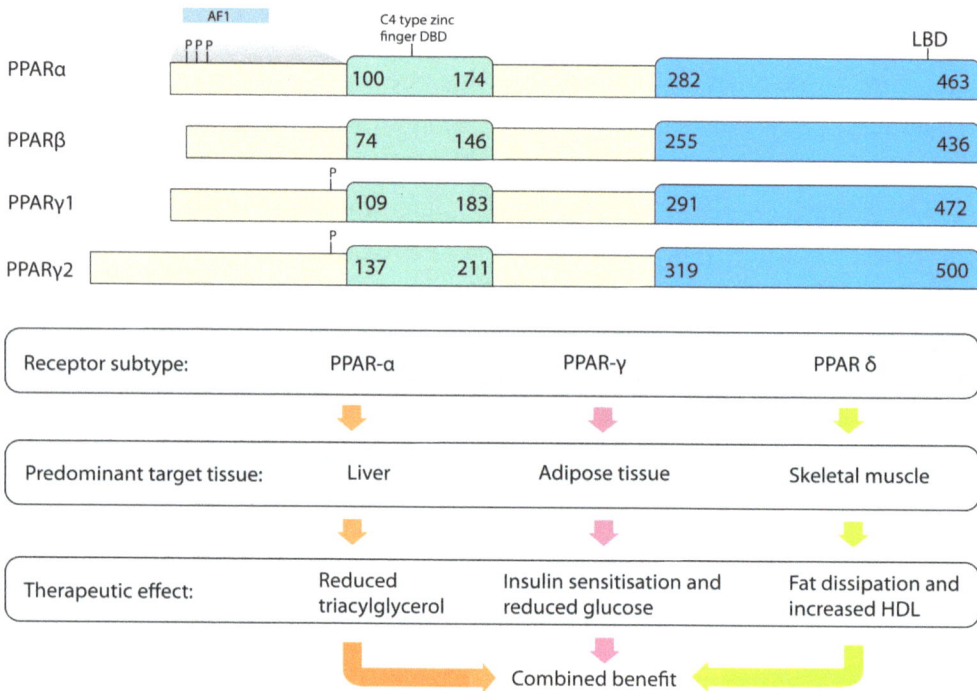

Figure 5: The structure and broad functions of the 3 subtypes of PPAR.

It is reasonable to speculate that alterations in the PPAR structure, either by polymorphisms, mutations or alternative splice sites can affect its function. This could have an impact on associated diseases that include T2DM, atherosclerosis and obesity. Dominant negative forms of PPARα and PPARγ have been identified. A truncated form of PPARα (PPARα$_{tr}$ lacks part of the D domain together with the entire E and F domains) due to alternative splicing and an early stop codon, functions as a dominant negative form [11]. PPARα variants containing different residues at positions 71, 123 and 444 acting as dominant negatives have been identified [12]. Reduced activation by ligands has been observed with the PPARα (L162V) polymorphism [13, 14]. Dominant negatives have been seen with PPARγ (substitutions at residues 290 and 467) resulting in severe insulin resistance [15].

PPAR Ligands

We have seen that certain agents induced peroxisome proliferation and it was subsequently discovered in 1990 that these agents bound PPARα. Since then other exogenous substances binding PPAR have been identified, some very specific for a subtype and others less so. Endogenous factors such as fatty acids and their metabolites were only identified later; thus, the receptors were classified as orphan receptors for a period of time. PPARα was the first of the group to be identified to be a target of endogenous fatty acids by Latruffe and Vamecq in 1997 [16]. Both exogenous and endogenous ligands of PPAR are presented in Tables **2** and **3** respectively. A basic summary of the binding and subsequent effects of PPAR activation is described in Fig. **6**. Similar to other nuclear receptors binding of ligands results in a three dimensional alteration of the receptor complex and in the case of PPAR activation many of the details have not been ascertained.

Interestingly ligands binding PPAR can have variable effects depending on different co-activators and co-repressors that they interact with. There is evidence that different co-activators are recruited depending on the specific ligand. Thus, different ligands even with similar binding affinity could lead to different levels of gene expression [17]. It is important to recognise this as exogenous drugs could exhibit within class variability of effects.

Fatty acids (especially saturated fatty acids) have been found to bind to PPARα. VLDL and LDL have been seen to activate PPARα in the presence of lipoprotein lipase suggesting that fatty acids and the esterified triacylglycerols are the natural ligands [18, 19]. Although metabolites of arachidonic acid (prostaglandins and leukotrienes) have been demonstrated to bind and activate PPARα there has been debate as to whether the tissue concentrations of these substances are sufficient to do so *in vivo*.

PPARγ has been seen to be activated by fatty acids (especially polyunsaturated) and eicosanoids. Eicosanoids are considered to play a part in the inflammatory processes. It has been identified that intracellular metabolism of fatty acids to eicosanoids by lipoxygenases enhances their ability to activate PPARγ. VLDL and LDL, especially with the addition of lipoprotein lipase, become activators of PPARγ as are oxidised LDL and phospholipids.

The endogenous ligands for PPARß/δ have not been established. The ligand binding pocket appears to be much larger than the other subtypes. Fatty acids and eicosanoids are seen to activate the system. Like the other subtypes, the addition of lipoprotein lipase to triglyceride rich VLDL and LDL enhances its activation.

PPAR upon activation binds to RXR and then interacts with PPRE found upstream of the target genes. Nuclear receptors binding RXR have been seen to bind to 2 copies (half sites) of DNA repeats (the sequence AGGTCA is similar for many receptors) separated by 1 to 6 nucleotides. It has been shown that the first half site (5' region) binds the PPAR while the second half site interacts with the RXR [20]. The structure of the PPRE is similar for all 3 PPAR subtypes. The PPRE is also recognised by certain other proteins and these can block the function of PPAR. Further, PPAR can also bind to factors other than RXR (*e.g.,* thyroid receptor); this too can result in inactivation. It has been observed that some of the effects of PPAR can be due to it interacting and blocking the DNA binding of other transcription factors. The actions of co-regulators are similar to that described for the generic nuclear receptor previously. Table **4** lists co-regulators identified for the various PPAR subtypes.

PPAR can also undergo kinase dependent activation, this being a ligand independent mechanism. Phosphorylation can affect DNA and ligand binding to the receptor as well as co-activator recruitment. It has been seen that MAPK, protein kinase A and C, AMP kinase and glycogen synthase kinase-3 have been seen to modulate the activity of PPAR although the effects can be distinct depending on the PPAR subtype. For example MAPK activates PPARα in hepatocytes, but inhibits PPARγ in adipocytes. In contrast protein kinase A phosphorylation activates all the PPAR subtypes.

Table 2: Exogenous ligands binding to PPAR. (References are listed in the PPAR Resource page (http://ppar.cas.psu.edu))

PPARα	PPARβ/δ	PPARγ
Fibrates	Valproic Acid	Thiazolidinediones
Phytanic Acid	Bezafibrate	Indomethacin
Indomethacin	Tetradecylthioacetic	Phthalates
Dehydroepiandrosterone (DHEA)	Sulindac Sulfide	Telmesartan
Phthalates	Benzanthracene	Bezafibrate
Valproic acid	Treprostinil Sodium	Glimepiride
Telmesartan		Diclofenac
Phenobarbital		DHEA
Epoxyisoprostane		Tolbutamide
		Gliclazide
		Glibenclamide
		Ibuprofen
		Atorvastatin
		Resveratrol

Table 3: Endogenous ligands binding to PPAR. (References are listed in the PPAR Resource page (http://ppar.cas.psu.edu))

Endogenous Ligands	PPARα	PPARβ/δ	PPARγ
LPL- treated VLDL	✓	✓	✓
VLDL	✓	✓	
OxVLDL		✓	
OxLDL			✓
Long-chain alkylamines	✓		
Saturated fatty acids	✓	✓	
Polyunsaturated fatty acids	✓	✓	✓
Prostaglandins	✓	✓	✓
Leukotriene B4	✓		

Table 4: Co-activators and co-repressors of PPAR are listed (References are listed in the PPAR Resource page (http://ppar.cas.psu.edu))

Cofactors	PPARα	PPAR β/δ	PPARγ
<u>Coactivators</u>			
Bifunctional enzyme (BFE)	✓		
BRG1-associated factor 60c (BAF60c)			✓
Constitutive coactivator of PPARγ (CCPG)			✓
CREBP- binding protein (CBP)	✓		✓
Hydrogen peroxide-inducible clone 5 protein (Hic5)			✓
LIM domain-only protein (LMO4)			✓
Lipin 1	✓		
Mediator subunit 1 (MED1)	✓	✓	✓
Mediator subunit 14 (MED14)			✓
Multiple Endocrine Neoplasia type 1 (MEN1)			✓
Multiprotein bridging factor 1 (MBF-1)			✓
Nuclear Receptor coactivators- 1 &2	✓	✓	✓
Nuclear Receptor coactivators- 3	✓	✓	
Nuclear Receptor coactivators- 4, 6 & 7			✓
p300	✓	✓	✓
Poly ADP-ribose polymerase 2 (PARP-2)			✓
Positive transcription elongation factor b complex (P-TEFb)			✓
PPARα-interacting complex 285 (PRIC285)	✓	✓	✓
PPARα-interacting complex 320 (PRIC320)	✓		
PPAR-interacting protein (PRIP)	✓		✓
PPARγ coactivator 1α (PGC-1α)	✓		✓
PPARγ coactivator 2 (PGC-2)			✓
PR domain containing 16 (PRDM16)	✓		✓
Protein arginine N-methyltransferase 2 (PRMT2)			✓
Steroid receptor coactivator-1 (SRC-1)	✓	✓	✓

Table 4: contd....

Steroid receptor coactivator-2 (SRC-2)	✓		✓
Steroid receptor coactivator-3 (SRC-3)	✓	✓	✓
Tat interactive protein (Tip60)			✓
Thyroid hormone receptor interacting protein 3 (TRIP3)			✓
Corepressor			
Insulin-like growth factor-binding protein-3 (IGFBP-3)			✓
Nuclear receptor corepressor 1 (NCoR)	✓	✓	✓
Receptor-interacting protein 140 (RIP140)	✓		✓
Scaffold attachment factor B1 (SAFB1)	✓	✓	✓
Silencing mediator of retinoid and thyroid receptors (SMRT)	✓	✓	✓
Tribbles homolog 3 (TRB3)			✓

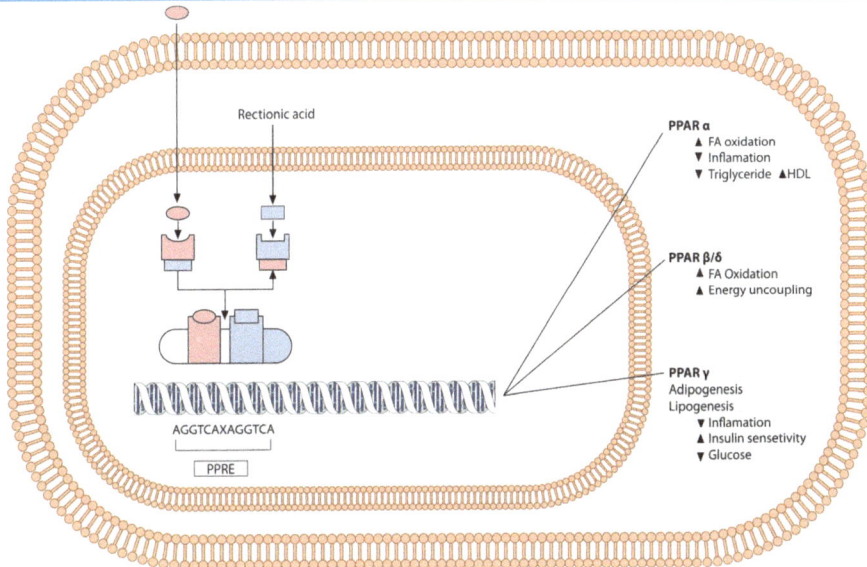

Figure 6: The mechanism of action of the PPAR subtypes and their effects on metabolism.

Functions of PPAR

The physiological role of PPAR is dependent on the target genes that are regulated. Although similarities in structure and activation are seen between the subtypes, differences in phenotypic expression are evident. For example PPARα activation modulates fatty acid oxidation whereas PPARγ activation results in adipogenesis and PPARβ/δ promote mitochondrial fatty acid oxidation, energy consumption, and thermogenesis. Therefore each PPAR subtype will be discussed separately.

PPARα Function

PPARα activation in the hepatocytes results in an increase in peroxisomal enzymes and some increase in mitochondrial oxidation of fatty acids. Expression of PPARα is altered by calorie status and stress levels of the individuals. Interestingly PPARα expression shows a diurnal variation which mirrors levels of glucocorticoids, a stress hormone. Insulin has been seen to reduce hepatic expression. PPARα also affects the transcription of genes coding for proteins influencing lipid metabolism. In the fasted state fatty acids provide a source of energy and concentrations increase. The catabolism of fatty acids appears to require a functional PPARα as this is absent in PPARα null mice where fasting leads to hypoglycaemia and elevated free fatty acids.

Dietary fatty acids also result in PPARα dependent peroxisomal and microsomal fatty acid oxidation. Feeding a high fat diet to PPARα null mice leads to hepatic fat accumulation. These mice compared to wild type controls have dereased glycogen storage and increased concentrations of serum cholesterol and HDL-C. These mice also have lower metabolic rates and greater adipose stores. PPARα also plays a role in lipogenesis and disabling the gene results in greater triglyceride accumulation in the liver. Treatment with PPARα agonists reduces this hepatic steatosis. PPARα is also involved in regulating lipoprotein metabolism. It is also thought to suppress inflammation *via* down regulating the expression of pro-inflammatory genes. This is of importance in clinical medicine as dyslipidaemia and inflammation are considered major risk factors for atherosclerosis.

Plasma VLDL levels are dependent on hepatic synthesis and peripheral tissue clearance by LPL. The liver produces VLDL continuously from endogenous synthesis, cytosolic triglyceride stores, plasma lipoproteins and circulating free fatty acids in order to provide fatty acids to peripheral tissues. VLDL metabolism together with factors affecting it is illustrated in Fig. **7**. Production of the VLDL begins with lipidation of apoprotein B100 by microsomal triglyceride transfer protein to a pre-VLDL particle. Further, lipidation results in VLDL2 and then VLDL1 particles. VLDL formation is stimulated by hepatic triglyceride (as a result of increased fatty acid availability; *e.g.,* increased lipogenesis, decreased fatty acid oxidation and increased hepatic fatty acid delivery), mediated by an increase of ApoB leading to more VLDL particles and lipidation. LPL activity is influenced by ApoC3, ApoA5, ANGPTL-3 and 4. VLDL, following this LPL mediated delipidation, becomes intermediate density lipoprotein and then LDL following further TG removal by hepatic lipase.

We have previously seen that PPARα increases fatty acid oxidation. Thus, it is considered that its stimulation decreases VLDL synthesis as reduced levels of fatty acids are available. However, PPARα has been seen to up-regulate the expression of several genes involved in fatty acid synthesis. Details of possible mechanisms are beyond this clinical review. PPARα stimulates LPL expression, although this may be indirectly *via* activators or inhibitors. It has been seen that the LPL inhibitor ApoC3 is down regulated by PPARα activation. In contrast ApoA5, an activator of LPL, is up regulated. Interestingly PPARα also exerts an inhibitory effect on LPL perhaps *via* increasing ANGPTL-4 levels. This may result in tissue specific differential effects and/or fine tune the predominant pro with some anti-lipolytic activity. Most of the above mechanistic studies have been based on mouse data. Obvious differences are present between the two models. PPARα is expressed in lower concentrations in the human liver compared to the mouse. However, it is clear that PPAR activation (using fenofibrate) in the human leads to lower TG levels with increased LPL activity. However, it is not clear whether VLDL production is decreased in humans. Small dense LDL was reduced by PPARα activation in both mouse and human studies.

Differences in HDL metabolism are seen between humans and mice. PPARα activation in the mouse results in decreased HDL-C while the reverse is seen in Fig. **8** illustrates some of the mechanisms by which HDL-C levels may be raised in humans. HDL levels are affected by nascent HDL synthesis in the liver and enterocytes, reverse cholesterol transport and rate of catabolism. Lipid poor ApoA1 takes up cholesterol *via* an ABCA1 mediated mechanism becoming nascent HDL. PPARα activation results in lower ApoA1 synthesis in mice while in humans ApoA1 (and ApoA2) levels are increased, more in the liver than in the enterocytes. Both ApoA1 and ApoA2 genes contain PPRE in the promoter regions. ABCA1 synthesis has been seen to be up-regulated by PPARα activation. The protein ABCG1 mediates cholesterol efflux into mature HDL as opposed to ApoA1 (carried out by ABCA1). Interestingly, PPARγ has been seen to increase ABCG1 levels in macrophages. The scavenger receptor SRB1 also enhances cholesterol efflux to mature HDL in macrophages and is up-regulated by PPARα. It is also thought that PPARα agonism may increase free cholesterol (cf. esterified cholesterol) which is available for efflux into HDL with esterification by lecithin cholesterol acyl transferase. Esterified cholesterol transfer in exchange for TG from other lipoproteins is mediated by CETP (not seen in mice), also up-regulated by PPARα. Cholesterol from HDL is also transferred to LDL by a CETP mediated process. Biliary cholesterol elimination may also be influenced by PPARα as they appear to induce ABCB4, a canalicular phospholipids translocator. Bile acid synthesis is also thought to be regulated by PPARα, although this aspect of data is often contradictory.

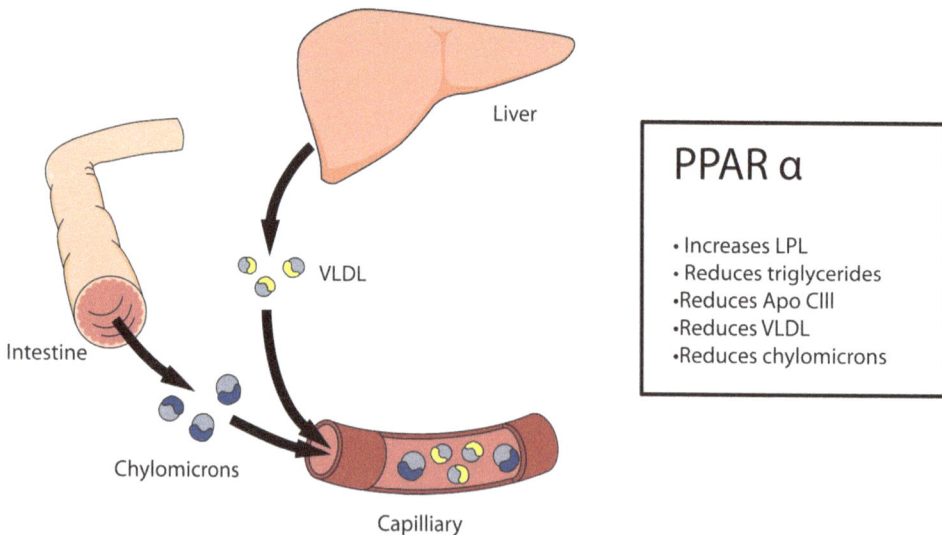

Figure 7: Effects of PPARα on VLDL and chylomicron metabolism.

PPARγ Function

PPARα and PPARγ expression is very different in both the fasting and fed states. While fasting decreases expression the fed state has an opposite effect. The role of

PPARγ appears to lie with fatty acid accumulation and storage in the adipocyte, cell differentiation and endothelial dysfunction, amongst other pleiotropic effects. Disruption of PPARγ in the mouse embryo results in non-viability associated with disrupted organogenesis. When pre-adipocytes *in vitro* enter the cell cycle the degree of PPARγ expression coincides with the increase in DNA synthesis. Several studies have suggested that PPARγ is involved in cell cycle withdrawal which may be due to suppressing synthesis of some forms of the transcription factor E2F. This results in cell differentiation.

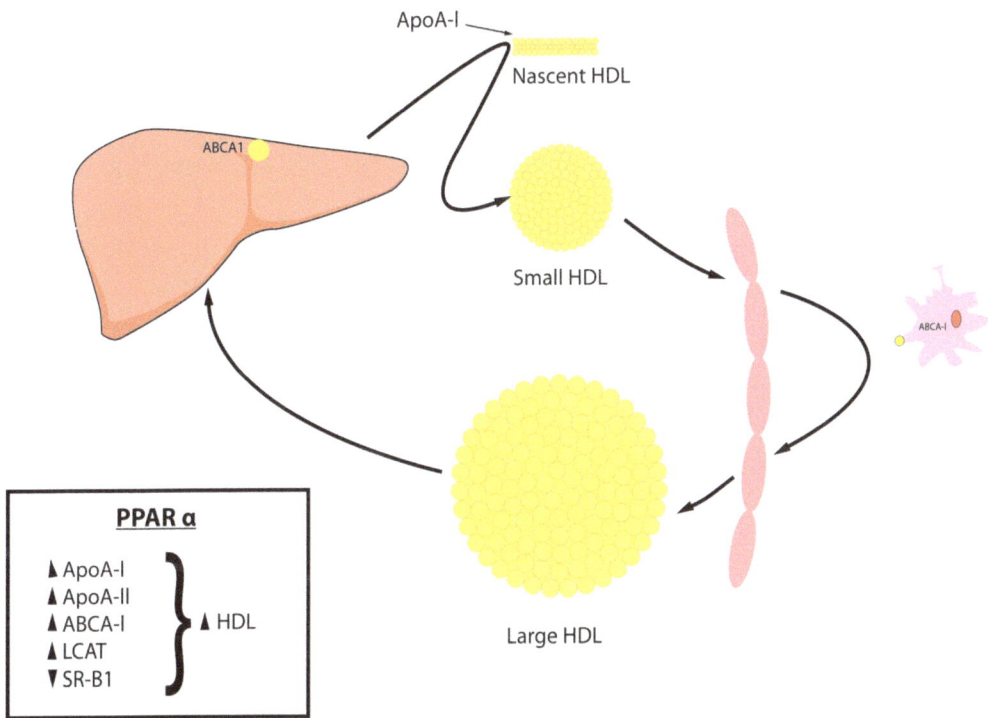

Figure 8: Effects of PPARα on HDL metabolism.

The function of PPARγ is influenced by both its regulation and how it regulates other genes. PPARγ expression is influenced by insulin, TNFα, SREBP1 and C/EBP. Inhibition of PPARγ by TNFα (at pre and post-translational points) may be associated with insulin resistance, atherosclerosis and inflammation. SREBP1, a member of a family of proteins involved in lipid synthesis, induces PPARγ transcription by up-regulating transcription. There appears to be cross talk between PPARγ and C/EBP (most genes where PPARγ bind also possessing

C/EBP binding sites), thus ensuring effective adipogenesis. C/EBP is induced by insulin and glucocorticoids, both hormones promoting adipogenesis. PPARγ regulates genes involved in increasing differentiation of adipocytes from the immature to the mature form able to accumulate fat. These include LPL, aP2 (a carrier protein of fatty acids) and CD36 (a fatty acid membrane binding protein and scavenger receptor) all of which are expressed in adipocytes and macrophages. Details of the possible mechanisms of adipogenesis and interactions with the above factors are beyond the scope of this chapter [21].

Apart from cell differentiation PPARγ has a role in glucose metabolism and reducing endothelial cell inflammation. These and other effects are shown in Fig. **9**. Reduction of plasma glucose may be due to an increase in glucose uptake and up-regulation of adiponectin and leptin expression. The Pro12Ala mutations in PPARγ are associated with insulin resistance and risk of T2DM. However, the Pro115Gln polymorphism does not increase the risk of T2DM or affect lipid metabolism. Interestingly, mouse studies have suggested that phosphorylation of PPARγ has been associated with insulin resistance and this phenomenon has been linked to decreases in adiponectin levels. There has been speculation that a conformational change to PPARγ may be induced by TZD that may prevent phosphorylation; this leading to improved insulin sensitivity. It has been suggested that adipose tissue PPARγ protects non-adipose tissue such as skeletal muscle from lipid overload, thus ensuring normal function. Reduced risk of atherosclerosis may be associated with the changes to pro-inflammatory cytokines such as TNFα and matrix metallopeptidase 9, reducing endothelial inflammation.

PPARß/δ Function

PPARß/δ is widely expressed, but especially in brain, adipose tissue and skin. However, they are less well understood than the other subtypes. Activation of PPAR ß/δ leads to increased synthesis of ADRP, fatty acid binding protein and cathepsin E. There are genes with response elements specific to PPARß/δ, although there are many that are also activated by the other PPAR subtypes.

PPARß/δ homozygous null mice show normal foetal development, but are smaller than their heterozygote and wild type counterparts in the neonatal and postnatal period. There have been suggestions that this is due to changes in adipocyte lipid stores. The PPARß/δ null mice have also shown myelination defects and impaired wound healing. There is evidence suggesting that activation of these nuclear receptors may regulate metabolic activity and inflammation in a range of tissues. There have been some ligands developed (not used in humans) which bind to PPARß/δ influencing cellular preference for either glucose or fatty acids. Thus, they are considered to be factors that may play a role in metabolic adaptation to changes such as diet. Studies using PPARß/δ ligands have demonstrated that lipoproteins are affected with significant decreases in TG and LDL-C and an increase in HDL-C in insulin resistant obese Rhesus monkeys [22]. This was accompanied by decreased levels of fasting insulin. Possible actions of PPARß/δ on various tissues influencing the metabolic syndrome are summarised in Fig. **10**. Some evidence that PPARß/δ may down regulate the Neimann-Pick like 1gene and reduce cholesterol absorption exists [23]. There is interest in its role in cancer development. The PPARß/δ gene was associated with the possible aetiology of adenomatous poliposis coli in mouse models, although contradictory data also exists [24-26].

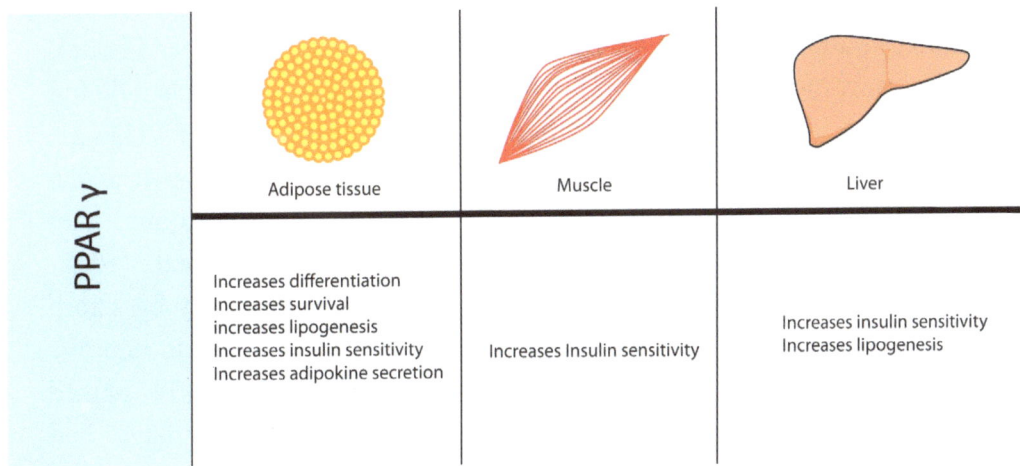

	Adipose tissue	Muscle	Liver
PPAR γ	Increases differentiation Increases survival increases lipogenesis Increases insulin sensitivity Increases adipokine secretion	Increases Insulin sensitivity	Increases insulin sensitivity Increases lipogenesis

Figure 9: Biological functions of PPARγ.

CLINICAL USE OF PPAR ACTIVATORS

The above sections set the scene for this clinical review of the use of PPAR activators in clinical medicine. Looking at the effects of PPAR activation it is clear that many systems will be affected. Potential benefits may be countered by deleterious effects and this has been the case with many of these agents, not surprising in view of the diverse pathways influenced. We will consider PPARα and PPARγ agonists initially, then move to the PPARα/γ dual activators and finally to the developmental work on the PPARß/δ agonists – GW501516 and L-165041.

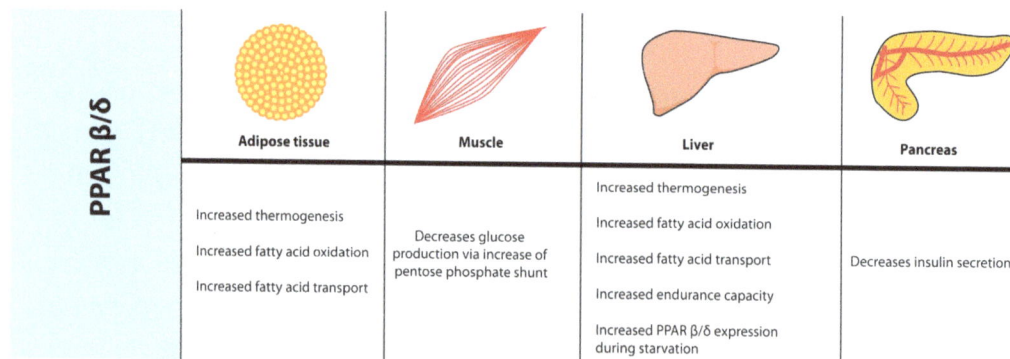

Figure 10: Targets of PPARß/δ activation and resulting effects.

Drugs Targeting PPARα

The evolution of a class of drug is well demonstrated when we consider fibrates. The initial observation that derivatives from dehydrocholic acid, phenylethyl acetic acid possessed lipid lowering properties was made in France in 1953 [27, 28] well before the discovery of PPAR. This was followed by research at the Imperial Chemical Industries Laboratories in England, identifying that oxyisobutyric acid derivatives had cholesterol lowering effects. Clofibrate (ethyl-α-4-chlorophenoxyisobutyrate) was discovered in the above laboratory by Thorp and Waring in1962 [1]. The most successful modification of clofibrate to improve upon its lipid lowering ability was the benzoyl derivative in 1974 named fenofibrate (initially referred to as procetofen). Since then we have had gemfibrozil, bezafibrate and ciprofibrate designed for use.

Development of fibrates as a class was hindered by the demonstration of associated murine hepatic carcinogenecity considered to be due to peroxisomal proliferation. It was in 1983 that Blümke *et al.* [29] demonstrated that humans were not endangered. Time ran out for clofibrate when the WHO cooperative trial on primary prevention of ischaemic heart disease showed a 47% increase in mortality, mainly non-cardiac, in the treatment arm. Further, there have been reports of amino transferase elevation and gall stone formation. In view of these, generic clofibrate, although still available, has virtually disappeared from the clinical radar. Structures of the fibrates are shown in Fig. **11**. They are well absorbed from the gastrointestinal tract and are highly bound to albumin.

Clofibrate and fenofibrate are pro-drugs that are hydrolysed to clofibric acid and fenofibric acid respectively. Gemfibrozil and bezafibrate are active compounds. Gemfibrozil is mostly metabolised to glucuronide conjugates whilst clofibrate, fenofibrate and bezafibrate are excreted unchanged or as glucuronide conjugates. There appears to be a lack of clarity as to the exact contribution of phase 1 metabolism and the role of cytochrome p450 in fibrates other than gemfibrozil. Gemfibrozil inhibits CYP2C8 which contributed to the metabolism of cerivastatin leading to an increase in the plasma concentration of the statin. The other fibrates have not demonstrated a similar increase in statin levels.

We have previously described the clinical effects of PPAR activation. Clinical use is influenced principally by safety, tolerability and evidence of benefit. The demise of clofibrate demonstrates this. We will now examine the position of the different fibrates at present.

Since the 4S study in 1994 [30], statins have been the mainstay of lipid lowering treatment. Statins inhibit hepatic cholesterol synthesis by competitively inhibiting the enzyme HMG CoA reductase leading to up-regulation of LDL receptors consequently increasing LDL uptake, in turn lowering plasma cholesterol. Since then more recent studies; TNT [31], REVERSAL [32] and PROVE-IT [33] have clearly demonstrated that 'lower is better' regarding cholesterol levels. These intervention trials have formed the basis of the latest BHS and JBS secondary prevention guidelines, BHS–IV [34] and JBS-2 [35], published in December 2005 and focused mainly on LDL-C. Regression studies such as ASTEROID [36] and SATURN [37] suggest that benefits may be possible if LDL-C levels are lowered sufficiently.

Despite optimal LDL-C reduction with statins the metabolic syndrome remains an underlying risk factor for cardiovascular disease [38, 39]. This may be due to low

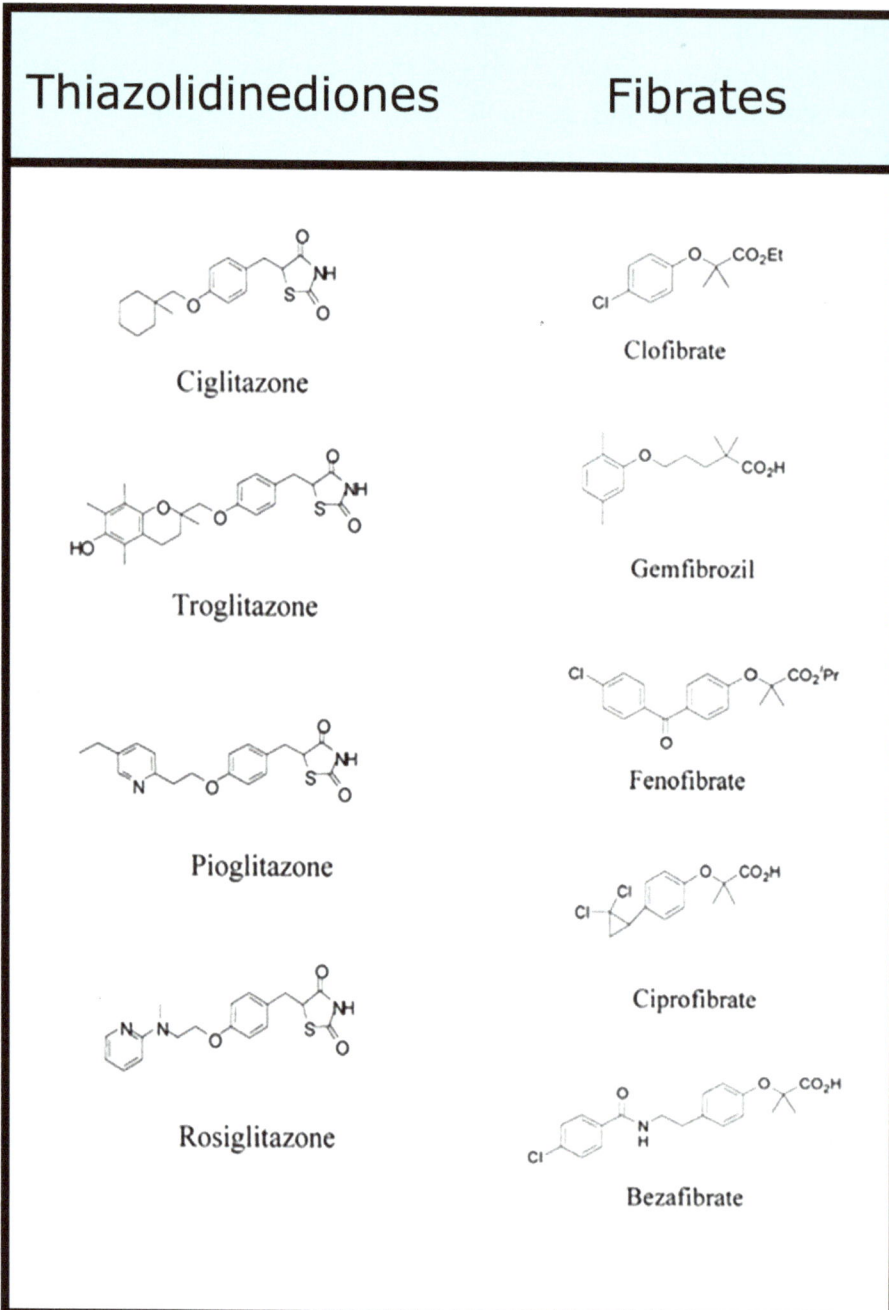

Figure 11: Molecular structures of thiazolidinedione and fibrate classes of drugs.

HDL-C and/or raised triglyceride, both of which are characteristics of the metabolic syndrome and are associated with cardiovascular disease. Definitions of the metabolic syndrome are presented in Fig. **12**. The underlying characteristics, while remaining the same, differ in emphasis and threshold. The association between cardiovascular disease and low HDL-C was initially observed nearly 60 years ago [40]. Since then evidence has accumulated confirming the relationship between HDL-C and cardiovascular disease (Framingham) [41]. This association is observed to be independent of LDL-C and exists even following reduction of LDL-C to low levels following statin treatment [42, 43]. Similarly, raised triglyceride levels have also been linked with cardiovascular disease.

a. **ATP3/NCEP definition** (Expert Panel on Detection, Evaluation, and Treatment of High Blood Cholesterol in Adults: Executive Summary of the Third Report of the National Cholesterol Education Program (NCEP) Expert Panel on Detection, Evaluation, and Treatment of High Blood Cholesterol in Adults (Adult Treatment Panel III). *JAMA* 285 : 2486 –2497,200)

A diagnosis of metabolic syndrome is made when three or more of the following are present:

- Abdominal obesity: waist circumference > 102 cm in men and > 88 cm in women
- Plasma triglyceride: > 1.6 mmol/L
- HDL-C: < 1.0 mmol/L in men and < 1.25 mmol/L in women
- Blood pressure: ≥ 130/≥85 mm Hg
- Fasting glucose: ≥ 110 mg/dl

b. **IDF classification** (IDF, 2005, Diabetes Voice, vol. 50, issue 3)

Metabolic syndrome is defined as:

High waist circumference
Plus any two of:
- Raised Triglyceride (>1.7 mmol/L)
- Low HDL (Men < 1.0; Women < 1.3 mmol/L)
- BP > 130/85 mmHg
- Fasting plasma glucose > 5.6 mmol/L

Figure 12: Characteristics defining the metabolic syndrome.

However, results from intervention studies raising HDL-C and lowering triglyceride using fibrates are less clear. While some studies have shown overall benefit, others have been negative in outcome. This is in contrast with the results observed following reduction in LDL-C with statin treatment.

We will briefly describe some of the notable trials with fibrates (Table **5**). Obviously there is selection bias, but we make no apology as clinical medicine is still to a large part subjective. However, to partly address this we will also describe a recently published meta-analysis.

The HHS was a randomised double blind study over 5 years comparing gemfibrozil (2051 men) 600 mg twice daily with placebo (2030 men). Inclusion criteria were men aged between 40 and 55 years with a non-HDL-C greater than 5.2 mmol/l [44]. A reduction in cardiovascular outcomes (after 5 years) of 34% (27.3 events per 1,000 individuals in the gemfibrozil arm compared to a corresponding figure of 41.4 in the placebo arm) was observed. A subgroup analysis of patients with raised triglyceride and low HDL-C showed the relative risk reduction to be greater. This provided an early indication regarding possible benefits of fibrates in patients with dyslipidaemia characterising the metabolic syndrome.The VAHIT study in 1999 again investigated the benefits of gemfibrozil (1,200 mg daily) *vs.* placebo, in patients (men under 74 years of age) with HDL-C below 1.0 mmol/l and interestingly a LDL-C below 3.6 mmol/l, thus aiming to select individuals with a lower LDL-C related risk [45]. It was a secondary prevention study with all 2531 patients having established coronary disease and the primary outcome was myocardial infarction or cardiovascular death. A relative risk reduction of 22% was associated with the treatment arm (21.7% event rate in the placebo controls against 17.3% in the gemfibrozil treated patients) over a median follow-up of 5.1 years. These two studies led to a positive outlook for fibrates as a class, although both were gemfibrozil trials. It appeared that cardiovascular risk due to HDL-C and TG, both seen as risk factors in population surveys, could be reduced by fibrates; a position of relative clarity.

A non-significant outcome was observed in BIPS in 2000 [46]. Bezafibrate 400 mg was compared to placebo in patients (men and women aged between 45 and 74 years) with established coronary heart disease. Inclusion criteria also included TC between 4.7 and 6.5 mmol/L and HDL-C ≤ 1.17 mmol/L MI, both fatal and non-fatal, and sudden death were lower in the patients treated with bezafibrate (13.6%) compared with the controls (15%), albeit not significantly.

T2DM is one of the factors associated with the metabolic syndrome. The FIELD study [47] investigated the effects of fenofibrate (4895 individuals) *vs.* placebo

(4900 individuals) in subjects aged between 50 and 75 years with T2DM not on statin therapy at entry. The cohort was a mix of primary (7664 individuals) and secondary prevention (2131 individuals). Inclusion criteria of the study were TC between 3.9 and 6.5 mmol/l, with either a TG/HDL-C ratio ≥ 4.0 or TG between 1.0 and 5.9 mmol/l. Patients with a plasma creatinine > 130 µmol/l were excluded. Over a mean follow-up of 5 years 5.2% of individuals in the fenofibrate arm and 5.9% in the control arm developed a coronary event, a non-significant risk reduction. The study was complicated by the fact that statins were added to 17% and 8% of the control and fenofibrate groups respectively in order to attain lipid targets. It was accepted that this could have masked a larger treatment benefit due to fenofibrate. Secondary analysis provided mixed findings. However, even in the initial publication there appeared to be a decrease in albuminuria progression and retinopathy needing laser treatment in the fenofibrate arm. We will discuss this later.

Fibrates now appeared to be at the cross roads regarding macrovascular benefit. The ACCORD-LIPID study was published in 2010 [48]. Statin therapy due to weight of evidence has assumed a front-line status in regards to cardiovascular risk reduction. Thus, the ACCORD-LIPID study was designed to determine the effects of fenofibrate as an add-on agent in individuals with T2DM. Fenofibrate or placebo was added to 5518 patients on open-label simvastatin with or without established cardiovascular disease. Glycated Hb was ≥ 7.5% (58 mmol/mol) and if cardiovascular disease was present, age at entry was restricted to 40 – 79 years. When two additional cardiovascular disease risk factors were present the age range for inclusion was 55 - 79 years. The LDL-C at entry was 1.55 – 4.65 mmol/l and HDL-C < 1.29 mmol/l for men and < 1.42 mmol/l for women. The triglyceride levels were < 8.5 mmol/l and < 4.5 mmol/l for those receiving and not receiving lipid therapy. The primary outcome was first occurrence of non-fatal myocardial infarction or stroke or death from cardiovascular causes. This figure was 2.4% in the control group and 2.2% in the fenofibrate treated patients over a mean follow-up of 4.7 years.

Thus, we have had 3 large non-gemfibrozil trials that did not show benefit. Many points were raised based on these trials at this stage. The crucial one was whether fibrates had any role at all, apart from in patients with marked hypertriglyceridaemia. Was there heterogeneity among fibrates with benefit specific to gemfibrozil, this raising a problem as adverse effects were much higher when gemfibrozil was combined with a statin? Another question was whether the trials were too wide, with only a subgroup of patients benefiting from fibrates. Sub-group analyses of the above RCTs could potentially provide a hint as to who these patients might be, although it would take another RCT with narrower inclusion criteria to establish the benefit.

Interestingly, subgroup analysis of all the above trials have suggested that cardiovascular benefits appear to be maximal in subjects with insulin resistance and other characteristics of the metabolic syndrome [49-52]. Bruckert *et al.* carried out an analysis of 5 large trials, selecting trial participants by cut-offs closest to that of the atherogenic dyslipidaemia (HDL-C < 0.91 mmol/L and TG > 2.2 mmol/l). The patients possessing these characteristics accounted for between 11% and 33% of the total cohort. This subgroup saw a significant reduction in cardiovascular risk of 28%. Importantly the complementary group only demonstrated a non-significant 6% risk reduction [53].

A meta-analysis of the outcome trials with fibrates over the past 40 years or so have been neatly summarised by Jun *et al.* in 2010, following the ACCORD-LIPID study. Meta-analyses pose dangers, but can be useful when reviewing large numbers of trials. Some of the benefits of meta-analyses include the ability to investigate variability in results, overcome small sample sizes of some of the individual trials and perhaps increase the precision of the findings. It can result in the generation of new hypotheses, this pointing to the next set of studies. There are limitations that affect most meta-analyses and caveats must be applied to their interpretation. Criteria for selection, the pooling of the different patient characteristics, variance in sample sizes, types of statistical analysis and measured outcomes have to be taken into account.

The meta-analysis by Jun *et al.* considered 18 randomised controlled trials using fibrates; outcomes included major cardiovascular events (MI, CVA), coronary events, coronary revascularisation, stroke, heart failure, cardiovascular deaths, new-onset albuminuria and adverse effects due to the drugs. Studies were identified from Medline, Embase and the Cochrane Library database with appropriate keywords and these were filtered down to 18 including 45,058 individuals. Significant decreases were observed in major cardiovascular events (Fig. **13**) and coronary events (Fig. **14**). No significant change in mortality, all cause or specific to cardiovascular disease was apparent. Jun *et al.* also carried out sub-group analyses demonstrating a greater effect in trials with higher mean triglyceride levels. There did not appear to be any difference between trials with regards baseline HDL-C concentrations or type of fibrate or dose used in the trials. A significant reduction in albuminuria progression was observed. This meta-analysis also highlighted that fibrates were associated with hypercreatininemia [54].

Table 5: Features of the large fibrate outcome trials described

Trial Characteristics	HHS	VA-HIT	BIP	FIELD	ACCORD
Drug	Gemfibrozil	Gemfibrozil	Bezafibrate	Fenofibrate	Fenofibrate
Dose	600 mg 2X/day	1200 mg/day	400 mg/day	200 mg/day	200 mg/day
Primary endpoint	MI (fatal and non-fatal), cardiac death	Combined incidence of nonfatal MI and death from CAD	MI (fatal and non-fatal), sudden death	CHD death, non-fatal MI	Non-fatal MI, non-fatal stroke, or CVD death
Mean duration of follow-up (years)	5	5	6	5	5
# of patients (total)	Fibrate= 2051 Placebo= 2030	Fibrate = 1264 Placebo = 1267	Fibrate = 1548 Placebo =1542	Fibrate =4895 Placebo=4900	Fibrate =2765 Placebo =2753
Effect on Lipid Levels (% change from baseline)	LDL-C: -10 TC: -11 TG: -43 HDL-C: +10	LDL-C: 0 TC: -4 TG: -31 HDL-C: +6	LDL-C: -6.5 TC: -4.5 TG: -21 HDL-C: +18	LDL-C: -12 TC: -11 TG: -29 HDL-C: +5	LDL-C: -19 TC: -14 TG: -22 HDL-C: +8.4
Outcomes	CHD: ↓ 34% Non-fatal MI: ↓37% Total mortality: no change	CHD and Non-fatal MI: ↓22% Total mortality: ↓ 11% (NS)	Fatal and nonfatal MI and sudden death: ↓ 9% (NS) Total mortality: no change	CHD and nonfatal MI: ↓11% (NS) ↑Total mortality: 19% (NS)	Nonfatal MI Nonfatal Stroke CVD Death: ↓8% (NS) Total mortality: ↓9 % (NS)

(adapted from Saha *et al.*, International Journal of Cardiology 2010; 141: 157-66).

The DAIS assessed the progression of coronary artery disease by angiography following treatment with micronised fenofibrate (207 patients) *vs.* placebo (211 patients). All patients had T2DM with reasonable glycaemia control (mean HbA1c was 7.5%, x mmol/mol) with at least one coronary lesion seen on angiography. Following 3 years of treatment the fenofibrate treated group demonstrated reduced angiographic progression of coronary artery disease [55].

The above described meta-analysis, subgroup analysis of the large fibrate trials described individually in addition to HHS and VA-HIT, and DAIS suggest that fibrates may be of benefit. We recently carried out a study on the effects of fibrates on HDL-C in a routine clinic setting and demonstrated the HDL-C increase was greater in patients with lower baseline HDL-C [56]– Fig. **15**. We did not observe any significant HDL-C change when baseline HDL-C was greater than 1.0 mmol/l. We also noted that the HDL-C increase was not as great in patients with T2DM. Interestingly the relationship between HDL-C change and pre-treatment levels was absent in patients already on statins. This finding would have major impact on treatment strategy if validated, as statins have rightly assumed a primary role in cardiovascular prevention.

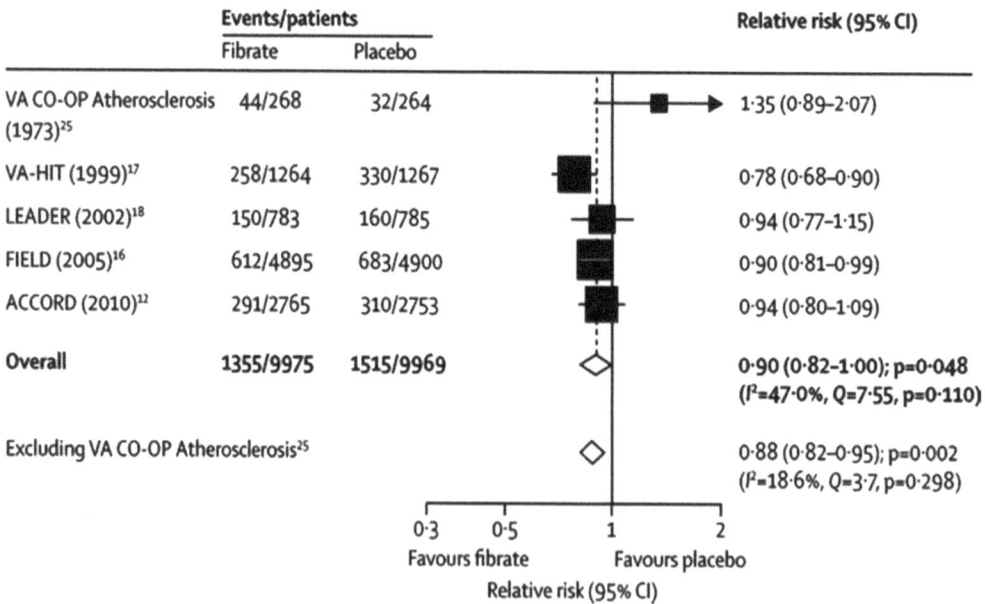

	Events/patients		Relative risk (95% CI)
	Fibrate	Placebo	
VA CO-OP Atherosclerosis (1973)[25]	44/268	32/264	1·35 (0·89–2·07)
VA-HIT (1999)[17]	258/1264	330/1267	0·78 (0·68–0·90)
LEADER (2002)[18]	150/783	160/785	0·94 (0·77–1·15)
FIELD (2005)[16]	612/4895	683/4900	0·90 (0·81–0·99)
ACCORD (2010)[12]	291/2765	310/2753	0·94 (0·80–1·09)
Overall	1355/9975	1515/9969	0·90 (0·82–1·00); p=0·048 (I^2=47·0%, Q=7·55, p=0·110)
Excluding VA CO-OP Atherosclerosis[25]			0·88 (0·82–0·95); p=0·002 (I^2=18·6%, Q=3·7, p=0·298)

0·3   0·5   1   2
Favours fibrate   Favours placebo
Relative risk (95% CI)

Figure 13: Effect of fibrates on major cardiovascular outcomes (Reproduced with permission from Jun *et al.* Lancet. 2010 May 29; 375(9729): 1875-84).

	Events/patients			Relative risk (95% CI)
	Fibrate	Placebo		
Newcastle-Tyne clofibrate trial (1971)[19]	121/244	130/253		1·01 (0·85–1·20)
IHD prevention clofibrate trial (1971)[20]	59/350	79/367		0·78 (0·58–1·06)
VA CO-OP Atherosclerosis (1973)[25]	8/268	9/264		0·88 (0·34–2·24)
Coronary Drug Project (1975)[24]	309/1103	839/2789		0·93 (0·83–1·04)
WHO CO-OP Trial (1978)[8]	167/5331	208/5296		0·80 (0·65–0·97)
Helsinki Heart (1987)[26]	56/2046	84/2035		0·66 (0·48–0·93)
Hanefeld et al (1991)[30]	32/379	31/382		1·04 (0·65–1·67)
BECAIT (1997)[29]	3/42	11/39		0·25 (0·08–0·84)
LOCAT (1997)[27]	7/197	7/198		1·01 (0·36–0·81)
SENDCAP (1998)[28]	6/81	17/83		0·36 (0·15–0·87)
VA-HIT (1999)[17]	219/1264	275/1267		0·80 (0·68–0·94)
BIP (2000)[23]	168/1548	189/1542		0·89 (0·73–1·08)
DAIS (2001)[22]	38/207	50/211		0·78 (0·53–1·13)
LEADER (2002)[18]	90/783	111/785		0·81 (0·63–1·05)
FIELD (2005)[16]	256/4895	288/4900		0·89 (0·76–1·05)
ACCORD (2010)[12]	332/2765	353/2753		0·94 (0·81–1·08)
Overall	1871/21503	2681/23164		0·87 (0·81–0·93); p<0·0001 (I^2=22·1%, Q=19·3, p=0·202)

0·2   0·5   1   2 2·5
Favours fibrate   Favours placebo
Relative risk (95% CI)

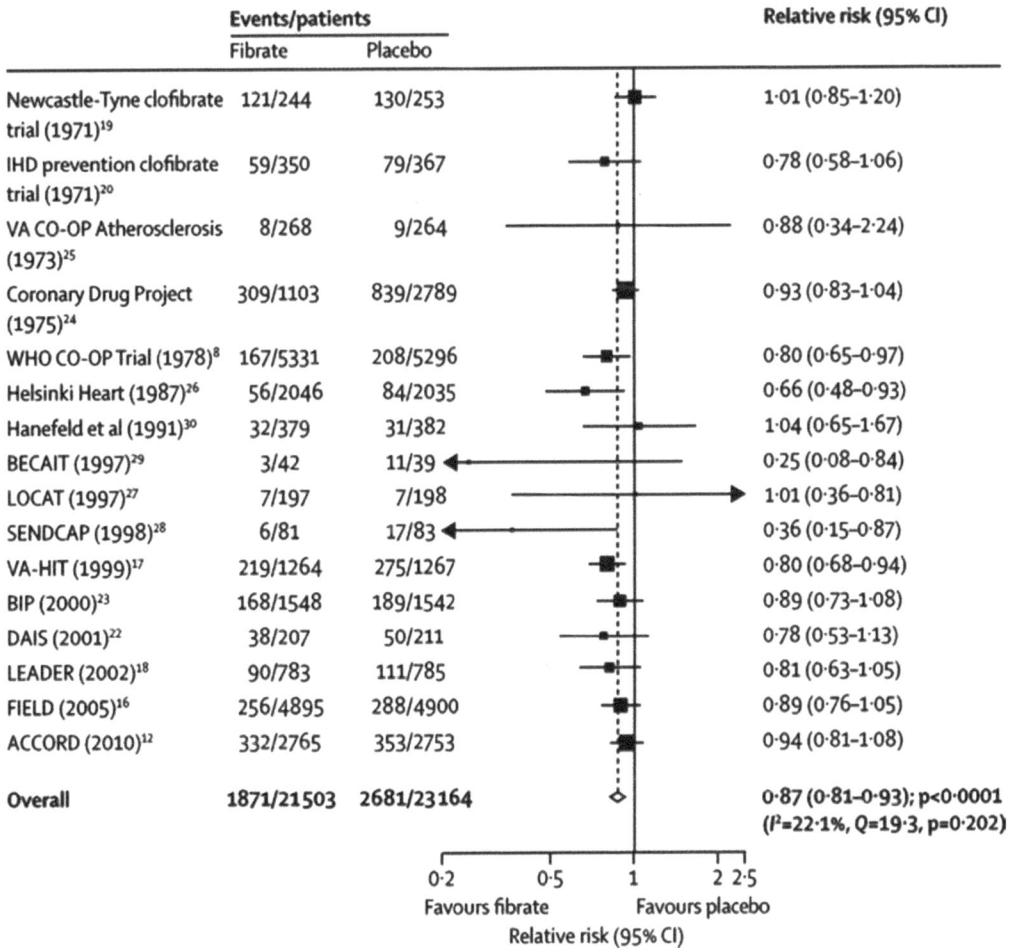

Figure 14: Effect of fibrates on coronary events (Reproduced with permission from Jun *et al.* Lancet. 2010 May 29; 375(9729): 1875-84).

We now examine heterogeneity between drugs within the class with regards outcome and adverse effects. The only 2 trials to show benefit in primary outcome used gemfibrozil. Is gemfibrozil different to the other fibrates?

Heterogeneity in drug interactions is observed when fibrates have been combined with statins. It has been seen that gemfibrozil has increased the plasma concentrations of statins. This was not observed in studies combining fenofibrate with statins. It has been demonstrated that gemfibrozil inhibits glucuronidation and it also moderately interacts with the CYP2C8 and CYP3A4; the enzyme systems metabolising simvastatin and atorvastatin. The area under the curve of statin concentrations increase 2-6 folds when gemfibrozil is co-administered [57].

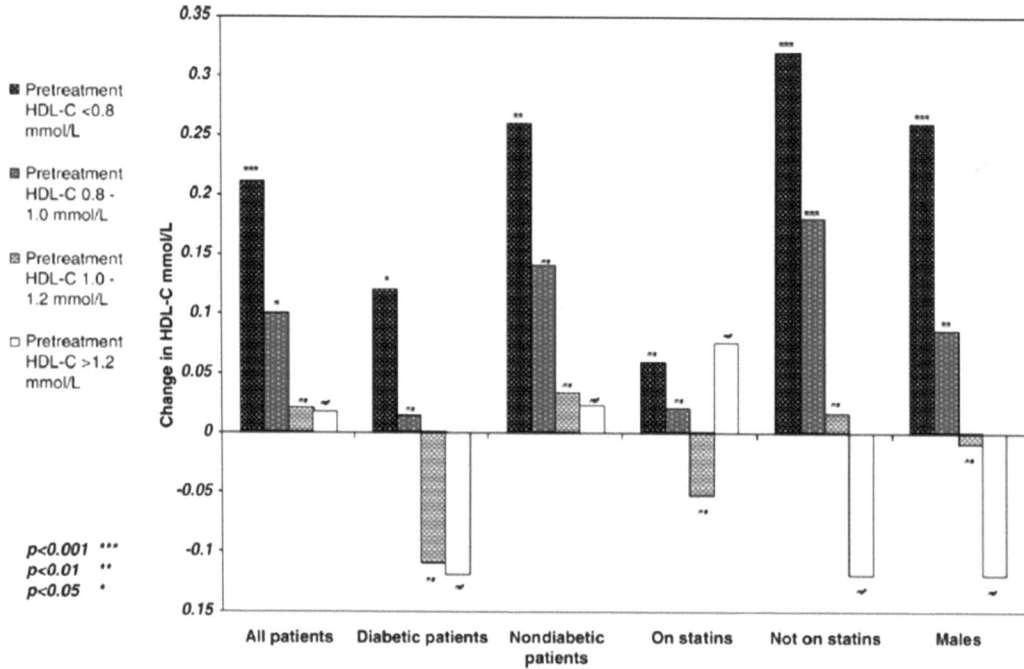

Figure 15: Association between HDL-C increase and baseline HDL-C levels (Ramachandran *et al.* Metab Syndr Relat Disord. 2012 Jun; 10(3): 189-94).

There appear to be further differences between gemfibrozil and the other fibrates. Homocysteine levels are increased by fibrates other than gemfibrozil [58-60]. Homocysteine has been seen to be associated with cardiovascular disease. Co-addition of folate to the fibrate prevents this increase in homocysteine [61] and has been seen to decrease lipoprotein oxidation. Thus, we can speculate whether elevation in homocysteine negates some of the fibrate associated benefit and whether this may be the reason for the gemfibrozil trials demonstrating better outcome. Thus, we ask the question as to whether folate should be given with a fibrate to improve benefit? However, no benefit in cardiovascular disease has been seen following folate administration. It has been suggested that the increase in homocysteine may be due to impaired renal function seen with fibrates other than gemfibrozil. However, some evidence points to the increased creatinine associated with fibrates to be non-renal in mechanism as GFR has not been seen to differ. We will describe changes in creatinine in greater detail later. Gemfibrozil, unlike other fibrates, does not down-regulate the renal cyclo-oxygenase enzyme system, this impairing the synthesis of vasodilatory prostaglandins. Although the implications of these phenomena are not well understood, it is important to be aware of within class differences.

Fibrates are contra-indicated in pregnancy, during breast feeding, acute liver disease, primary Biliary cirrhosis, gallbladder disease and renal dysfunction (including dialysis). Elevated liver function tests have been associated with fibrates and we would recommend baseline checks with regular monitoring. It is interesting that NASH/NAFLD is associated with many of the patients requiring fibrates therapy. Mechanism of action suggests that fibrates may improve NASH/NAFLD, although there is a lack of evidence. We have observed improvements in LFT with fibrates in our clinic setting (EAS 2012: abstract number 1153), not on the internet as yet) with the greatest reductions seen in patients with higher baseline concentrations of GGT, ALT and ALP. Unfortunately, as it was in a routine clinical setting, we did not carry out liver biopsies which may have confirmed the underlying reason for the elevated baseline LFT [62].

Elevated creatinine has been reported, although the mechanisms are unclear. We have recently published the scale of this, converted to eGFR (Fig. **16**) which forms the cornerstone of current management of kidney disease in the United Kingdom [63]. Our findings were comparable with that observed in the FIELD and ACCORD-LIPID studies. The increase in creatinine is reversible and awareness is important, especially in patients with T2DM where metformin may be discontinued and the patient subject to unnecessary investigations to determine the cause of the worsening kidney function.

Muscle toxicity in combination with statins has not been significantly increased in trials (FIELD, ACCORD-LIPID) not using gemfibrozil. In view of the interactions outlined above, the combination of gemfibrozil and statins is not recommended. Although not undergoing metabolism *via* the cytochrome p450 system, changes in anticoagulation status may be observed in patients on warfarin. The mechanism may be due to mild inhibition of CYP2C9 [64] or displacement of warfarin from protein binding. Increased monitoring of INR until it is stabilised, following either initiation or changes in fibrate dose, is recommended.

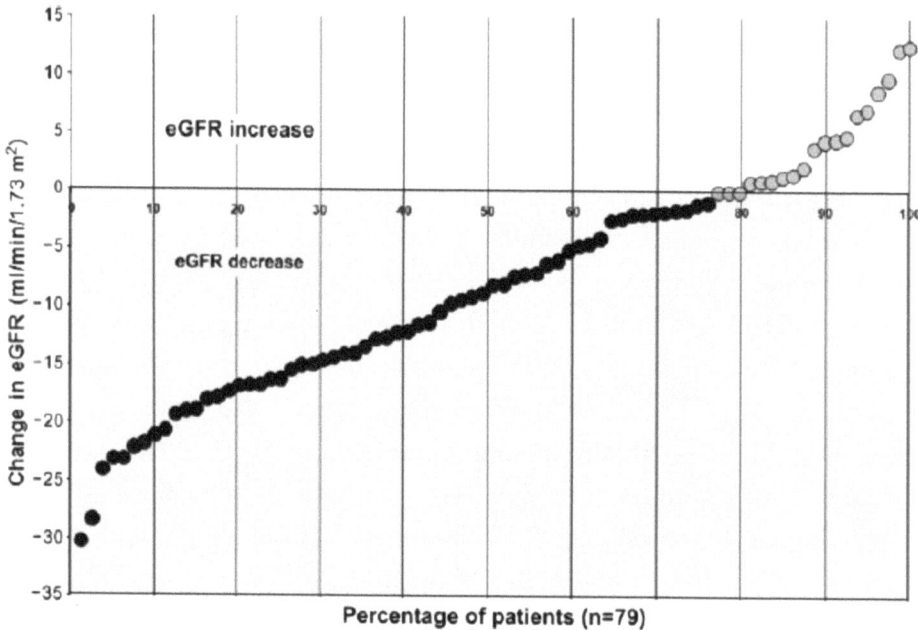

Figure 16: Changes in eGFR in patients treated with fibrates (Reproduced with permission, Abbas A *et al.*, Postgrad Med J. 2012).

NICE guidelines in the United Kingdom suggest that fibrates should only be considered in patients intolerant to statins in both primary and secondary prevention. In our view this is unfortunate as it can give the impression to non-specialists that stains and fibrates are similar. The guidelines ignore the underlying differences in their action on lipoprotein metabolism. We consider it far better if medical practitioners make a decision based on available evidence and lipid results than guidelines aiming for simplicity. The ESC/EAS guidelines focus more on individual care based on disease mechanisms and evidence [65]. They recommend consideration of fibrate therapy in patients with mixed dyslipidaemia and hypertriglyceridaemia and we would recommend these if guidance is required regards fibrate therapy.

There has been considerable interest in the ability of fenofibrate to reduce the rates of microvascular complications. The FIELD study demonstrated that fenofibrate significantly reduced the rate of progression of albuminuria [47]. Further, more patients in the treatment arm did not progress, or even regressed, compared to the control arm. The FIELD washout sub-study also showed less progression of albuminuria, with more patients not deteriorating or even regressing after 5 years of washout [66]. Lower creatinine values were observed in patients who were on fenofibrate during the trial. However, it must be noted that a reduction of albuminuria and the increased number of patients whose albuminuria did not progress or regressed was not detected in the patients in FIELD recruited in Helsinki; this may be due to lower patient numbers [67]. The ACCORD-LIPID trial also showed that adding fenofibrate to simvastatin was associated with significantly reduced development of both microalbuminria and macroalbuminuria [68]. The FIELD study also demonstrated a significant reduction in patients requiring laser treatment for either macular oedema or proliferative retinopathy [69]. Progression of diabetic retinopathy was reduced in the ACCORD Eye study by both intensive glycaemia control and by the addition of fenofibrate to simvastatin, but this improvement was not noted with intensive blood pressure control [70].

It must be stated that the above microvascular complication rates were not included as primary end points. Thus, we urgently require studies with microvascular complications as primary end-points to help determine the baseline characteristics of those patients who may experience maximum benefit.

Drugs Targeting PPARγ

This class of drugs, primarily PPARγ activators, was born out of attempts to create more potent fibrates in Japan in the 1970's. Takeda laboratories synthesized many analogues of clofibrate and observed that in addition to lipid lowering properties many also possessed hypoglycaemic effects in diabetic mice. This led to the discovery of ciglitazone in 1982. The Sankyo Company created troglitazone in 1988 and this drug improved both insulin resistance and reduced hepatic glucose synthesis, and was the first thiazolidinedione approved for clinical use in 1997. However, reports of hepatotoxicity led to its withdrawal in 2000 and probably delayed the launch of other agents within the class. Two of these agents were rosiglitazone and pioglitazone, developed by SmithKline in 1988 and

Takeda in 1990 respectively. Both were approved for clinical use in the USA in 1999. By 2011 rosiglitazone had been largely withdrawn due to an increase in the risk of cardiac disease outweighing benefit. Pioglitazone currently is at the crossroads following suspension of its use in France and Germany due to fears of association with bladder cancer.

As there are only 2 thiazolidinediones currently available (rosiglitazone is still available in the USA to patients previously treated and subject to restrictions) we will consider them separately. We have seen previously that these drugs activate PPARγ receptors and reduce hyperinsulinaemia and plasma glucose levels. Details of all the trials with both agents are presented on the "trials-results centre" web page [71]. Rosiglitazone is primarily metabolised by the cytochrome p450 enzymes (by 2C8 (mostly) and 2C9) and 99% bound to plasma proteins, mainly albumin. It has a 100 fold higher affinity to bind to PPARγ receptors than pioglitazone [72].

It was reported by Lupil *et al.* that rosiglitazone prolonged ß cell life by preventing apoptosis caused by free fatty acids [73]. Although there were no issues about its effects on improving glycaemia [74], the resulting changes to the lipid profile were of concern when compared to pioglitazone [75]. Total LDL and VLDL particles (including all sub-fractions) were seen to be increased. The ApoB1 lipoproteins were shown to increase [76]. Thus, there were early concerns about potential cardiovascular risk outcomes based on risk factors.

The DREAM study was reported in 2006 to assess the ability of rosiglitazone to prevent high risk patients with impaired fasting glucose and/or impaired glucose tolerance from developing T2DM compared to placebo [77]. Over a mean follow-up of 3 years it was seen that 11.6% and 26% of individuals in the rosiglitazone and placebo arms respectively met the primary outcome of incident diabetes or death. Our view even at the time was that this trial was meaningless as diabetes was a condition diagnosed by the degree of glycaemia and any drug decreasing glucose concentrations would have led to a reduction of the condition. Interestingly, it did not show any association between rosiglitazone and cardiovascular deaths. We will discuss the development of heart failure, as with other adverse effects common to both rosiglitazone and pioglitazone, later.

In 2007 Nissen and Wolski carried out a meta-analysis on the cardiovascular morbidity and mortality following rosiglitazone therapy [78]. Selection criteria for studies included study duration > 24 weeks, the presence of a control group not receiving rosiglitazone and availability of outcome data on myocardial infarctions and cardiovascular mortality. Data from 42 studies reported increased rates of both outcomes with the odds ratio for myocardial infarction and death from cardiovascular causes being 1.43 (p=0.02) and 1.64 (p=0.06) respectively. Following this publication the US Food and Drug Administration reviewed the available data [79] and concluded that the issue of cardiovascular disease remained unclear and inconclusive following data submitted by GlaxoSmithKline the manufacturers of rosiglitazone. Pooled data from 42 trials showed that rosiglitazone treated patients compared with comparators were associated with increased rates of myocardial infarction with a significant hazard ratio of 1.31. We have seen that the DREAM study [77] did not show any relationship between cardiovascular deaths and rosiglitazone, The ADOPT study [80] published in 2006 did not suggest any increase in ischaemic disease with rosiglitazone compared to metformin or glyburide. Interim data from RECORD [81], an open label study evaluating non-inferiority of rosiglitazone against controls with a primary outcome including cardiovascular deaths and hospitalisation, on-going at the time was also considered. This study was published in 2009 and stated that rosiglitazone did not increase cardiovascular morbidity or mortality compared with metformin or sulfonylurea [82]. Further, Diamond *et al.* in 2007 analysed the same studies as in the Nissen meta-analysis, using different statistical methodology and did not show a significant increase in myocardial infarction or death from cardiovascular causes [83].

Investigation regarding cardiovascular disease continued in the USA with inconclusive outcomes. However, the use of rosiglitazone is severely curtailed and is available only through a risk evaluation and mitigation strategy program. In Europe, following evaluation of the data and disclosures of concerns going as far back as 2000, the European Medicines Agency recommended suspension of rosiglitazone. It is food for thought that although no individual RCT demonstrated increased myocardial infarction rates with rosiglitazone its use was curtailed on the back of meta-analyses.

Thus, at present pioglitazone alone represents the PPARγ class and even this agent is under threat of suspension. Pioglitazone is rapidly absorbed with extensive protein binding (>99%) mainly to albumin. It undergoes hepatic hydroxylation and oxidation with CYP2C8, CYP3A4 and CYP1A1 involved. About 15 – 30% is excreted in the urine as metabolites. The remaining fraction is excreted unchanged *via* bile.

Pioglitazone was not affected by the controversy that dogged rosiglitazone in 2007. In contrast the PROactive study in 2005 demonstrated a significant benefit in death from any cause, non-fatal myocardial infarction or stroke (this was a secondary end point, the primary end point being the secondary end point as well as acute coronary syndrome, leg amputation, coronary revascularisation or revascularisation of the leg) in patients with T2DM who had evidence of macrovascular disease [84]. During the mean study period of 34.5 months 301 of the 2605 patients in the pioglitazone arm and 358 of the 2633 placebo control patients developed the above mentioned secondary end points.

A meta-analysis of ischaemic cardiovascular events in outcome trials using pioglitazone in 2007 by Lincoff *et al.* [85] showed a very different outcome. There were 19 trials included in this meta-analysis with 16,390 patients (8554 in the pioglitazone arm and 7836 in the control arms), although nearly a third of patients were from the PROactive study. Death, myocardial infarction and strokes were significantly reduced in the patients receiving pioglitazone; HR: 0.82 (95% CI: 0.72-0.94, p=0.005). The HR values for PROactive and the remaining trials were 0.84 (95% CI: 0.72-0.98) and 0.75 (95% CI: 0.55-1.02) respectively. In light of the above findings Lincoff *et al.* speculated that the differences in cardiovascular events may be due to pioglitazone having a more favourable effect on lipids. Both thiazolidinediones had similar effects on glycaemia control.

Since the initial observation by Caldwell in 2001 there have been further reports that improvements in hepatic steatosis may also be associated with both rosiglitazone and pioglitazone treatment [86-88]. Argo *et al.* demonstrated that there was sustained benefit in non-alcoholic steatohepatitis even after treatment discontinuation [89].

Recently pioglitazone has been suspended in France and Germany following association with bladder cancer. Increased bladder cancer was noted in the PRO-active study in 2005 with 14 and 6 cases in the pioglitazone and placebo arms respectively. Of these, 11 were excluded after consulting with external experts as they occurred within the first year of randomisation and it was thought that it could not have been drug related. This left 6 and 3 cases in the pioglitazone and placebo groups respectively. The report also stated that there were more risk factors associated with bladder cancer in the pioglitazone group and it was unlikely that the cancer was associated with pioglitazone.

Lewis *et al.* published an interim report of a 10 year observational cohort KPNC study [90] and found that pioglitazone exposure led to an increased risk of bladder cancer, reaching statistical significance only when drug exposure was greater than 24 months (HR=1.4, 95% CI: 1.03-2.0).

Mamtani *et al.* studied the general practitioner database (Health Improvement Network) between 2000 and 2010 in the United Kingdom and compared diabetic patients on thiazolidinediones with those on other agents [91]. The risk of bladder cancer increased with time elapsed since commencement of thiazolidinedione treatment irrespective of the treatment duration. Similar patterns were observed for rosiglitazone and pioglitazone.

Recently Azoulay *et al.* [92] surveyed patients with T2DM in over 600 general practices contributing to the general practitioner database, who were commenced on oral hypoglycaemic agents between 1988 and 2009. Pioglitazone exposure was associated with bladder cancer, the risk being highest in patients with more than 24 months exposure. Interestingly, unlike Mamtani *et al.* they did not see an increase in risk associated with rosiglitazone and concluded that it was a drug specific effect as opposed to a class effect.

With the weight of ever increasing evidence it will be interesting to see if other countries follow the action taken by France and Germany. In our clinical practice the role of thiazolidinediones had diminished even prior to the association with bladder cancer being reported. Our practice has been influenced by other effects such as weight gain, fluid retention and fracture risk.

Weight gain is a recognised side effect of thiazolinedione treatment affecting between 1-10% of patients [93]. An average weight gain of 3-4kg is seen during the initial six months of treatment; it has been seen that the rate of weight gain decreases and stabilises after the first 6-12 months of treatment. Weight gained with thiazolidinedione monotherapy is greater than with metformin, but comparable with that of sulphonylurea and insulin treatment [94]. The effect is greater in combination with insulin or sulphonylureas, and interestigly is reduced or absent in combination with metformin [93]. Disciplined calorie control was shown to curtail this weight gain [95]. Further, weight gain did not appear to have an effect on glycaemic control sustainability over the duration of thiazolidinedione treatment [96]. There have been studies investigating possible mechanisms for the paradox of improving insulin sensitivity and weight gain. Some evidence exists that a favourable shift of fat distribution from visceral to subcutaneous adipose depots may result in improvement of hepatic and peripheral tissue sensitivity to insulin. Despite the above evidence, in our clinical practice patient acceptability is influenced by this potential weight gain.

Fluid retention was recognized as an adverse effect of thiazolidinedione therapy even back in 1999 [97]. The incidence ranged from 7% in monotherapy and 15% in combination with insulin. Presentation was often seen as peripheral oedema progressing to pulmonay oedema and cardiac failure, often resistant to loop diuretics. Due to potential decompensation of heart failure caution was advised in patients with established cardiac failure (NYHA grade 3 and 4). Even in healthy volunteers a 6-7% increase in plasma volume is seen (4-5). Although the mechanisms have not been fully understood, both increased fluid reabsorption *via* renal collecting duct and increased vascular permeability in adipose tissue have been shown to be involved [98].

All the large prospective RCTs have demonstrated this phenomenon. The PROactive study demonstrated significantly increased heart failure, both needing and not needing hospital admission. However, no change in heart failure related mortality was seen. In this context it must be noted that patients with established heart failure (NYHA grades 2 and above) were excluded from the study. The RECORD study also reported heart failure causing hospital admission or death was significantly increased in the rosiglitazone group. Patients with a history of heart failure or those on treatment for it were excluded. Thus, fluid retention worsening cardiac failure appears to be a class effect, demonstrated in virtually every outcome study recording adverse effects.

The first reports of macular oedema in patients treated with rosiglitazone [99] appeared in 2005 which reversed on reduction of the dose. This was followed by a survey of 30 patients (17 on pioglitazone, 11 on rosiglitazone and the remaining 2 were on both at different times) by Ryan *et al.* They concluded that both drugs appeared to be associated with reversible macular oedema in some patients who demonstrated fluid retention [100]. A prospective cohort study of 170,000 diabetic patients by Fong and Contreras [101] in 2009 confirmed this finding that the thiazolidinedione drugs were associated with macular oedema (OR: 2.6, 95% CI: 2.4-3.0). Idris *et al.* [102] after assessing relevant data in the Health Care Network database in the United Kingdom noted a 3 fold risk of macular oedema, although the absolute risk was below 1% due to the low underlying risk of the condition. Thus, a baseline visual acuity test should be carried out and in the event of it worsening iatrogenic causes should be considered.

The ADOPT study in 2006 first brought to our attention the increased fracture risk associated with thiazolidinediones. This was not surprising as pre-clinical studies demonstrated activation of PPARγ inhibited bone formation perhaps by diverting mesenchymal stem cells to the adipocytic rather than to the osteogenic lineage within the bone. It was speculated that thiazolidinediones may increase bone resorption by stimulating osteoclastic activity. Numerous studies in humans have demonstrated decreased bone turnover, accelerated bone loss and impaired bone mineral density both in healthy volunteers and in patients with type 2 diabetes. Following on from the ADOPT study, other studies have shown that women are at particular risk of non-vertebral fractures [103] with a near doubling of hip fractures [104].

Loke *et al.* carried out a meta-analysis of 10 RCTs and 2 observational studies with 13,715 and 31,679 individuals respectively, all patients being exposed to thiazolidinediones for greater than 12 months [105]. The thiazolidinediones were associated with increased fractures in all the RCTs and observational studies. This increased risk was seen in women (OR: 2.23, 95% CI: 1.65-3.01) as opposed to men (OR: 1.00, 95% CI: 0.73-1.39). Bone mineral density in women on thiazolidinediones was also significantly lower in the lumbar spine and hip. Further, work needs to be carried out to determine the causative mechanisms of this gender related fracture risk.

Drugs Targeting Multiple PPARs

Thus far we have dealt with drugs targeting specific PPAR receptors. There are instances, however, where established drugs may have an effect *via* multiple PPAR pathways simultaneously. Furthermore the Glitizars are a group of new drugs that have been designed to bind to both PPARα and γ. What follows is a description of the observed heterogeneity when drugs act on more than one PPAR receptor.

Paradoxical Effect on HDL Cholesterol

This rare phenomenon is presented separately as we have considerable clinical experience and have observed it with fibrate monotherapy as well as when a thiazolidinedione is added to a fibrate. It was initially observed by Chandler *et al.* in 1994 when HDL-C was observed to decrease from 0.9 mmol/l to 0.18 when ciprofibrate was initiated [106]. A report by Ebcioglu *et al.* in 2003 documented this paradoxical change in 2 patients when rosiglitazone was added to fenofibrate [107]. We published a case series of this phenomenon observed in our clinic which hinted at heterogeneity in the response pattern [108]. We were subsequently first to report that pioglitazone was also associated with a paradoxical HDL-C decrease [109]. Keidar *et al.* [110] reported this phenomenon as rare with incidences of 0.02% and 1.39% with fibrates and a combination of fibrates and glitazones respectively. We think this might be an underestimate as we have identified more than 20 patients demonstrating this decrease in our centre. The question remains as to what defines the paradoxical HDL-C decrease; we have used a 50% reduction in our series. The causative mechanism has not been determined and may prove to be difficult due to its relative rarity. However, it may provide some useful clues as to PPARα activation/antagonism that may be extended to other areas of PPAR mechanisms. We do think that even though thiazolidinediones have been implicated, it is likely to be a PPARα based mechanism as most patients showed continued improvement in glycaemia control.

The PPARγ activators are known to cross react with the PPARα receptor. Interestingly we have observed presentation heterogeneity and some of these are;

Fibrates

1. With only one fibrate.

2. With more than one fibrate.

Thiazolidinediones

3. With only one when combined with fibrate.

4. With more than one combined with fibrate.

Dose Change

5. When dose of fibrate or thiazolidinedione was increased.

We have also reported one patient whose HDL-C decreased by more than 50% with simvastatin and atorvastatin monotherapy [111]. There was no decrease in HDL-C when pravastatin and rosuvastatin were tried. Once again no explanation was possible with the dearth of knowledge, but we are aware of a link between statins and PPAR activation. We again bring to the attention of readers that our retrospective study of patients showed that concurrent statin treatment affected the pattern of HDL-C change that we observed with fibrate monotherapy [56].

Statins and PPAR

Statins are HMG-CoA reductase inhibitors that reduce LDL-C levels. This inhibition reduces intracellular cholesterol concentration and increases a proteolytic activation of the sterol responsive element-binding protein and this transcription factor in turn increases genes coding for the LDL receptor, lipoprotein lipase and cholesterol 7α hydroxylase [112-114]. It has been reported that statins can regulate PPARα transcription and PPARα mRNA expression *in vitro* in human endothelial cells and HepG2cells as well as in mouse studies [115-117]. PPARγ activation has also been seen with statins [118]. The exact mechanisms of this activation are still unclear. There is speculation that some of the pleiotropic anti-atherogenic effects of statins may be *via* PPAR activation. These may include anti-inflammatory and anti-oxidant effects [119]. These are illustrated in Fig. **17**.

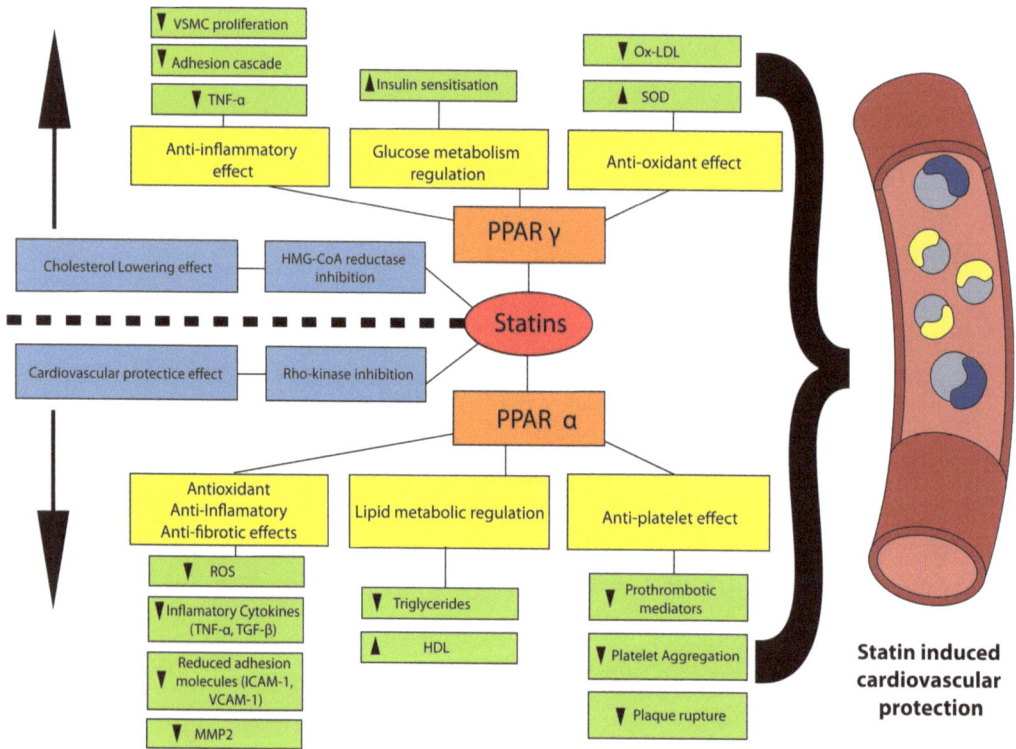

Figure 17: Some of the potential anti-atherogenic pleiotropic effects of statins are shown in this figure.

Drugs Binding PPARα and γ - Glitazars

The glitazars are dual action PPARα and γ agonists that were designed to improve both lipid profile and glycaemic control and also reduce the pill burden and improve compliance. The idea was promising as patients with T2DM often exhibit the atherogenic lipid profile that fibrates were effective against. Many compounds entered the clinical trial phase, but most were discontinued at that stage due to safety concerns. We have seen within class variation in both PPARα and γ. Both, types (fibrates and thiazolidinediones) have shown benefit and harm. Thus, designing a new dual agonist against this background was going to be a challenge with both PPAR sub types regulating expression of a number of genes.

As there were concerns, especially regarding rodent bladder cancer, the US FDA requested a 2 year period prior to phase 3 trials to estimate toxicity. Only muraglitazar and tessaglitazar proceeded into phase 3 trials. However, both were discontinued in 2006. Muraglitazar and tessaglitazar were associated with heart failure and reversible hypercreatininaemia respectively. These effects have been evident with thiazolidinediones and fibrates as we have seen previously and should not have been surprising. Since the demise of these 2 drugs we have not had any further in the class. We would argue that when dealing with systems as complex as PPAR up-regulation to try and develop dual agonists was always a step too far.

Drugs Binding PPARß/δ

We do not have any PPARß/δ agonists licensed for use in clinical practice. Compound GW 501516 has been best studied as a treatment for obesity, dyslipidaemia and hyperglycaemia. It has been shown to alter the metabolic preference towards fat utilisation instead of carbohydrate or protein. It has also been associated with increased muscle mass. Other compounds such as KD3010, CER-002, NNC61-5920 and GW610742 are being studied in various disease models. It is premature to review these as they have not reached clinical practice. It is essential that a detailed study of safety is carried out in view of our experience with the other PPAR agonists.

CONCLUSIONS

We have tried to write a balanced review of nuclear receptors and their therapeutic agonists. We consider that these drugs are agents which assume an illusion of complexity, due to the variation of the human genome. The poor safety data on thiazolidinediones has meant that their use has significantly declined. It is essential that they are not forgotten, for as in the case of fibrates, they may rise again on the back of advanced pharmacogenomics.

Our current methods of evaluating benefits and risks in these drugs may be too basic. The randomised controlled trial that is favoured by the medical establishment is a blunt instrument. It may be true that this is the best available technique at the moment and we have to accept that. But this should not stop us from realising its failings and trying to develop more sophisticated methodology that takes into account the complexity of the human being. Only then will we gain a better insight into the workings of drugs such as the PPAR activators. We hope that readers are able to reflect on the issues we have presented and perhaps even expand it to other areas that also deserve a similar approach.

ACKNOWLEDGEMENTS

We should like to acknowledge the scientists working in this area of interest who have made this chapter possible. We would also like to thank Nathaniel J Douglas (KraftNclick) for the illustrations.

CONFLICT OF INTEREST

The authors confirm that this chapter contents have no conflict of interest.

ABBREVIATIONS

ABC	=	ATP-binding cassette (A1, B4, G1)
ADRP	=	Adipocyte differentiated related protein
AF-1	=	Activated function (1, 2)
ALP	=	Alkaline phosphatase
ALT	=	Alanine transaminase
AMP kinase	=	Adenosine monophosphate-activated protein kinase
ANGPTL	=	Angiopoietin like protein
aP2	=	Adipocyte Protein 2

Apo	=	Apoprotein (B, C3, A1, A2, A5)
ATP3	=	Adult Treatment Panel III
BHS	=	British Hypertension Society
CCAAT-C/EBP	=	CCAAT/enhancer binding protein
CD36	=	Cluster of Differentiation 36
CETP	=	Cholesterol ester transfer protein
CVA	=	Cerebro-Vascular Accident
DBD	=	DNA Binding Domain
D-box	=	Distal box
DNA	=	Deoxyribonucleic acid
eGFR	=	Estimated Glomerular Filtration Rate
ERAP160	=	Oestrogen Receptor-Associated Protein 160
ESC	=	EAS guidelines focus
FFA	=	Free Fatty Acid
GGT	=	Gamma-glutamyltransferase
H_2O_2	=	Hydrogen peroxide
Hb	=	Haemoglobin
HDL	=	High Density Lipoprotein
HDL-C	=	High Density Lipoprotein Cholesterol
HMG CoA	=	3-hydroxy-3-methyl-glutaryl-CoA reductase
HR Values	=	Hazard Ratio Values
HRE	=	Hormone response elements
IDF	=	International Diabetes Federation
INR	=	International Normalized Ratio

JBS	=	Joint British Societies' guidelines
LBD	=	Ligand Binding Domain
LCAT	=	Lecithin Cholesterol Acyl Transferase.
LDL	=	Low Density Lipoprotein
LDL-C	=	Low Density Lipoprotein Cholesterol
LFT	=	Liver Function Test
LPL	=	Lipoprotein Lipase
MAPK	=	Mitogen Activated Protein Kinase
MI	=	Myocardial Infarction
mRNA	=	Messenger Ribonucleic Acid
NAFLD	=	Non-Alcoholic Fatty Liver Disease
NASH	=	Non-Alcoholic Steatohepatitis
NCEP	=	National Cholesterol Education Program
NYHA	=	New York Heart Association
P-box	=	Proximal Box
PPAR	=	Peroxisome Proliferator-Activated Receptor
PPARß/δ	=	Peroxisome Proliferator-Activated Receptor Beta/Delta
PPARα	=	Peroxisome Proliferator-Activated Receptor Alpha
PPARγ	=	Peroxisome Proliferator-Activated Receptor Gamma
PPRE	=	Peroxisome Proliferator Response Element
RA	=	Rheumatoid Arthritis
RCTs	=	Randomized Controlled Trials
Redox	=	Reduction Oxidation
RNA	=	Ribonucleic acid

RXR   = Retinoid-X-Receptors

SR-B1   = Scavenger receptor class B member 1

SREBP1   = Sterol response element binding protein 1

T2DM   = Type 2 Diabetes

TG   = Triglyceride

TNFα   = Tumour necrosis factor α

tRNA   = Transfer Ribonucleic acid

TZD   = Thiazolidinedione

VLDL   = Very Low Density Lipoprotein

ABBREVIATED TRIAL TITLES

4S   = Scandinavian Simvastatin Survival Study

ACCORD Lipid   = Action to Control Cardiovascular Risk in Diabetes Lipid Trial

ADOPT   = A Diabetes Outcome Progression Trial

ASTEROID   = A Study to evaluate the Effect of Rosuvastatin on Intravascular ultrasound-Derived coronary atheroma burden

BIPS   = The Bezafibrate Infarction Prevention Study

DAIS   = Diabetes Atherosclerosis Intervention Study

DREAM   = Diabetes Reduction Assessment with Ramipril and Rosiglitazone Medication

FIELD   = Fenofibrate Intervention and Event Lowering in Diabetes

HHS   = Helsinki Heart Study

KPNC = Kaiser Permanente Northern California's

PROactive = PROspective Pioglit Azone Clinical Trial in Macro Vascular Events

PROVE-IT = The Pravastatin or Atorvastatin Evaluation and Infection Therapy

RECORD = Rosiglitazone Evaluated for Cardiovascular Outcomes in Oral Agent Combination Therapy for Type 2 Diabetes

REVERSAL = Reversal of Atherosclerosis with Aggressive Lipid Lowering

SATURN = Study of Coronary Atheroma by in Travascular Ultrasound: Effect of Rosuvastatin *vs.* Atorvastatin

TNT = Treating to New Targets

VAHIT = Veterans Affairs High-Density Lipoprotein Cholesterol Intervention Trial

REFERENCES

[1] Thorp JM, Waring WS. Modification of metabolism and distribution of lipids by ethyl chlorophenoxyisobutyrate. Nature. 1962; 194: 948-949.

[2] Crick, F. On protein synthesis. Symposia of the Society for Experimental Biology. 1958; 12: 138–163.

[3] Elwood V. Jensen. http://www.annualreviews.org/doi/pdf/10.1146/annurev-physiol-020911-153327.

[4] Hollenberg SM, Weinberger C, Ong ES, *et al.* Primary structure and expression of a functional human glucocorticoid receptor cDNA. Nature. 1985; 318: 635-41.

[5] Committee NRN. A unified nomenclature system for the nuclear receptor superfamily. Cell. 1999; 97: 164-3.

[6] Issemann I, Green S. Activation of a member of the steroid hormone receptor superfamily by peroxisome proliferators. Nature. 1990; 347: 645-50.

[7] Dreyer C, Krey G, Keller H, *et al.* Control of the peroxisomal beta-oxidation pathway by a novel family of nuclear hormone receptors. Cell. 1992; 68: 879-87.

[8]   Escher P, Wahli W. "Peroxisome proliferator-activated receptors: insight into multiple cellular functions, " Mutation Research 2000; vol. 448, no. 2, pp. 121–138.

[9]   Rhodin J. "Correlation of ultrastructural organization and function in normal and experimentally changed proximal tubule cells of the mouse kidney". Doctorate Thesis. KarolinskaInstitutet, Stockholm. 1954.

[10]   deDuve C, Baudhuin P. Peroxisomes (microbodies and related particles). Physiol Rev. 1966; 46, 323-357.

[11]   Gervois P, Torra IP, Chinetti G, *et al.* A truncated human peroxisome proliferator-activated receptor α splice var- iant with dominant negative activity. Molecular Endocrinology. 1999; 13: 1535-49.

[12]   Roberts RA, James NH, Woodyatt NJ, *et al.* Evidence for the suppression of apoptosis by the peroxisome proliferator activated receptor alpha (PPAR alpha). Carcinogenesis. 1998; 19(1): 43-8.

[13]   Sapone A, Peters JM, Sakai S, *et al.* The humanperoxisomeproliferator-activatedreceptor alpha gene: identification and functionalcharacterization of twonaturalallelicvariants. Pharmacogenetics. 2000; 10: 321-33.

[14]   Vohl MC, Lepage P, Gaudet D, *et al.* Molecular scanning of the human PPARα gene: association of the L162v mutation with hyperapobetalipoproteinemia. JLipidRes. 2000; 41: 945-52.

[15]   Barroso I, Gurnell M, Crowley VEF *et al.* Dominant negative mutations in human PPARγ associated with severe insulin resistance, diabetes mellitus and hypertension. Nature. 1999; 402: 880-3.

[16]   Latruffe N. , Vamecq J. Peroxisome proliferators and peroxisome proliferator activated receptors (PPARs) as regulators of lipid metabolism. Biochimie 1997; 79: 81-94.

[17]   Sporn M. B. , Suh N. , Mangelsdorf D. J. Prospects for prevention and treatment of cancer with selective PPARgamma modulators (SPARMs) Trends in Molecular Medicine. 2001; 7: 395–400.

[18]   Chawla, A. , C. H. Lee, Y. Barak, *et al.* PPARdelta is a very low-density lipoprotein sensor in macrophages. Proc. Natl. Acad. Sci. USA. 2003. 100: 1268–1273.

[19]   Ziouzenkova O, Perrey S, Asatryan L, *et al.* Lipolysis of triglyceride-rich lipoproteins generates PPAR ligands: evidence for an antiinflammatory role for lipoprotein lipase. ProcNatlAcadSci USA. 2003; 100: 2730-2735.

[20]   Juge-Aubry C, Pernin A, Favez T, *et al.* DNA binding properties of peroxisome proliferator-activated receptor subtypes on various natural peroxisome proliferator response elements. Importance of the 5'-flanking region. J BiolChem 1997; 272: 25252-9.

[21]   Lowe C. E. , O'Rahilly S. , Rochford J. J. Adipogenesis at a glance. J Cell Sci. 2011; Aug 15; 124(Pt 16): 2681-6.

[22]   Oliver WR Jr, Shenk JL, Snaith MR, *et al.* A selective peroxisome proliferator-activated receptor delta agonist promotes reverse cholesterol transport. Proc Natl Acad Sci U S A. 2001; Apr 24; 98(9): 5306-11. Epub 2001 Apr 17.

[23]   van der Veen J. N. , Kruit J. K. , Havinga R. , *et al.* Reduced cholesterol absorption upon PPARdelta activation coincides with decreased intestinal expression of NPC1L1. J. Lipid Res. 2005; 46: 526–534.

[24]   He TC, Chan Ta, *et al.* PPARdelta is an APC-regulated target of nonsteroidal anti-inflammatory drugs. Cell. 1999; 99: 335-45

[25]   Reed K R, Sansom O J, Hayes A J. *et al.* PPARdelta status and Apc-mediated tumourigenesis in the mouse intestine. Oncogene. 2004; 23(55)8992-8996.

[26]   Gupta RA, Wang D, Katkuri S, *et al.* Activation of nuclear hormone receptor peroxisome proliferator-activated receptor-δ accelerates intestinal adenoma growth. Nature Medicine. 2004; 10(3): 245-247.

[27]   Mathivat A, Cottet J. Clinical trials on the hypocholesterolemia-producing effect of 2-phenylbutyric acid. Bull MemSoc Med Hop Paris. 1953; 69: 1030-1048.

[28]   Cottet J, Redel J, Krumm-Heller C, *et al.* Hypocholesterolemic property of sodium phenylethylacetate (22 TH) in the rat. Bull Acad Natl Med. 1953; 137: 441-442.

[29]   Blümcke S, Schwartzkopff W, Lobeck H, *et al.* Influence of fenofibrate on cellular and subcellular liver structure in hyperlipidemic patients. Atherosclerosis. 1983 Jan; 46(1): 105-16.

[30]   Randomised trial of cholesterol lowering in 4444 patients with coronary heart disease: the Scandinavian Simvastatin Survival Study (4S) Lancet. 1994 Nov 19; 344(8934): 1383-9.

[31]   Waters DD, Guyton JR, Herrington DM, *et al.* Treating to New Targets (TNT) Study: does lowering low-density lipoprotein cholesterol levels below currently recommended guidelines yield incremental clinical benefit? Am J Cardiol. 2004 Jan 15; 93(2): 154-8.

[32]   Nissen SE, Tuzcu EM, Schoenhagen P, *et al.* Effect of intensive compared with moderate lipid-lowering therapy on progression of coronary atherosclerosis: a randomized controlled trial. JAMA. 2004 Mar 3; 291(9): 1071-80.

[33]   Cannon CP, McCabe CH, BelderR, *et al.* Design of the Pravastatin or Atorvastatin Evaluation and Infection Therapy (PROVE IT)-TIMI 22 trial. Am J Cardiol 2002; 89: 860-1.

[34]   British Hypertension Society guidelines (BHS-IV). Williams B, Poulter NR, Brown MJ, *et al.* British hypertension society guidelines for hypertension management 2004 (BHS-IV): summary. BMJ 2004; 328: 634-640.

[35]   JBS 2: Joint British Societies' guidelines on prevention of cardiovascular disease in clinical practice. Heart. 2005 December; 91(Suppl 5): v1-v52.

[36]   Nissen SE, Nicholls SJ, Sipahi I, *et al.* Effect of very high-intensity statin therapy on regression of coronary atherosclerosis: the ASTEROID trial. JAMA. 2006 Apr 5; 295(13): 1556-65.

[37]   Nicholls SJ, Borgman M, Nissen SE, *et al.* Impact of statins on progression of atherosclerosis: rationale and design of SATURN Study of Coronary Atheroma by InTravascular Ultrasound: effect of Rosuvastatin *vs.* Atorvastatin. Curr Med Res Opin. 2011 Jun; 27(6): 1119-29.

[38]   Pyörälä K, Ballantyne CM, Gumbiner B, *et al.* Scandinavian Simvastatin Survival Study (4S). Reduction of cardiovascular events by simvastatin in non-diabetic coronary heart disease patients with and without the metabolic syndrome: subgroup analyses of the Scandinavian Simvastatin Survival Study (4S). Diabetes Care. 2004 Jul; 27(7): 1735-1740.

[39]   Girman CJ, Rhodes T, Mercuri M, *et al.* 4S Group and the AFCAPS/TexCAPS Research Group. The metabolic syndrome and risk of major coronary events in the Scandinavian

Simvastatin Survival Study (4S) and the Air Force/Texas Coronary Atherosclerosis Prevention Study (AFCAPS/TexCAPS). Am J Cardiol. 2004 Jan 15; 93(2): 136-41.

[40] Barr DP, Russ EM, Eder HA. Protein-lipid relationships in human plasma. Am J Med 1951; 11: 480-5.

[41] Thomas R. Dawber, M. D. , Gilcin F. Meadors, M. D. , M. P. H. , and Felix E. Moore, Jr. , National Heart Institute, National Institutes of Health, Public Health Service, Federal Security Agency, Washington, D. C. , Epidemiological Approaches to Heart Disease: The Framingham Study Presented at a Joint Session of the Epidemiology, Health Officers, Medical Care, and Statistics Sections of the American Public Health Association, at the Seventy-eighth Annual Meeting in St. Louis, Mo. , November 3, 1950.

[42] Heart Protection Study Collaborative Group. MRC/BHF Heart Protection Study of cholesterol lowering with simvastatin in 20, 536 high-risk individuals: A randomised placebo-controlled trial. Lancet. 2002, 360: 7-22.

[43] Barter P, Gotto AM, LaRosa JC, Treating to New Targets Investigators. HDL cholesterol, very low levels of LDL cholesterol, and cardiovascular events. N Engl J Med. 2007 Sep 27; 357(13): 1.

[44] Frick MH, Elo O, Haapa K, Heinonen OP, *et al.* Helsinki Heart Study: primary-prevention trial with gemfibrozil in middle-aged men with dyslipidemia. Safety of treatment, changes in risk factors, and incidence of coronary heart disease. N Engl J Med. 1987 Nov 12; 317(20): 1237-45.

[45] Rubins HB, Robins SJ, Collins D, *et al.* Gemfibrozil for the secondary prevention of coronary heart disease in men with low levels of high-density lipoprotein cholesterol. Veterans Affairs High-Density Lipoprotein Cholesterol Intervention Trial Study Group. N Engl J Med. 1999 Aug 5; 341(6): 410-8.

[46] Secondary prevention by raising HDL cholesterol and reducing triglycerides in patients with coronary artery disease: the Bezafibrate Infarction Prevention (BIP) study. Circulation. 2000 Jul 4; 102(1): 21-7.

[47] Keech A, Simes RJ, Barter P, *et al.* ; FIELD Study Investigators. Effects of long-term fenofibrate therapy on cardiovascular events in 9795 people with type 2 diabetes mellitus (the FIELD Study): randomised controlled trial. Lancet 2005; 366: 1849–1861.

[48] Abate N, Chandalia M. The ACCORD LIPID trial. MetabSyndrRelatDisord. 2010 Oct; 8(5): 373-4.

[49] Tenkanen L, Mänttäri M, Manninen V. Some coronary risk factors related to the insulin resistance syndrome and treatment with gemfibrozil. Experience from the Helsinki Heart Study. Circulation. 1995 Oct 1; 92(7): 1779-85.

[50] Rubins HB, Robins SJ, Collins D, *et al.* Diabetes, plasma insulin, and cardiovascular disease: subgroup analysis from the Department of Veterans Affairs high-density lipoprotein intervention trial (VA-HIT). Arch Intern Med. 2002 Dec 9-23; 162(22): 2597-604.

[51] Tenenbaum A, Motro M, Fisman EZ, *et al.* Bezafibrate for the secondary prevention of myocardial infarction in patients with metabolic syndrome. Arch Intern Med. 2005 May 23; 165(10): 1154-60.

[52] Scott R, O'Brien R, Fulcher G; Fenofibrate Intervention and Event Lowering in Diabetes (FIELD) Study Investigators. Effects of fenofibrate treatment on cardiovascular disease risk

in 9, 795 individuals with type 2 diabetes and various components of the metabolic syndrome: the Fenofibrate Intervention and Event Lowering in Diabetes (FIELD) study. Diabetes Care. 2009 Mar; 32(3): 493-8. Epub 2008 Nov 4.

[53] Bruckert E, Labreuche J, Deplanque D, etal. Fibrates effect on cardiovascular risk is greater in patients with high triglyceride levels or atherogenicdyslipidemia profile: a systematic review and meta-analysis. J CardiovascPharmacol. 2011 Feb; 57(2): 267-72.

[54] Jun M, Foote C, Lv J, *et al.* Effects of fibrates on cardiovascular outcomes: a systematic review and meta-analysis. Lancet. 2010 May 29; 375(9729): 1875-84.

[55] McLaughlin PR, Gladstone P. Diabetes Atherosclerosis Intervention Study (DAIS): quantitative coronary angiographic analysis of coronary artery atherosclerosis. CathetCardiovascDiagn. 1998 Jul; 44(3): 249-56.

[56] Ramachandran S, Abbas A, Saraf S, *et al.* Significant increase in high-density lipoprotein cholesterol with fibrates is associated with low pretreatment high-density lipoprotein cholesterol: findings from an outpatient clinic setting. MetabSyndrRelatDisord. 2012 Jun; 10(3): 189-94. Epub 2012 Jan 27.

[57] Davidson MH. Statin/fibrate combination in patients with metabolic syndrome or diabetes: evaluating the risks of pharmacokinetic drug interactions. ExpertOpin Drug Saf. 2006 Jan; 5(1): 145-56.

[58] Giral P, Bruckert E, Jacob N, Chapman MJ, Foglietti MJ, Turpin G. Homocysteine and lipid lowering agents. A comparison between atorvastatin and fenofibrate in patients with mixed hyperlipidemia. Atherosclerosis. 2001 Feb 1; 154(2): 421-7.

[59] Westphal S, Dierkes J, Luley C. Effects of fenofibrate and gemfibrozil on plasma homocysteine. Lancet. 2001; 358: 39–40.

[60] Dierkes J, Westphal S, Kunstmann S, *et al.* Vitamin supplementation can markedly reduce the homocysteine elevation induced by fenofibrate. Atherosclerosis. 2001; 158: 161–4.

[61] Mayer O. , Šimonetal J. Folate Co-Administration Improves the Effectiveness ofFenofibrate to Decrease the Lipoprotein Oxidation andEndothelial Dysfunction Surrogates. Physiol. Res. 55: 475-481, 2006.

[62] Bhartia M, Gandhi N, Lenton R, *et al.* Liver function tests changes following fibrate treatment in patienst with metabolic syndrome http://eas. ekonnect. co/EAS_279/poster_30486/program. aspx

[63] Abbas A, Saraf S, Ramachandran S, Raju J, Ramachandran S. Fibrates and estimated glomerular filtration rate: observations from an outpatient clinic setting and clinical implicationsPostgrad Med J. 2012 Mar 29.

[64] Miller DB, Spence JD. Clinical pharmacokinetics of fibric acid derivatives (fibrates). ClinPharmacokinet. 1998 Feb; 34(2): 155-62.

[65] ESC/EAS Guidelines for the management of dyslipidaemias: the Task Force for the management of dyslipidaemias of the European Society of Cardiology (ESC) and the European Atherosclerosis Society (EAS). Eur Heart J. 2011 Jul; 32(14): 1769-818. Epub 2011 Jun 28.

[66] Davis TM, Ting R, Best JD, Donoghoe MW, Drury PL, Effects of fenofibrate on renal function in patients with type 2 diabetes mellitus: the Fenofibrate Intervention and Event

Lowering in Diabetes (FIELD) Study. Diabetologia. 2011 Feb; 54(2): 280-90. Epub 2010 Nov 4.

[67] Forsblom C, Hiukka A, Leinonen ES, Effects of long-term fenofibrate treatment on markers of renal function in type 2 diabetes: the FIELD Helsinki substudyDiabetes Care. 2010 Feb; 33(2): 215-20. Epub 2009 Oct.

[68] ACCORD Study Group, Effects of combination lipid therapy in type 2 diabetes mellitus. N Engl J Med. 2010 Apr 29; 362(17): 1563-74. Epub 2010 Mar 14.

[69] Keech AC, Mitchell P, Summanen PA, O'Day J, Davis TM. Effect of fenofibrate on the need for laser treatment for diabetic retinopathy (FIELD study): a randomised controlled trial. Lancet. 2007 Nov 17; 370(9600): 1687-97. Epub 2007 Nov 7.

[70] ACCORD Study Group; ACCORD Eye Study Group, Effects of medical therapies on retinopathy progression in type 2 diabetes. N Engl J Med. 2010 Jul 15; 363(3): 233-44. Epub 2010 Jun 29.

[71] http://www.trialresultscenter.org/ultimate3/cadre.asp?pathologie=20&mecanisme=120&question=416&classe=glitazones

[72] Cantello BCC, Cawthorne MA, Haigh D, *et al.* The synthesis of BRL 49653: a novel and potent antihyperglycaemic agent. Bioorg Med Chem Lett. 1997; 4: 1181–4.

[73] Lupi R, Dotta F, Marselli L *et al.* Prolonged exposure to free fatty acids has cytostatic and pro-apoptotic effects on human pancreatic islets: evidence that beta-cell death is caspase mediated, partially dependent on ceramide pathway, and Bcl-2 regulated. Diabetes. 2002 May; 51(5): 1437-42.

[74] Raskin P, Rappaport EB, Cole ST, Yan Y, Patwardhan R, Freed MI. Rosiglitazone short-term monotherapy lowers fasting and post-prandial glucose in patients with type II diabetes. Diabetologia. 2000 Mar; 43(3): 278-84.

[75] Deeg MA, Buse JB, Goldberg RB, *et al.* Pioglitazone and Rosiglitazone Have Different Effects on Serum Lipoprotein Particle Concentrations and Sizes in PatientsWith Type 2 Diabetes and Dyslipidemia. Diabetes Care. 2007; 30(10): 2458-64

[76] Freed MI, Ratner R, Marcovina SM, Kreider MM, Biswas N, Cohen BR, Brunzell JD; Rosiglitazone Study 108 investigators. Effects of rosiglitazone alone and in combination with atorvastatin on the metabolic abnormalities in type 2 diabetes mellitus. Am J Cardiol. 2002 Nov 1; 90(9): 947-52.

[77] DREAM (Diabetes REduction Assessment with ramipril and rosiglitazone Medication) Trial Investigators, Gerstein HC, Yusuf S, Bosch J, Pogue J, Sheridan P, Dinccag N, Hanefeld M, Hoogwerf B, Laakso M, Mohan V, Shaw J, Zinman B, Holman RR. Effect of rosiglitazone on the frequency of diabetes in patients with impaired glucose tolerance or impaired fasting glucose: a randomised controlled trial. Lancet. 2006 Sep 23; 368(9541): 1096-105.

[78] Niessen S, Wolski K. Effect of Rosiglitazone on the risk of myocardial infarction and death from cardiovascular causes. NEJM. 2007; 356: 24: 2457–41.

[79] (http://www.fda.gov/Drugs/DrugSafety/PostmarketDrugSafetyInformationforPatientsandProviders/ucm143460. htm)

[80] Kahn SE, Haffner SM, Heise MA, *et al.* , for the ADOPT study group. Glycemic durability of rosiglitazone, metformin, or glyburide monotherapy. N Engl J Med 2006; 355: 2427–43.

[81] Home PD, Pocock SJ, Beck-Nielsen H, *et al.* Rosiglitazone Evaluated for Cardiovascular Outcomes—An Interim Analysis. NEJM. 2007; 357(1): 28–8.

[82] Home PD, Pocock SJ, Beck-Nielsen H, *et al.* Rosiglitazone evaluated for cardiovascular outcomes in oral agent combination therapy for type 2 diabetes (RECORD): a multicentre, randomised, open-label trial. Lancet 2009; 373: 2125–35; Published Online June 5, 2009 DOI: 10. 1016/S0140-6736(09)60953-3.

[83] Diamond GA, Bax L, Kaul S. Uncertain effects of rosiglitazone on the risk for myocardial infarction and cardiovascular death. Ann Intern Med. 2007 Oct 16; 147(8): 578-81. Epub2007 Aug 6.

[84] Dormandy JA, Charbonnel B, Eckland DJA, *et al.* Secondary prevention of macrovascular events in patients with type 2 diabetes in the PROactiveSutdy (PROspectivepioglitAzone Clinical Trial In macroVascular Events): a randomised controlled trial. Lancet. 2005; 366(9493): 1279-1289.

[85] Lincoff AM, Wolski K, Nicholls SJ, Nissen SE. Pioglitazone and risk of cardiovascular events in patients with type 2 diabetes mellitus: a meta-analysis of randomized trials. JAMA. 2007 Sep 12; 298(10): 1180-8.

[86] Caldwell SH, Hespenheide EE, Redick JA, Iezzoni JC, Battle EH, Sheppard BL. A pilot study of a thiazolidinedione, troglitazone, in nonalcoholicsteatohepatitis. Am J Gastroenterol. 2001 Feb; 96(2): 519-25.

[87] Neuschwander-Tetri BA, Brunt EM, Wehmeier KR, Oliver D, Bacon BR. Improvednonalcoholicsteatohepatitis after 48 weeks of treatment with the PPAR-gamma ligand rosiglitazone. Hepatology. 2003 Oct; 38(4): 1008-17.

[88] Promrat K, Lutchman G, Uwaifo GI, Freedman RJ, Soza A, Heller T, Doo E, Ghany M, Premkumar A, Park Y, Liang TJ, Yanovski JA, Kleiner DE, Hoofnagle JH. A pilot study of pioglitazone treatment for nonalcoholicsteatohepatitis. Hepatology. 2004 Jan; 39(1): 188-96.

[89] Argo CK, Iezzoni JC, Al-Osaimi AM, Caldwell SH. Thiazolidinediones for the treatment in NASH: sustained benefit after drug discontinuation. J Clin Gastroenterol. 2009 Jul; 43(6): 565-8.

[90] Lewis JD, Ferrara A, Peng T, Hedderson M, Bilker WB, Quesenberry CP Jr, Vaughn DJ, Nessel L, Selby J, Strom BL. Risk of bladder cancer among diabetic patients treated with pioglitazone: interim report of a longitudinal cohort study. Diabetes Care. 2011 Apr; 34(4): 916-22.

[91] RonacMamtani, Kevin Haynes, Warren B. Bilker *et al.* Long-term therapy with thiazolidinediones and the risk of bladder cancer: A cohort study. J Clin Oncol 30, 2012 (suppl; abstr 1503).

[92] Azoulay L, Yin H, Filion KB, Assayag J, Majdan A, Pollak MN, Suissa S. The use of pioglitazone and the risk of bladder cancer in people with type 2 diabetes: nested case-control study. BMJ. 2012 May 30; 344: e3645. doi: 10. 1136/bmj. e3645.

[93] Wilding J. Thiazolidinediones, insulin resistance and obesity: Finding a balance. Int J Clin Pract. 2006 Oct; 60(10): 1272-80.

[94] Barnett A, Allsworth J, Jameson K, Mann R. A review of the effects of antihyperglycaemic agents on body weight: the potential of incretin targeted therapies. Curr Med Res Opin. 2007 Jul; 23(7): 1493-507.

[95]   Fonseca V. Effect of thiazolidinediones on body weight in patients with diabetes mellitus. Am J Med. 2003 Dec 8; 115Suppl 8A: 42S-48S.

[96]   Tan MH, Baksi A, Krahulec B, Kubalski P, Stankiewicz A, Urquhart R, Edwards G, Johns D; GLAL Study Group. Comparison of pioglitazone and gliclazide in sustaining glycemic control over 2 years in patients with type 2 diabetes. Diabetes Care. 2005 Mar; 28(3): 544-50.

[97]   Gorson DM. Significant weight gain with rezulin therapy. Arch Intern Med. 1999 Jan 11; 159(1): 99.

[98]   Yang T, Soodvilai S. Renal and vascular mechanisms of thiazolidinedione-induced fluid retention. PPAR Res. 2008; 2008: 943614.

[99]   Colucciello M. Vision loss due to macular edema induced by rosiglitazone treatment of diabetes mellitus. ArchOphthalmol. 2005 Sep; 123(9): 1273-5.

[100]   Ryan EH (Jr), Han DP, Ramsay RC, Cantrill HL, Bennett SR, Dev S, Williams DF. Diabetic macular edema associated with glitazone use. Retina. 2006 May-Jun; 26(5): 562-70.

[101]   Fong DS, Contreras R. Glitazone use associated with diabetic macular edema. Am J Ophthalmol. 2009 Apr; 147(4): 583-586. e1. Epub 2009 Feb 1.

[102]   http://diabetes. diabetesjournals. org/content/60/Supplement_1/A1. full. pdf)

[103]   Bonds DE, Larson JC, Schwartz AV, Strotmeyer ES, Robbins J, Rodriguez BL, Johnson KC, Margolis KL. Risk of fracture in women with type 2 diabetes: the Women's Health Initiative Observational Study. J Clin Endocrinol Metab. 2006 Sep; 91(9): 3404-10. Epub 2006 Jun 27.

[104]   Janghorbani M, Van Dam RM, Willett WC, Hu FB. Systematic review of type 1 and type 2 diabetes mellitus and risk of fracture. Am J Epidemiol. 2007 Sep 1; 166(5): 495-505. Epub 2007 Jun 16.

[105]   Loke YK, Singh S, Furberg CD. Long-term use of thiazolidinediones and fractures in type 2 diabetes: a meta-analysis. CMAJ. 2009 Jan 6; 180(1): 32-9. Epub 2008 Dec 10.

[106]   Chandler HA, BatchelorAJ. Ciprofibrate and lipid profile. Lancet. 1994 Jul 9; 344(8915): 128-9.

[107]   Ebcioglu Z, Morgan J, Carey C, Capuzzi D Paradoxical lowering of high-density lipoprotein cholesterol level in 2 patients receiving fenofibrate and a thiazolidinedione. Ann Intern Med. 2003 Nov 4; 139(9): W80.

[108]   Shetty C, Balasubramani M, Capps N, Milles J, Ramachandran S. Paradoxical HDL-C reduction during rosiglitazone and fibrate treatment. Diabet Med. 2007 Jan; 24(1): 94-7.

[109]   Saraf S, Nishtala S, Paretti H. *et al.* Paradoxical fall in HDL cholesterol observed in a patient treated with rosiglitazone and pioglitazone. doi: 10. 1177/1474651409341325 British Journal of Diabetes & Vascular Disease July/August 2009 vol. 9 no. 4 186-189.

[110]   Keidar S, Guttmann H, Stam T, Fishman I, Shapira C. High incidence of reduced plasma HDL cholesterol in diabetic patients treated with rosiglitazone and fibrate. Pharmaco epidemiol Drug Saf. 2007 Nov; 16(11): 1192-4.

[111]   Ramachandran S, Saraf S, Shetty C, Capps N, Bailey C. Paradoxical decrease in HDL-cholesterol and apolipoprotein A1 with simvastatin and atorvastatin in a patient with type 2 diabetes. Ann Clin Biochem. 2011 Jan; 48(Pt 1): 75-8. Epub 2010 Nov 29.

[112] Brown MS, Goldstein JL (1999). A proteolytic pathway that controls the cholesterol content of membranes, cells, and blood. Proc Natl Acad Sci USA 96 (20): 11041–8. DOI: 10. 1073/pnas. 96. 20. 11041.

[113] Fan P, Zhang B, Kuroki S, Saku K. Pitavastatin, a potent hydroxymethylglutaryl coenzyme a reductase inhibitor, increases cholesterol 7 alpha-hydroxylase gene expression in HepG2 cells. Circ J. 2004 Nov; 68(11): 1061-6.

[114] Saiki A, Murano T, Watanabe F, Oyama T, Miyashita Y, Shirai K. Pitavastatin enhanced lipoprotein lipase expression in 3T3-L1 preadipocytes. J Atheroscler Thromb. 2005; 12(3): 163-8.

[115] Seo M, Inoue I, Ikeda M, Nakano T, Takahashi S, Katayama S, Komoda T. Statins Activate Human PPARalpha Promoter and Increase PPAR alpha mRNA Expression and Activation in HepG2 Cells. PPAR Res. 2008; 2008: 316306. Epub 2008 Dec 24.

[116] Planavila A, Laguna JC, Vázquez-Carrera M. Atorvastatin improves peroxisome proliferator-activated receptor signaling in cardiac hypertrophy by preventing nuclear factor-kappa B activation. Biochim Biophys Acta. 2005 Feb 21; 1687(1-3): 76-83.

[117] Zapolska-Downar D, Siennicka A, Kaczmarczyk M, Kołodziej B, Naruszewicz M. Simvastatin modulates TNFalpha-induced adhesion molecules expression in human endothelial cells. Life Sci. 2004 Jul 30; 75(11): 1287-302.

[118] Yano M, Matsumura T, Senokuchi T, Ishii N, Murata Y, Taketa K, Motoshima H, Taguchi T, Sonoda K, Kukidome D, Takuwa Y, Kawada T, Brownlee M, Nishikawa T, Araki E. Statins activate peroxisome proliferator-activated receptor gamma through extracellular signal-regulated kinase 1/2 and p38 mitogen-activated protein kinase-dependent cyclooxygenase-2 expression in macrophages. Circ Res. 2007 May 25; 100(10): 1442-51. Epub 2007 Apr 26.

[119] Balakumar P, Mahadevan N. Interplay between statins and PPARs in improving cardiovascular outcomes: a double-edged sword? Br J Pharmacol. 2012 Jan; 165(2): 373-9. doi: 10. 1111/j. 1476-5381. 2011. 01597.

Send Orders for Reprints to reprints@benthamscience.net

CHAPTER 6

Topical Drugs for Diabetic Complications Current and Future

Ryuhei Sano and Tomohiko Sasase*

Central Pharmaceutical Research Institute, Japan Tobacco Inc., Takatsuki, 569-1125 Osaka, Japan

Abstract: Diabetes is a major metabolic disease, and the number of patients is still growing in both industrialized and developing countries. Drugs to reduce blood glucose levels by several mechanisms have been launched in past decades, and the options for treating hyperglycemia have been extended. However, it is still difficult to control blood glucose level strictly, and moreover, current therapies using hypoglycemic agents have not shown sufficient effect to prevent the development of microvascular complications that result in impaired quality of life (QOL) of patients. Therefore treatments for these diabetic complications are urgently needed. Drugs for diabetic peripheral neuropathy currently in use are mainly for management of diabetic neuropathic pain, and do not cure the underlying cause of the disease. In diabetic retinopathy, intravitreal injection of anti-VEGF agents is a current approach, but the administration route is insufficient. Diabetic nephropathy is treated with angiotensin-I converting enzyme (ACE) inhibitors and angiotensin-II receptor blockers (ARBs), which were originally prescribed for hypertension. These situations indicate the lack of effective target-oriented medication for diabetic compliations, and drugs with novel mechanism should be developed. In this chapter, we summarize drugs for three major diabetic complications currently in use and in clinical trial phases. It is hoped this will help the reader to understand the trend of current pharmacological approaches and the tide of next-generation drugs in this field.

Keywords: ACE inhibitor, analgesia, Anti-VEGF, ARB, ARI, blood circulation disorder, clinical trials, diabetes, diabetic complications, diabetic macular edema, diabetic nephropathy, diabetic neuropathy, diabetic retinopathy, metabolic disorder, microalbuminuria, microvascular complications, new drugs, painful diabetic neuropathy, RAS inhibitor.

INTRODUCTION

Diabetes mellitus is one of the most common metabolic diseases, and the number of patients is estimated at approximately 350 million worldwide [1]. More than half of all diabetics have one or more diabetic microvascular complications, such as diabetic peripheral neuropathy (DPN), diabetic retinopathy (DR), or diabetic

---

*Address correspondence to Tomohiko Sasase: Central Pharmaceutical Research Institute, Japan Tobacco Inc., Takatsuki, Osaka, Japan; Tel: +81-72-681-9700; Fax: +81-72-681-9722; E-mail: tomohiko.sasase@jt.com

nephropathy (DN). The Diabetes Control and Complications Trial (DCCT), the U.K. Prospective Diabetes Study (UKPDS), and the Kumamoto Study have established that intensive glycemic control on both type 1 and type 2 diabetes patients can delay the onset and progression of vascular complications [2-4]. However, it is difficult to control blood glucose level strictly. In DCCT, only 44% of the patients receiving intensive therapy achieved the goal of HbA1c value of 6.05% or less [3]. Furthermore, current therapies using blood glucose-lowering drugs have not shown sufficient effect to prevent the development of vascular complications. Therefore, original approaches to prevent diabetic complications are urgently needed. Although several drugs for these diabetic complications have been developed and launched, many people are still suffering from these severe diseases and endure impaired quality of life (QOL). In this chapter, current drugs and drugs under development for these diabetic complications are overviewed.

DIABETIC PERIPHERAL NEUROPATHY

Diabetic neuropathy occurs in the relatively early stage of diabetes mellitus, especially in peripheral nerves (diabetic peripheral neuropathy; DPN). In general, small fibers, which are autonomic nerves that transmit cold sense, burning sensation, and pain, are affected at the onset of diabetic neuropathy, and large fibers, which are sensory nerves, are gradually damaged. Damage to autonomic nerves causes gastrointestinal disorder, dyshidrosis, orthostatic hypotension, erectile dysfunction, *etc*.

To improve pain and numbness, which may result from ischemia, hypoxia, or nutritional deficiency in peripheral nerves due to microangiopathy, anticonvulsants and antidepressants are effective and conventionally used. Duloxetine hydrochloride and pregabalin are currently used as the standard treatment worldwide, but it has been essentially difficult to develop any follow-on candidates. As neuropathy proceeds to hypoaesthesia, patients become less sensitive to pain and heat and more vulnerable to foot injury and burn. In this stage, injury results in ulceration due to trophic nerve degeneration and microangiopathy. As the disease becomes worse, the tissue becomes gangrenous, requiring foot amputation to avoid death. In the treatment of diabetic foot ulcer, burden is reduced by depressurization using shoes designed to disperse pressure on the wound. At further stage, artificial skin grafting is effective for chronic wound unresponsive to conventional therapy. In recent years, anti-nerve growth factor antibodies, vascular endothelial growth factors (VEGFs), and plasmid preparations designed to increase the expression of genes coding for VEGF-A had been studied, but development of these substances has since been discontinued or suspended.

SYMPTOMATIC TREATMENT OF PAINFUL DIABETIC NEUROPATHY

Anticonvulsants

Ca^{2+} Channel a$_2$δ Ligands

Pregabalin

Pregabalin (Lyrica$^{®}$; Pfizer) is a γ-aminobutyric acid (GABA) derivative. It produces analgesia by strongly binding to the α$_2$δ subunit in voltage-gated Ca^{2+} channels in the overexcited excitatory nervous system to inhibit influx of Ca^{2+} into presynaptic excitatory neurons and thus inhibit release of various excitatory neurotransmitters such as glutamate. It is similar in structure to GABA, but does not bind to the GABA receptor, and does not exhibit the acute effect on the metabolism or the uptake of GABA. It has also been shown to have no activity at the binding site of various excitatory amino acid receptors such as NMDA, AMPA, kainic acid, and glycine receptors, as well as molecules that act on voltage-gated Ca^{2+} channels, Na^{+} channels, Cl^{-} channels, and K^{+} channels (except for gabapentin). In clinical studies, it has been shown to be as effective as tricyclic antidepressants (TCAs), with fewer anticholinergic side effects on the autonomic nervous system such as thirst and constipation.

In an 8-week phase III study in 146 patients with painful diabetic neuropathy, a significant improvement was observed in the mean endpoint pain score based on the daily patient diary (11-point numerical pain rating scale) at a dose of 100 mg x 3/day [5]. Pregabalin was initially approved for the treatment of peripheral neuropathic pain in Europe in 2004 and for the treatment of postherpetic neuralgia (PHN) and painful diabetic neuropathy in the US in 2004. It was also additionally approved for the treatment of central neuropathic pain in Europe in 2006. As of 2010, it has been approved for the treatment of neuropathic pain, *etc.*, in 110 countries and regions, including the US, EU, Australia, and Canada.

Gabapentin Enacarbil

Gabapentin enacarbil (Horizant®; GlaxoSmithKline/XenoPort) is a prodrug of gabapentin used as an anticonvulsant as well as a treatment for neurogenic pain, with the same mechanism of action as pregabalin [6]. Gabapentin has the following pharmacokinetic problems: 1) difficult to obtain dose-dependent blood concentrations due to saturation of upper gastrointestinal absorption at high doses; 2) large variation in blood concentrations resulting from large individual variation in the expression of L-type amino acid transporter 1 (LAT1), which is responsible for absorption of gabapentin; and 3) short half-life in blood, requiring frequent doses to maintain blood concentrations. Gabapentin enacarbil is designed to alleviate these problems by increasing the permeability. This improvement contributes to minimum individual difference in absorption of gabapentin and stable efficacy [7]. Gabapentin enacarbil has also been shown to be effective for PHN [8] and restless legs syndrome (RLS) [9, 10]. It was approved by the FDA for the treatment of RLS in 2011 (phase III in Europe). For PHN, an application for approval was submitted to the FDA in 2011 (phase II in Europe). However, no significant improvement was observed in the Pain Intensity-Numerical Rating Scale (PI-NRS) after 14-week treatment as compared with placebo control in a phase II study on diabetic neuropathic pain, leading to discontinuation of its development for the treatment of painful diabetic neuropathy.

In addition, a phase II study of DS-5565 (Daiichi Sankyo) for the treatment of DPN (*vs.* pregabalin, primary endpoint: change from baseline in the average daily pain score, 400 subjects, 5 weeks) is ongoing in the US.

GABA Transaminase Inhibitors

Valproate

Valproate (Depakote®; Sanofi, Depakene®; Abbott Laboratories) is a histone deacetylase 1 (HDAC1) inhibitor. The main pharmacological effect is inhibition of central GABA transaminase to inhibit inactivation of GABA and increase the GABA concentration. GABA acts on the $GABA_A$ receptor to open Cl^- channels and induce hyperpolarization in nerve cells. Valproate is used as a broad-spectrum anticonvulsant, because it also blocks voltage-gated Na^+ channels and T type Ca^{2+} channels. It is also used as a second-line treatment for neurogenic pain.

In clinical studies of valproate in patients with painful diabetic neuropathy, it was reported that the pain score was improved, but the nerve conduction velocity (NCV) was not [11, 12]. With a narrow therapeutic margin, it has a risk of liver disorder and teratogenicity at high doses.

Na$^+$ Channel Blockers

Carbamazepine

Damage to peripheral nerves under hyperglycemic or chronic ischemic conditions induces ectopic expression of adrenaline receptors on the axonal membrane. Binding of catecholamine in blood or from sympathetic nerve endings to the receptor leads to opening of Na$^+$ channels and ectopic impulses, which are the source of pain. In addition, a tactile stimulus is erroneously recognized as pain if an abnormal short circuit is formed between large myelinated fibers that transmit tactile impulses and small pain fibers (ephapse formation). This is called allodynia.

Carbamazepine, a specific medicine for temporal lobe epileptic seizure, blocks Na$^+$ channels in nerve cells to inhibit upstroke of the membrane action potential and thereby reduce peripheral pain impulse propagation. Therefore, it is also used in the treatment of painful diabetic neuropathy to inhibit abnormal excitement of peripheral nerve axons. It was reported to be significantly more analgesic than placebo in a clinical study of Tegretol® (Novartis Pharma) in painful diabetic neuropathy [13], and was often used in combination with TCAs. However, it has been replaced with new drugs such as pregabalin, because it is associated with a high incidence of toxic drug eruption and requires cumbersome blood concentration monitoring.

Others

Topiramate

Topiramate (Topamax®; Janssen Pharmaceuticals, Topina®; Kyowa Hakko Kirin) is an anticonvulsant that has been used in the treatment of epilepsy (as combination therapy) and migraine worldwide since 1995. The antiepileptic effect of topiramate may be based on multiple mechanisms of action: inhibition of

Voltage-gated Na^+ channels, inhibition of voltage-gated L type Ca^{2+} channels, inhibition of AMPA/kainite-type glutamate receptor function, enhancement of $GABA_A$ receptor function in the presence of GABA, and inhibition of carbonic anhydrase. In addition, Qsymia™ (Vivus), a formulation combining topiramate with anoretic phentamine, was approved by the FDA as an anti-obesity drug in 2012.

Several clinical studies of topiramate have been conducted for the treatment of painful diabetic neuropathy. In a study in 323 subjects, it was reported that a significant improvement was achieved in the pain visual analog (PVA) score with weight reduction after 12-week treatment at a dose of 400 mg [14].

Antidepressants

Tricyclic Antidepressants (TCAs)

Serotonin (5-HT), a neurotransmitter in the brain, small intestine, and platelets, is involved in emotional control (elevation and depression), sleep-wake, and eating behavior. Noradrenaline (NA), an excitatory neurotransmitter in sympathetic nerve endings and the central nervous system, is secreted in the brain in response to stress such as pain and fear to trigger a fight-or-flight response. If 5-HT or NA released from nerve cells in the brain cannot bind to the receptor on next nerve cells, therefore, depressive symptoms occur. In the central nervous system, 5-HT and NA activate the descending pain modulatory system. Accordingly, TCAs may inhibit reuptake of 5-HT and NA to enhance descending pain modulation and thereby produce analgesia. TCAs have strong effects by inhibiting reuptake of both 5-HT and NA, but also have side effects such as thirst, dysuria, constipation, and sleepiness by nonspecifically acting on the receptor of other neurotransmitters such as acetylcholine and histamine.

Several meta-analyses have shown that TCAs, which activate the central pain modulatory system, are the most effective for painful diabetic neuropathy [15-17]. In recent years, it has been pointed out that not only the efficacy for pain, but also side effects should be evaluated to assess the usefulness [17]. While TCAs are relatively superior in both efficacy and safety, some recommend secondary amines such as nortriptyline, which are associated with lower incidences of side effects such as sleepiness and light-headedness, rather than tertiary amines such as amitriptyline [17].

Amitriptyline

Amitriptyline (Elavil®; Merck, Laroxyl®; Roche, Saroten®; Bayer, Tryptizol®; Merck, Lentizol®; Pfizer) is a first-generation TCA with strong sedative effects. It inhibits reuptake of 5-HT more potently than reuptake of NA in nerve endings [18]. It is effective for depressed patients with a chief complaint of anxiety, irritation, and/or sleeplessness. It is also effective for neurogenic pain (especially with burning pain and/or paresthetic pain). Administration for 2 to 3 weeks is required to induce the onset of antidepressant effects, but 1 week or less to induce analgesic effects. Anticholinergic side effects such as thirst and constipation may sometimes limit its use. The FDA approved amitriptyline as an antidepressant in 1961. It has been reported to reduce diabetic neuropathic pain [19] and is recommended by the ADA for the treatment of painful diabetic neuropathy.

Nortriptyline

Nortriptyline (Aventyl®; Eli Lilly, Allegron®; Pfizer, Noritren®; H. Lundbeck) is a demethylated metabolite of amitriptyline and has less sedative effect than the parent drug. It inhibits reuptake of NA more potently than reuptake of 5-HT in nerve endings [18]. It antagonizes the effect of reserpine and inhibits reserpine-induced hypothermia [20]. It is effective for depressed patients with a chief complaint of avolition, constricted affect, and/or apathy, because it enhances motivation. It was approved by the FDA as an antidepressant in 1964. It is also effective for neurogenic pain (especially with burning pain and/or paresthetic pain) and used to treat painful diabetic neuropathy on an off-label basis [21, 22].

Serotonin Noradrenaline Reuptake Inhibitors (SNRIs)

SNRIs have antidepressant effects by binding to 5-HT transporter and NA transporter to inhibit reuptake of 5-HT and NA in the brain and thus increase 5-HT and NA in the synaptic cleft. Therefore, SNRIs are used as antidepressants like TCAs, selective serotonin reuptake inhibitors (SSRIs), and monoamine oxidase inhibitors (MAOIs), but have fewer side effects due to more specific inhibition of reuptake than TCAs.

It has long been said that pain and depression have a common psycho pharmacological profile, and it is known that compounds that inhibit reuptake of 5-HT and/or NA can reduce chronic pain and physical symptoms. In particular, it is suggested that NA plays a more important role in pain relief than 5-HT. Therefore, SNRIs that inhibit reuptake of both 5-HT and NA are now commercially available for the treatment of diabetic neuropathic pain as well.

Duloxetine

During 12-week treatment with duloxetine hydrochloride (Cymbalta®; Eli Lilly) in non-depressive patients with diabetic neuropathic pain, a rapid improvement was observed in the mean weekly score of the 24-hour average pain severity, with significant differences from the placebo group starting at week 1 of treatment [23]. It was also reported in a long-term treatment study that Brief Pain Inventory (BPI) 24-hour average pain was improved over 26 weeks in patients with diabetic neuropathic pain [24]. It has been reported that the analgesic efficacy of duloxetine was similar to that of amitriptyline, a TCA [25]. It was originally launched for the treatment of depression in the West, but was additionally approved for the treatment of diabetic neuropathic pain in Europe in 2004. In Japan, it was introduced for the treatment of depression in 2010, and was additionally approved for the treatment of diabetic neuropathic pain in 2012.

Bicifadine

Bicifadine (XTL Biopharmaceuticals) inhibits reuptake of NA potently, with a 5-HT/NA inhibitory profile different from that of conventional SNRIs [26]. It was effective in a clinical study on postoperative pain [27], but not in a phase III study on chronic low back pain. In a phase IIb study in patients with diabetic neuropathic pain, 14-week treatment failed to achieve the primary endpoint, defined as improvement in the 24-hour pain rate on an 11-point Pain Intensity Numeric Rating Scale (LIKERT scale), leading to discontinuation of development in 2007.

Desvenlafaxine

Desvenlafaxine (Pristiq®; Pfizer) is a major active metabolite of venlafaxine hydrochloride (Effexor®; Pfizer), a medicine for major depressive disorder, and was approved for the treatment of major depressive disorder in the US in 2008. Desvenlafaxine succinate sustained release (DVS SR; Pfizer) is a sustained-release formulation of desvenlafaxine. In addition to major depressive disorder, two phase III studies (primary endpoint: change from baseline in the mean pain score on the numeric rating scale (NRS)) were conducted for the treatment of painful diabetic neuropathy, but both studies were discontinued.

In addition, SEP-228432 (Dainippon Sumitomo Pharma), a new triple reuptake inhibitor (TRI) that inhibits reuptake of not only 5-HT and NE, but also dopamine, is under development for the treatment of neuropathic pain and depression, and a phase I study is ongoing.

Anxiolytics

Benzodiazepine Receptor Agonists

Benzodiazepine inhibits anxiety and excitation by stimulating central GABA receptors and thereby inhibiting neurotransmission in the central nervous system. It is also used to treat insomnia, because it causes drowsiness by inhibiting anxiety and excitement. Benzodiazepines are mainly used as hypnotics or anxiolytics (minor tranquilizers). Benzodiazepine receptors are classified into central benzodiazepine receptors (CBRs) and peripheral benzodiazepine receptors (PBRs). At present, however, it is shown that PBRs are also found in the central nervous system. PBRs are found in microglial cells in the brain and have been reported to increase in neuropsychiatric disorders involving microglial activation in the brain such as Alzheimer's disease.

SSAR180575

In an animal study, it was reported that PBR expression was increased in dorsal root ganglia (DRG) in a rat model of neurogenic pain [28]. SSAR180575 (Sanofi), a PBR agonist, stimulated nerve regeneration and increased the survival of nerve cells by up to 72% in a model of facial nerve disorder. It also improved function as measured by blink reflex [29]. A phase II study (primary endpoint: rate of epidermal nerve fiber regeneration following denervation of the epidermal layer with capsaicin, 309 subjects, 6 months) was conducted for the treatment of painful diabetic neuropathy, but development was discontinued.

Opioids

Opioids act on opioid receptors (μ, δ, κ, *etc.*) to exhibit morphine-like pharmacological effects. Full agonists for the major target receptor for analgesia are primarily used to treat cancerous pain and serious injury, because they can produce potent analgesia by binding to occupiable sites on receptors. While neurogenic pain is generally resistant to opioids, some opioids have been reported to be effective for PHN and painful diabetic neuropathy and are used.

Oxycodone

Oxycodone hydrochloride (OxyContin®; Purdue Pharma) is an alkaloid analgesic that acts on the μ receptor to produce analgesia. It is more bioavailable and 1.5 to 2 times more analgesic than morphine at the equivalent dose. It is used for analgesia in cancer patients with moderate to severe pain. Like morphine, it has side effects such as feeling of sickness, constipation, sedation, and respiratory depression, but has a lower incidence of nausea and delirium than morphine.

In a clinical study on moderate to severe diabetic neuropathic pain, it was reported that it reduced pain. There have been several reports, including one from a study in 159 patients that 6-week treatment resulted in a significant improvement in the overall average daily pain intensity during study days 28 to 42 [30].

Tramadol

Tramadol hydrochloride (Ultram®; Janssen Pharmaceuticals) is a phenol ether analgesic that has combined effects of μ agonist and TCAs, with moderate affinity for the μ receptor, but little affinity for the δ or κ receptor. Compared with other opioids, it has fewer side effects, especially on the gastrointestinal motility. It is an "atypical" opioid, with the effects of SNRIs, and has been shown to be effective for diabetic neuropathic pain with analgesia resulting from activation of the descending inhibitory system. In a study in 131 subjects with diabetic neuropathic pain, it was reported that 42-day treatment resulted in a significant improvement in the mean pain intensity scores [31]. In a 6-month long-term study in 117 subjects, improvements were observed in self-administered pain intensity scores and pain relief scores [32].

In a phase III study of a formulation combining tramadol hydrochloride and acetaminophen (Ultracet®; Janssen Pharmaceuticals) for the treatment of diabetic neuropathic pain, 66-day treatment resulted in a significant improvement in the change in the mean of average daily pain scores from baseline to final week in 313 subjects [33].

Tapentadol

Tapentadol hydrochloride (Nucynta®; Janssen Pharmaceuticals) is a central synthetic analgesic with two different mechanisms of action: agonism of the μ opioid receptor and inhibition of NA reuptake. In diabetic animals, it improves thermal hyperalgesia at doses from 0.1 to 1 mg/kg *i.v.* This effect is more potent than that of morphine, and unlike morphine, does not affect normal animals [34]. In a phase III study of Nucynta® ER (extended release) for the treatment of diabetic neuropathic pain, the optimal dose was individually determined after 3-week treatment Nucynta® ER (100 to 250 mg twice daily) for 588 subjects who had previously received an analgesic for at least 3 months, and 395 subjects with improvement in neuropathy were randomized to receive the active drug or placebo for 12 weeks. After treatment, at least 30% improvement was observed in the mean change from baseline in average pain intensity on the NRS 11-point scale in 53.6% (105 of 196 patients) in the Nucynta® ER group [35].

Nucynta® was launched for the treatment of acute pain in the US in 2009. It was additionally approved for the treatment of chronic pain in patients aged 18 years and older in 2010 and launched in Germany. Nucynta® ER, an extended-release formulation approved in 2011, was additionally approved by the FDA for the treatment of diabetic neuropathic pain in 2012.

Drugs with Other Mechanisms

Nav1.7/1.8 Blockers

Voltage-gated Na^+ channels (VGSCs) are seven-transmembrane proteins abundantly expressed in central nerve cells and DRGs. The fast inward Na^+ currents generated during polarization of VGSCs through the relevant cell membrane give rise to the action potential and act at the threshold potential to determine the sensitivity of sensory neurons. Nerve damage results in various changes in different Na^+ channels. As the α subunit of VGSCs in mammals, a total of 9 subtypes, from Nav1.1 to Nav1.9, have been identified. Among them, the most important roles are played by tetrodotoxin (TTX)-resistant Nav1.8 in excitation propagation for nociceptive pain of nociceptive DRG neurons and TTX-sensitive Nav1.7 in excitation propagation for inflammatory pain [36, 37].

Nav1.7/1.8 blockers are expected to be effective for peripheral neuropathic pain based on distribution of the relevant channels. They are also expected to be highly safe without affecting the central nervous system or cardiac system, unlike non-selective Na^+ channel blockers and non-selective Ca^{2+} channel blockers, existing treatments for neuropathic pain. A-803467, a Nav1.8 blocker, has been reported to inhibit conduction of polymodal nociceptive C-fibers in painful diabetic neuropathy [38].

At present, phase II studies are ongoing for Nav1.7 blockers CNV 1014802 (Convergence Pharmaceuticals) and PF-05089771 (Pfizer/Icagen), as well as Nav1.8 blocker developing by Pfizer. In addition, Nav1.7/1.8 blockers XEN402 (Xenon Pharmaceuticals/Teva Pharmaceutical Industries) and DSP-2230 (Dainippon Sumitomo Pharma) are in phase II and phase I of development for the treatment of PHN and neuropathic pain, respectively.

Nicotinic α4β2 Agonists

Neuronal nicotinic acetylcholine receptors (NNRs) are members of the ligand-gated ion channel family and found throughout the central and peripheral nervous systems. NNRs consist of α (α2-10) and β (β2-4) subunits, and many subtypes with different functions depending on the composition have been identified [39]. While α4β2 and α7 NNR subtypes are found in the pain pathway in the central nervous system, α3 and β4 subtypes are found in the autonomic nervous system.

Sofinicline

Sofinicline (ABT-498; Abbott Laboratories/NeuroSearch), an α4β2 subunit agonist, reduced mechanical hyperalgesia in an animal model of diabetes [40]. Subsequently, two phase II studies (primary endpoint: weekly mean of 24h average pain score, 404 subjects, 8 weeks) were conducted for sofinicline with improved specificity for α4β2 for the treatment of diabetic neuropathic pain, but failed to demonstrate the efficacy, leading to discontinuation of development in 2009 [41]. Sofinicline is currently under development for the treatment of attention deficit/hyperactivity disorder (ADHD), for which a phase II study has been completed.

α2A-adrenoceptor Agonists

Adrenaline receptors are G protein-coupled (muscarinic) receptors that are activated by catecholamines and mainly found in heart and smooth muscles. Adrenaline receptors are classified into α1 (A, B, D), α2 (A, B, D), and β (1, 2, 3) subtypes, and the α2-adrenoceptor is involved in various neural actions, including platelet aggregation and inhibition of lipolysis. α2-adrenoceptor agonists have been clinically used as central antihypertensive agents, but are also found throughout the central nervous system. In particular, the α2A-adrenoceptor is involved in inhibition of dorsal horn neurons in the descending pain modulatory system, and the α2A-adrenoceptor level in the central nervous system has been reported to decrease in diabetes [42]. In addition, the α2A-adrenoceptor may mediate analgesia produced by N_2O [43].

Clonidine

Clonidine has been used as a central antihypertensive agent since the 1970s. In the 1990s, it was reported to inhibit hyperalgesia in diabetic animals [44] and to be effective for some patients with diabetic neuropathic pain [45]. In 2009, a phase II study of ARC-4558 (topical clonidine gel; Arcion Therapeutics) was conducted for the treatment of diabetic neuropathic pain in the US and demonstrated the efficacy (primary endpoint: mean change from baseline in 0-10 numerical pain rating scale (NPRS) score, 179 subjects, 12 weeks) [46].

Rezatomidine

For rezatomidine (AGN-203818; ACADIA/Allergan), an α2A-adrenoceptor agonist, a phase II study (primary endpoint: reduction in daily pain score, 330 subjects, 4 weeks) was conducted for the treatment of chronic pain, including DPN, and completed in 2007.

Vanilloid VR1 Receptor (TRPV1) Agonists

Capsaicin

Capsaicin, a major ingredient of chili peppers, is not only pungent, but also causes pain. The capsaicin receptor was initially named vanilloid receptor subtype 1 (VR1) [47], but is now called TRPV1, as the first molecule of the TRPV subfamily of the TRP ion channel superfamily [48]. The TRPV1 gene is specifically expressed in unmyelinated C fibers in sensory nerves and induces delayed pain such as sensation of warmth, itching, and burning pain if stimulated. Capsaicin, which not only causes pain, but also produces analgesia, is used as an analgesic. This effect is understood as a result of desensitization, that is, sensory nerve terminals chronically or repeatedly exposed to capsaicin become unresponsive to other nociceptive stimuli.

Qutenza® (Astellas Pharma/NeurogesX) was approved by the FDA for the treatment of PHN in 2009 and for the treatment of peripheral neuropathic pain in nondiabetic patients in the EU. Burning pain on application is a known side effect. Currently, a phase III study of Qutenza® (8% capsaicin) patch is ongoing for the treatment of DPN in the West (primary endpoint: percentage change in health-related quality of life (HRQOL) total score as assessed by the disease-specific Norfolk scale). NeurogesX is also developing NGX-1998, a high-concentration liquid formulation of capsaicin.

NMDA Receptor Antagonists

The N-methyl-D-aspartic acid (NMDA) receptor is a glutamate receptor subtype. It is an ion channel-coupled receptor found throughout the body, including areas of the central nervous system such as hippocampus, and binds to glutamate to transmit cations. At the resting membrane potential, Mg^{2+} bound to the NMDA receptor near the channel pore inhibits activation of the receptor. Membrane depolarization results in dissociation of Mg^{2+} from the NMDA receptor, and the receptor exhibits ion channel activity. Release of glutamate, an excitatory neurotransmitter, followed by its activation of the dorsal horn NMDA receptor is involved in the onset of neuropathic pain such as central sensitization and wind-up phenomenon, which plays an important role in transmission of nociceptive messages such as pain. The NMDA receptor also reverses the resistance to opioids to enhance analgesia. While noncompetitive NMDA receptor antagonists are used for anesthesia, improvement of cerebral circulation, and treatment of Parkinson's disease, Alzheimer's disease, and cough, competitive antagonists are not suitable for clinical use, as they have many side effects.

Dextromethorphan

Dextromethorphan is a noncompetitive NMDA receptor antagonist that inhibits cough reflex by inhibiting the medullary cough center and used as a central nonnarcotic antitussive. While it was shown that dextromethorphan hydrobromide hydrate reduced pain in a 6-week clinical study in 13 patients with diabetic neuropathic pain [49], there was a report that its safety or efficacy was doubtful [50]. Administration at a 3-fold higher dose or more is required for neuropathic pain than for cough, and attention should be paid to side effects such as clouding of consciousness and mental confusion.

Neurodex (AVP-923; Avanir Pharma) is a formulation combining dextromethorphan hydrobromide and low-dose quinidine sulfate (CYP2D6 enzyme inhibitor), which serves to increase the bioavailability of dextromethorphan. It was launched for the treatment of pseudobulbar affect (PBA) in 2011. A phase III study was conducted for the treatment of DPN (primary endpoint: four pain rating scales applied daily using patient diaries, 379 subjects, 13 weeks), and it was reported in 2012 that it reduced the severity and frequency of leg pain [51].

Lacosamide

Lacosamide (Vimpat®; UCB/Schwarz Biosciences), an NMDA receptor antagonist, was approved as an adjunctive treatment of partial-onset seizure in epilepsy patients aged 17 years and older in the West in 2008. In an animal study, it was reported that it inhibited hyperalgesia to cold, thermal, and mechanical stimuli in an animal model of diabetes [52]. However, a phase III study intended for additional approval for the treatment of diabetic neuropathic pain in Europe failed to produce convincing evidence of efficacy (primary endpoint: average daily 11-point NPRS score from baseline to the last 4 weeks, 246 subjects, 6-week titration period followed by 12-week treatment) [53], leading to withdrawal of the application for approval in Europe.

NP-1

NP-1 (4% amitriptyline/2% ketamine, AmiKet™; Epicept) topical cream is a formulation combining amitriptyline (TCA) and ketamine (NMDA antagonist used as an anesthetic). In a phase IIb study on use of NP-1 for the treatment of diabetic neuropathic pain (primary endpoint: change from baseline in average pain, 215 subjects, 4 weeks), which was completed in 2008, it did not reduce pain significantly compared with placebo.

AMPA Receptor Antagonists

Like the NMDA receptor, the α-amino-3-hydroxy-5-methyl-4-isoxazole propionic acid (AMPA) receptor is a glutamate receptor subtype. It is mainly found in excitatory neurons and involved in central nervous system diseases characterized by overexcited neurons, such as epilepsy, neurodegenerative disease, movement disorder, and pain, by transducing signals from neurotransmitter glutamate to the brain. Excessive influx of Ca^{2+} into nerve cells due to glutamate-induced neuronal hyperexcitability results in abnormal activation of various enzymes and thus cellular dysfunction. Since the NMDA receptor is usually inactive, regular glutamate excitatory synaptic transmission in the central nervous system may be primarily mediated by the AMPA receptor. The AMPA receptor, which responds to ligand binding rapidly in milliseconds, is also responsible for rapid neurotransmission.

Perampanel

Perampanel (Fycompa®; Eisai), an AMPA receptor antagonist, was approved for the treatment of partial seizure in epilepsy patients aged 12 years and older in Europe in 2012 (currently under application in the US and phase III in Japan). In the West, a phase II/III study was also conducted for the treatment of diabetic neuropathic pain and PHN (primary endpoint: McGill Pain Questionnaire (SF-MPQ), 262 subjects, 16 weeks), but development for the treatment of these diseases was discontinued.

Antiarrhythmics

Mexiletine

Like lidocaine (Xylocaine®; AstraZeneca) used as a local anesthetic, mexiletine hydrochloride (Mexitil®; Boehringer Ingelheim) is a Na^+ channel blocker classified as a Vaughan-Williams Class Ib antiarrhythmic agent. In peripheral neuropathic pain, damaged nerves cause more irritation because of abnormal Na^+ channels formed as a result of quantitative and qualitative changes. Mexiletine blocks these Na^+ channels without affecting normal neurotransmission, inhibiting nerve irritability. It also prevents excessive action potential by inhibiting the dorsal horn neuron activity, which is evoked by stimuli from C fibers, and thereby preventing firing in spinal DRG.

Several clinical studies have been conducted to evaluate the efficacy of mexiletine for diabetic neuropathic pain, but the efficacy is still controversial. In a study involving a total of 216 subjects, for instance, the visual analog scale (VAS) for pain/discomfort was scored during daytime and nighttime on a daily basis, showing that pain was reduced in the high dose (675 mg/day) group, but without correlation between the efficacy and blood mexiletine concentration [54]. In Japan, mexiletine is approved for the treatment of DPN, for which it improved subjective symptoms (spontaneous pain and numbness) in 77 of 168 patients (45.8%), in addition to tachyarrhythmia. Lidocaine liniment has been shown to be effective for neuropathic pain, but no evidence for diabetic neuropathic pain has been presented.

PDE IV Inhibitors

Ibudilast

Ibudilast (Ketas®; MediciNova) is a phosphodiesterase IV (PDE IV) inhibitor that has been commercially available for the treatment of bronchial asthma and cerebrovascular disturbances since 1989 in Japan and several Asian regions [55]. It may inhibit PDE IV to enhance the antiplatelet effect of endothelium-derived nitric oxide (NO), resulting in peripheral vascular dilatation. In a single-dose study of ibudilast (10 mg) in 41 patients with type 2 diabetes, the cross-sectional area and blood flow index of the dorsal pedis artery and dermal microcirculatory blood volume were significantly increased 1 hour post-dose compared with the control group (elastase 1800 U), showing an improvement in leg microcirculation in diabetic patients [56].

On the other hand, glial cells abundant in the brain and central nervous system are involved in chronic pain by inducing cytokines, *etc.*, which are necessary for healing, at the site of pain. Persistent intense pain leads to overactivity of glial cells, which results in unexplained chronic pain. Ibudilast, which inhibits overactivity of glial cells [57], is also approved for the treatment of neurogenic pain in Japan and the US. At present, a phase Ib/IIa study has been completed for the treatment of DPN.

Anti-NGF Antibodies

A clinical study on a nerve growth factor (NGF) for the treatment of amyotrophic lateral sclerosis, *etc.*, was conducted, but without success. In this study, pain was reported as a side effect, revealing that the NGF enhances pain sensitivity, and the anti-NGF antibody produces analgesia [58].

Tanezumab

Tanezumab (PF-4383119/RN624; Pfizer/Rinat Neuroscience) is a humanized anti-NGF antibody. Many patients treated with tanezumab underwent joint replacement surgery due to rapid progression of osteoarthritis (OA) in a phase III study on tanezumab for the treatment of OA [59], and products being developed by other companies had a similar risk, leading to discontinuation of development in 2010 upon request of the FDA. A phase II study for the treatment of diabetic neuropathic pain was also discontinued, in 2010. Based on unmet needs for patients unresponsive to analgesics, resumption of a clinical trial only in refractory OA has been supported by the Arthritis Advisory Committee of the FDA since 2012. As it is shown that combined use of tanezumab and NSAIDs is a risk factor for worsening of OA, it may be approved as the first biologics for the treatment of chronic pain.

Fulranumab

A phase II study of fulranumab (Janssen Pharmaceuticals) was conducted for the treatment of painful diabetic neuropathy, PHN, and traumatic neuralgia. The results obtained before a clinical hold on anti-NGF antibodies showed that 12-week treatment of painful diabetic neuropathy with fulranumab at a dose of 10 mg resulted in a significant improvement in the pain score (change from baseline in the mean 7-day pain score at week 12 of treatment). On the other hand, it did not significantly reduce pain in PHN or PTN, suggesting that the degree of involvement of NGF in neuropathic pain varies depending on the disease. Development is expected to proceed once the clinical hold is lifted. In addition, REGN475 (Regeneron Pharmaceuticals) is under development for the treatment of OA, *etc.*

Cellular Metabolism Activators

Actovegin®

Actovegin® (Takeda Pharmaceutical) is a deproteinized calf blood extract and contains various low-molecular substances (nucleotides, nucleozides, glycolipids, oligo-peptides, amino acids, *etc.*). It enhances tissue function [60] and increases glucose transport and oxidation [61, 62] by promoting mitochondrial respiration and increasing ATP production. It is therefore used for cerebral circulatory disorders, peripheral circulation disorders, skin grafting, burn injuries, wounds, *etc.*, in Europe and Asia. In a phase III study (primary endpoint: total symptom score (TSS) of the lower limbs and vibration perception threshold (VPT), 567 subjects, intravenous administration at 8 mg/mL x 250 mL for 20 days followed by oral administration at 600 mg x 3 for 140 days) in type 2 diabetic patients with DPN, it was reported that it improved the neurological symptom score [63].

Etiology-Based Drugs

Causes of DPN include increased metabolism of polyols, abnormal protein kinase C (PKC) activity, increased oxidative stress, and increased glycation, which reflect chronic hyperglycemic conditions and may result in functional and morphological abnormalities of neural cells, vascular endothelial cells, and smooth muscle cells (decreased nerve blood flow). It has been shown that these metabolic abnormalities are deeply involved in neurodegeneration by inducing neurotrophic factor deficiency in the perineural tissue. In brief, the etiology of DPN essentially consists of 1) metabolic abnormality, 2) blood circulation disorder, and 3) neurotrophic factor deficiency, and complex interactions of these elements result in the onset and development of neuropathy. Therefore, effective treatment with drugs targeting these causes is needed.

Drugs Correct Metabolic Disorder

Aldose Reductase Inhibitors (ARIs)

Under hyperglycemic conditions, intracellular glucose concentrations increase in the optic lens, retina, renal glomeruli, and nerves because these tissues uptake glucose in an insulin-independent manner. Under hyperglycemic conditions, glucose is taken up in excess of the capacity of the glycolytic pathway, requiring glucose consumption in other pathways. One of these pathways is a polyol pathway. Glucose is converted into sorbitol by aldose reductase (AR). However, sorbitol is also neurotoxic that increases intracellular osmotic pressure and inhibits the uptake of myoinositol and taurine, which are important for intercellular communication. In addition, increased AR activity results in excessive consumption of its coenzyme NADPH. NADPH is an important coenzyme for NOS and glutathione reductase.

Many pharmaceutical companies have embarked on the development of ARIs, such as sorbinil (Pfizer), zopolrestat (Pfizer), zenarestat (Astellas Pharma/Pfizer), and fidarestat (Sanwa Kagaku Kenkyusho), but epalrestat is the only product on the market now.

Epalrestat

Epalrestat (Kinedak[®]; Ono Pharmaceutical) is the only clinically available ARI, and used only in Japan because its usefulness has not been established on a worldwide basis. Its efficacy has been reported not only in cellular experiments using nerve cells [64] and vascular smooth muscle cells [65], and in a study in diabetic animals [66], but also clinically. In a phase IV study (ARI-Diabetes Complications Trial; ADCT) reported in 2006, the effect of epalrestat on neurological function was evaluated over 3 years in 594 diabetic patients with mild neuropathy, showing that it significantly inhibited delay of motor nerve conduction velocity (MNCV) and prolongation of F-wave minimum latency in the median nerve, with greater efficacy in patients with better glycemic control and milder neuropathy [67]. Similar findings were reported with respect to subjective symptoms of neuropathy [68], suggesting the importance of early ARI treatment and good glycemic control.

Ranirestat

Ranirestat (AS-3201; Dainippon Sumitomo Pharma), which is being studied in an ongoing clinical study, was reported to significantly decrease the sorbitol level in the sural nerve in a clinical study [69], with high affinity for human AR. In a phase IIb study in 285 patients with DPN in Japan, 52-week treatment with ranirestat at doses from 10 to 40 mg once daily resulted in a significant improvement in the primary endpoint (sum of changes in sensory nerve conduction velocity (SNCV) and MNCV) at week 4 of post-treatment follow-up, compared with the placebo group. The tibial MNCV was significantly increased in the 40 mg group compared with the placebo group. In the modified Toronto Clinical Neuropathy Score (mTCNS), which consists of sensory scores for the assessment of positive symptoms such as spontaneous pain and numbness, as well as negative symptoms manifested as objective test abnormalities such as decreased vibratory sense, no significant differences were observed between the active drug and placebo groups, because the placebo effect was unexpectedly high. However, the efficacy was suggested in an analysis excluding patients who had fluctuating symptoms during follow-up. Similar results were reported in a clinical study in the US and Canada [70].

Eisai acquired the exclusive rights to develop, manufacture, and distribute ranirestat worldwide except in Japan in 2005. It is conducting a phase III study in the US, Canada, and Europe, which will be completed in 2013.

In addition, BNV-222 (choline diepalrestat; Bionevia Pharmaceuticals) was shown to be effective in a preclinical study. QR-333 (Quigley Pharma), a formulation (liniment) combining quercetin, a flavonoid with an inhibitory effect on AR, ascorbyl palmitate, and vitamin D3, was also shown to be effective for DPN. In a 12-week phase IIb study in 140 patients with DPN, it improved the sural NCV, but did not improve pain, symptoms, QOL, or sleep, leading to suspension of development in 2011.

PKCβ Inhibitors

Ruboxistaurin

In diabetes, excess blood glucose is metabolized through the collateral pathways rather than the glycolytic system, resulting in various biochemical abnormalities.

For instance, diacylglycerol (DG) production is increased, and PKC is activated. In fact, increase in DG levels and activation of PKCβ in the vascular tissue (retina, heart, aorta, and renal glomerulus) were reported in diabetes, and PKCβ inhibitors were shown to be effective for retinopathy, nephropathy, and neuropathy in diabetic animals [71, 72].

Ruboxistaurin mesylate (Arxxant®; Eli Lilly), a PKCβ inhibitor, was reported to improve neurological symptoms and skin blood circulation in a clinical study [73]. In a phase II study in 256 patients with DPN, however, 1-year treatment resulted in an improvement in neurological symptoms and vibratory sense in patients with mild disease, but no improvement in the vibratory sense and Neuropathy Total Symptom Score-6 (NTSS-6) [74]. In a phase III study in approximately 400 patients with DPN, it was also ineffective for the NTSS-6, leading to discontinuation of development. In addition, JTT-010 (Japan Tobacco) improved DPN in a preclinical study [75], but development was discontinued.

Advanced Glycation Endproducts (AGE) Inhibitors

Abnormal signaling resulting from production of AGEs by nonenzymatic glycation reaction and binding of AGEs to the receptor for AGE (RAGE) is an important cause of DPN [76]. RAGE is expressed in perineural and endoneurial endothelial cells and Schwann cells in rat peripheral nerves. It was reported that diabetes was induced, but neuropathy was not in RAGE knockout mice [77], indicating that drugs that act on the AGEs-RAGE may be useful in the treatment of neuropathy. However, many of AGE inhibitors have been developed for the treatment of diabetic nephropathy or diabetic retinopathy, and the clinical efficacy for DPN has not been shown.

Aminoguanidine

Aminoguanidine (Pimagedine; Synvista Therapeutics) is an AGE production inhibitor that inhibits glycation reaction by binding the amino group to the intramolecular carbonyl group of Amadori compounds in the early glycation reaction and 3-deoxyglucosone (3-DG) in the advanced glycation reaction to prevent further reaction. It has been shown to inhibit AGE production by glycation reaction and protein cross-linking and polymerization *in vitro*, and has been reported to inhibit neurological dysfunction and morphological changes without lowering blood glucose levels in diabetic animals [78, 79]. Clinically, studies were conducted for the treatment of diabetic nephropathy, but failed to demonstrate the efficacy [80, 81]. Side effects such as anemia, liver disorder, and vitamin B6 deficiency have been reported.

OPB-9195

OPB-9195 (Otsuka Pharmaceutical) is an AGE production inhibitor with a thiazoline structure. It improves diabetic nephropathy at a 10- to 20-fold lower dose than aminoguanidine, and has been shown to inhibit diabetic nephropathy and TGF-β production in diabetic animals [82]. As for DPN, it has been reported to improve the NCV in diabetic animals [83]. Development was discontinued in the preparatory stage of a clinical study.

Alagebrium

Alagebrium chloride (ALT-711; Synvista Therapeutics) is an AGE crosslink breaker developed for the treatment of geriatric diseases that breaks protein cross-linking formed by glycation reaction. Metal chelation, antioxidation, and capture of carbonyl compounds may be involved in this action. Long-term coexistence of glucose and collagen or elastin in living organisms results in protein cross-linking, causing glycation/degeneration (protein aging) in the cardiovascular system, blood vessels, and skin. Unlike glycation inhibitors, which are expected to prevent degeneration, ALT-711 may also be useful in the treatment of diseases involving advanced tissue degeneration. In diabetic animals, it prevented collagen cross-linking in the tail and left ventricle, decreased AGEs in the heart muscle and N-ε-(carboxymethyl) lysine (CML) in the kidney, and improved arteriosclerosis, myocardial abnormalities, and nephropathy [84, 85]. Clinical studies were conducted for the treatment of heart disease, hypertension, diabetic nephropathy, and erectile dysfunction, but development was discontinued.

There are several known forms of RAGE due to alternative splicing. Endogenous secretory RAGE (esRAGE) is secreted in the absence of the C-terminal transmembrane domain of full-length membrane-bound RAGE as a result of alternative splicing. esRAGE may bind extracellularly to AGE to inhibit binding of AGE to transmembrane RAGE on the cell surface and thereby inhibit the cellular effects of AGE [86]. However, it has been reported that the blood esRAGE level was not correlated with neuropathy in type 2 diabetic patients, indicating that application to DPN is doubtful [87]. Full-length membrane-bound RAGE is shed by matrix metalloproteinase 9 (MMP9), disintegrin, and metalloproteinase 10 (ADAM10) above the cell membrane to form soluble RAGE (sRAGE). Enhancement of the shedding mechanism leads to additive inhibition of AGEs-RAGE system by not only decreasing full-length membrane-bound RAGE, which induces intracellular signaling, but also increasing sRAGE, which acts as a decoy receptor. It is hoped that this mechanism will be fully clarified and drugs that enhance the shedding will be developed.

Antioxidants

Progression of oxidative stress not only causes cell damage by fractionating and/or oxidizing DNA, proteins, and lipids, but also impairs neurological function by inhibiting NO production and decreasing blood flow. Diabetic patients are more vulnerable to oxidative stress, and the importance of oxidative stress as a cause of DPN is of high interest. Therefore, the usefulness of various drugs with antioxidant properties has been studied, but there have been few reports showing the clinical efficacy for DPN.

α-Lipoic acid (Thioic acid)

α-Lipoic acid (Thioctacid®; Meda Pharma), which has carboxyl groups and cyclic disulfide in its molecule, has antioxidant properties, reactivates antioxidants, and enhances energy production. It was launched for the treatment of DPN in Germany in 1955. It has been reported that it improved the NCV and nerve blood flow in diabetic animals after intraperitoneal administration 5 times a week for 1 month [88]. In four clinical studies in Germany, α-Lipoic acid was administered to a total of 1258 diabetic patients by intravenous infusion at doses from 100 to 1200 mg/day for 3 weeks, resulting in an improvement in the subjective symptom score (primary endpoint: Total Symptom Score (TSS) in the feet on a daily basis) at doses of 600 mg/day or more [89].

Vitamin E (α-Tocopherol)

It has been reported that DPN was severer in diabetic animals fed with vitamin E-deficient diet than in those fed with normal diet [90]. A clinical study of vitamin E (α-tocopherol) was conducted in 21 patients with DPN. Improvement was achieved in several primary endpoints (median and peroneal NCVs) in 11 patients given vitamin E at a dose of 900 mg for 6 months compared with 10 patients in the placebo group [91].

Poly (ADP-Ribose) Polymerase (PARP) Inhibitors

PARP is a protein involved in DNA repair and apoptosis. Normal cells use PARP in self-repair. Chronic stress, such as hyperglycemic conditions, results in formation of radicals, which severely damage DNA and thereby requires excessive activation of PARP for DNA repair. The resultant depletion of energy such as NAD^+ or high-energy phosphates and activation of PKC affect gene expression in various tissues. It has been reported that NAD^+ and ATP in nerve cells rapidly decreased in models of radical-induced neuronal death and focal brain ischemia, and these decreases and cell death were inhibited in studies of a PARP inhibitor and in PARP-1 knockout mice [92, 93].

PARP is overexpressed in the nerve tissue in diabetic animals, and PJ34 (Inotek Pharmaceuticals), a PARP inhibitor, improved the sciatic NCV after 2-week treatment [94]. GPI-15427 (Eisai) also improved the NCV and hypoesthesia/nervousness in diabetic animals after 4-week treatment [95]. While several compounds have been developed as anticancer drugs, no clinical study for the treatment of DPN has been reported.

Transketolase Activators

Benfotiamine

Benfotiamine is a lipophilic derivative of vitamin B1 (thiamine) and a long-acting vitamin B1 compound. After gastrointestinal absorption, benfotiamine is metabolized into thiamine and shows approximately 5 times more bioavailablity than thiamine itself. In diabetic microvascular complication, it may activate transketolase in the pentose phosphate pathway to inhibit the polyol metabolic pathway, PKC pathway, and glycation reaction, which cause the diabetic complications.

In diabetic animals given benfotiamine mixed in food, the MNCV as well as CML and 3-DG levels in the brachial nerve were significantly decreased, showing that it not only improved DPN, but also inhibited AGE accumulation [96]. It significantly prevented DPN without affecting blood glucose levels after administration at a daily dose of 40 mg for 3 weeks in a clinical study (BEDIP study) in 40 diabetic patients [97] and significantly improved the neuropathy symptom score (NSS) in a 6-week phase III study (BENDIP study) in 165 patients [98].

Rho Kinase Inhibitors

Rho kinase is an intracellular serine-threonine protein kinase identified as a target protein for Rho, a small GTP-binding protein [99-101], and has been shown to be involved in physiological functions such as contraction of smooth muscle cells, morphological control and migration of various cells, and gene expression control [102]. It is also deeply involved in a variety of diseases, including cardiovascular diseases, and it is suggested that Rho kinase inhibitors may be useful for the treatment of these diseases. It has also been reported that Rho kinase is activated in diabetes. Rho kinase inhibits insulin signaling by phosphorylating the insulin receptor substrate (IRS-1) [103]. Long-term treatment with fasudil hydrochloride (Eril®; Asahi Kasei, indicated for cerebrovascular spasm after surgery for subarachnoid hemorrhage), a Rho kinase inhibitor, improved insulin resistance in diabetic rats [104]. Rho kinase is also expressed in Schwann cells in peripheral nerves [105], and fasudil improved painful diabetic neuropathy in animal models [106]. In addition, fasudil has been reported to induce erection in a NO-independent manner [107].

SAR-407899

For SAR-407899 (Sanofi), an orally-bioavailable Rho kinase inhibitor [108], a phase I study was conducted for the treatment of DPN, but development was suspended. Based on a report that like fasudil, it induced erection in diabetic animals [109], a phase II study was conducted for the treatment of erectile dysfunction in France (primary endpoint: duration of penile rigidity during sexual stimulation, 24 subjects), but development was discontinued in 2009.

Other Categories

C-Peptide

C-peptide is a component of proinsulin, an insulin precursor, and a peptide with 31 amino-acid residues. When insulin is synthesized, insulin and C-peptide are separated at a 1:1 ratio. In recent years, the physiological effects of C-peptide, which was traditionally considered to be biologically inactive, have been clarified [110, 111]. While no details such as the presence or absence of the receptor are known, a NOS inhibitor decreased C-peptide-induced improvement in nerve function and blood flow in diabetic rats, suggesting that C-peptide may improve nerve function through the vascular system [112]. It has also been reported to activate Na^+/K^+-ATPase [113] and increase nerve growth factors [114]. Replacement with C-peptide can be expected to be effective for type 1 diabetes, which is associated with decreased secretion of not only insulin but also C-peptide. On the other hand, type 2 diabetes, in which C-peptide-induced release of ATP from platelets is decreased, might be resistant to C-peptide [115].

In a study of 6-month treatment with C-peptide in type 1 diabetic patients, it was reported that it improved the NCV and sensory assessment, with higher efficacy in patients with less severe diabetic neuropathic pain [116]. In this study, however, C-peptide was subcutaneously administered 4 times daily because the half-life in blood is approximately 10-30 minutes. PEGylated C-peptide (Ersatta™; Cebix) is more stable with a half-life of 1 hour to 3 days in blood and can be administered using a once-weekly regimen. Currently, a phase IIb study in DPN in type 1 diabetic patients (primary endpoint: SNCV, 240 subjects, 12 months) is ongoing in the US.

Acetyl-*L*-carnitine

L-carnitine (levocarnitine), an amino acid derivative, is a carrier molecule that transports long-chain fatty acids into the mitochondria in cells. Since fatty acids are important energy substrates for the tissues such as skeletal muscle and heart muscle, *L*-carnitine deficiency may result in myopathy, hypoglycemia or cardiomyopathy due to impaired muscle metabolism. Carnitine deficiency results from congenital metabolic disorders, acquired medical conditions such as liver cirrhosis, and medical practices such as long-term dialysis and pharmacotherapy. Levocarnitine is used to treat carnitine deficiency.

It has also been reported that acetyl-*L*-carnitine, a metabolite of *L*-carnitine, improved DPN. In animal studies, it was reported that it improved the NCV [117, 118]. In clinical studies as well, the efficacy was demonstrated for axonal degeneration, neuroregeneration, *etc.* [119, 120], including improvement in the NCV [121] and VAS [122] in diabetic patients after 12-month treatment.

Drugs for Blood Circulation Disorders

Angiogenic Factors

Angiogenesis, a physiological phenomenon of vascular network formation involving new blood vessels from existing blood vessels, is a good therapeutic target for hypoperfusion due to vascular abnormality or ischemia. Improved blood circulation resulting from increased angiogenesis is expected to provide a cure to DPN and diabetic foot ulcers. Known angiogenic factors include VEGF [123], hepatocyte growth factor (HGF) [124], and basic fibroblast growth factor (bFGF) [125], which have been reported to improve DPN in animal models.

Recombinant Human VEGF$_{165}$

VEGF are produced by macrophages and tumor cells, and signaling through VEGF receptors (VEGFR) plays an important role in angiogenesis. Telbermin (recombinant human VEGF$_{165}$; Genentech) is developed for the treatment of chronic neuropathic diabetic foot ulcers. In the US, the safety of 6-week topical treatment was shown in a phase I study [126], but the efficacy endpoint (skin re-epithelialization without drainage or dressing requirements) was not achieved after 12-week treatment in a phase II study, leading to discontinuation of development in 2007.

SB-509

SB-509 (Sangamo BioSciences) is an injectable formulation of plasmid DNA that encodes a zinc finger DNA-binding protein (ZFP) transcription factor and designed to up-regulate the expression of the gene encoding VEGF-A. In a preclinical study, SB-509 improved the MNCV/SNCV by increasing the expression of VEGF-A after intramuscular injection into the lower limb in a rat model of diabetes [127]. In the US, the efficacy was not shown for the primary endpoint regarding the sural NCV or the secondary endpoint regarding the neuropathy impairment score in the lower limb (NIS-LL) or intraepidermal nerve fiber density (IENFD) after 180-day treatment in a phase IIb study, leading to discontinuation of development in 2011.

Hepatocyte Growth Factor (HGF)

HGF, which was initially found as a hepatocyte-specific growth factor, has been reported to induce angiogenesis, increase and protect vascular endothelial cells and nerve cells. Attempts are being made to use these effects for therapeutic purposes. HGF promotes the growth of vascular endothelial cells through c-Met, an HGF receptor expressed on the endothelial cells [128]. It has been reported that neuropathy improved in diabetic animals after HGF gene transfection [124, 129]. Clinically, gene therapy for severe leg ischemia with no alternative treatment available has been reported, but with no statistically significant effect, and no studies have been conducted in DPN.

Recombinant Human Basic Fibroblast Growth Factor (bFGF)

bFGF (FGF-2) acts through the FGF receptor to promote the growth of fibrocytes, vascular endothelial cells, and epidermal cells, as well as benign granulation and epithelialization at the wound site, which shorten the time to healing. It also improves the quality of wound healing (QOWL) by improving the quality of scars [130]. It has been reported that bFGF with cross-linked gelatin hydrogel improved the NCV, nerve blood flow, and thermohypesthesia after intramuscular injection into the lower limb in diabetic animals [125]. A clinical study of recombinant human bFGF (Trafermin, Fiblast®; Kaken), a drug for decubitus and skin ulcer, is ongoing for the treatment of diabetic foot ulcers as a phase III study (primary endpoint: wound closure rate of diabetic foot ulcers, approximately 210 subjects, 12 weeks) in Europe.

Recombinant Human Platelet-Derived Growth Factor (PDGF-BB)

Becaplermin (Regranex® Gel; Healthoint Biotherapeutics), a recombinant human PDGF-BB, has biological activities similar to those of endogenous PDGF, including promotion of cell growth and vascular formation involved in wound repair. After topical application, becaplermin penetrates the subcutaneous tissue to fasten healing by promoting angiogenesis and granulation and improving blood flow. In a US phase III study (primary endpoint: complete healing rate of ulcers, 382 subjects, at least 8 weeks), ulcers completely healed in 50% of patients after application at a dose of 100 μg/g (35% in the placebo group), with a 32% reduction in time to complete healing (83 days *vs.* 127 days) [131]. The FDA approved becaplermin for the treatment of lower extremity diabetic neuropathic ulcers (with adequate arterial perfusion) in 1997. However, it is not approved for lower extremity diabetic neuropathic ulcers that do not reach the subcutaneous tissue through the dermis (stage I or II, IAET staging classification) or ischemic diabetic ulcers.

On the other hand, a retrospective cohort study on cancer occurrence and death was conducted in 1622 becaplermin-treated patients and 2809 control patients, showing that the cancer mortality increased in those who used at least 3 tubes of becaplermin [132]. In addition, malignancies distant from the application site have been reported in both clinical studies and post-marketing use. Consequently, the FDA announced addition of boxed warning to the product label of Regranex® Gel in 2008 that it should be used with caution in patients with known malignancy (of all kinds) only when benefits outweigh risks.

Besides these drugs, treatments designed to increase angiogenic factors indirectly are used. Autologous platelet-rich plasma (PRP) is concentrated plasma prepared by centrifugation of whole blood from the patient. In the treatment with autologous platelet gel (APG), PRP from the patient's blood is artificially activated, that is, a variety of growth factors (PDGF-BB, TGF-β1, VEGF, EGF, IGF-1, *etc.*) are massively released from α-granules of platelets to promote healing of chronic wounds such as nonhealing diabetic foot ulcers. APG treatment, in which autologous growth factors are topically applied with the involvement of patients only in blood collection, is shown to be effective.

In addition, cell transplantation is being studied for more effective regenerative medicine. Transplantation of endothelial progenitor cells (EPCs) into the ischemic region is clinically evaluated in the cardiovascular field [133] and has been shown to improve DPN in diabetic animals [134]. Peripheral blood mononuclear cells (PBMNCs) and bone marrow mononuclear cells (BMMNCs) include EPCs, hematopoietic stem cells, and mesenchymal stem cells, and secrete many cytokines. These cells are expected to serve as cytokine cocktail by secreting a variety of cytokines after transplantation and have been reported to improve DPN through VEGF after transplantation into the hindlimb skeletal muscle in diabetic animals [135]. Highly proliferative, cytokine-secreting mesenchymal stem cells (MSCs) that can differentiate into mesenchymal cells may be suitable for transplantation. In diabetic animals, transplantation of MSCs resulted in increased local production of VEGF and bFGF as well as improvement in the NCV and blood flow [136]. Embryonic stem (ES) cells and induced pluripotent stem (iPS) cells are also expected to play a central role in the next generation of regenerative medicine.

Low-Molecular-Weight Heparin (LMWH)

Since the discovery of LMWH in 1976, studies have been conducted with focus on molecular weight-related differences in the inhibitory activity of heparin against coagulation factors to inhibit blood coagulation without stimulating bleeding [137]. LMWH is a fraction with a molecular weight of 1000-10000 Da (average molecular weight: 4000-5000 Da) resulting from gel filtration of enzymatically or chemically treated heparin. LMWH inhibits factor Xa, but hardly inhibits thrombin, with a high anti-factor Xa/antithrombin ratio, indicating that it marginally stimulates bleeding [138]. LMWH, which is generally used in the treatment of deep vein thrombosis, pulmonary embolism, and disseminated intravascular coagulation (DIC) as well as for the prevention of blood coagulation during extracorporeal blood circulation, has been reported to improve diabetic foot ulcers by improving microcirculation.

Dalteparin

Dalteparin sodium (Fragmin[®]; Pfizer) is a LMWH with an average molecular weight of 6000 Da and an anti-factor Xa/antithrombin ratio of 2.5. In a study of dalteparin sodium subcutaneously given to 10 patients with diabetic foot ulcers at a dose of 2500 units once daily for 8 weeks, ulcers improved in 8 patients, including 4 patients with complete healing [139]. In a clinical study in 87 diabetic patients with chronic foot ulcers, ulcers significantly improved with fewer cases of lower limb amputation in the dalteparin sodium group (subcutaneous administration at a dose of 5000 units once daily for up to 6 months), compared with the control group [140]. The oxygen supply to the skin also improved [141]. A phase II/III study in neuroischaemic foot ulcers in diabetic patients with peripheral arterial occlusive disease (PAOD) and DPN (primary endpoint: percentage of subjects with at least 50% improvement in the ulcer area, 276 subjects, 6 months) was completed in 2010.

Bemiparin

Bemiparin sodium (Hibor®; Rovi Pharmaceuticals Laboratories) is a second-generation LMWH with an average molecular weight of 3600 Da and an anti-factor Xa/antithrombin ratio of 9.7, and called ultra-LMWH [142, 143]. It has lower antithrombin activity than the preexisting LMWHs, with a more specific anti-factor Xa effect. It was launched as an antithrombotic in Spain and Britain in 1998. In a clinical study of bemiparin sodium given to 70 patients with diabetic foot ulcers at a dose of 3500 units for 10 days and then at a dose of 2500 units for up to 3 months in combination with usual treatment, the percentage of subjects with improvement in the ulcer area was significantly higher compared to placebo group [144]. A phase III study in diabetic foot ulcers (primary endpoint: percentage of subjects with at least 50% improvement in the ulcer area, 329 subjects, 3 months) in Europe was completed in 2010.

Drugs Supplementing Nerve Growth Factor

Neurotrophic Factor and Its Enhancers

Neurotrophic factors (NFs) are a family of proteins involved in survival and maintenance of physiological functions of neural cells. NFs are also involved in brain development and synaptic plasticity by regulating the growth of nerve cells through protein synthesis and neurotransmitter production. NFs can be a therapeutic target for neurodegenerative diseases such as Huntington's disease and

Alzheimer's disease. Some NFs have been reported to be involved in DPN. It has been reported that replacement with NGF [145], neurotrophin-3 (NT-3) [146], neuritin [147], ciliary neurotrophic factor (CNTF) [148], and glial cell line-derived neurotrophic factor (GDNF) [149], which were decreased in the nervous tissue and surrounding tissues in diabetic animals, improved DPN. Drugs that stimulate NF production or act on downstream signaling of NFs to enhance their physiological functions are under development.

Recombinant Human NGF

NGF is involved in survival and maintenance of smaller sensory nerve fibers in the peripheral nervous system. Recombinant human NGF (rhNGF) has been shown to improve sensory nerve function in an animal study [145] and a phase II study in patients with DPN [150]. In 2000, however, it was reported that no improvement was observed in the lower extremity neuropathy score, the primary endpoint in a 48-week phase III study in 1019 patients with DPN [151].

Coleneuramide

Coleneuramide (MCC-257; Mitsubishi Tanabe Pharma) is a sialic acid derivative with a cholesterol skeleton. It stimulated the secretion of NGF to improve the NCV in diabetic animals [152]. It also enhanced NGF-induced improvement in survival of cultured DRG and PC12 cells [152]. A phase II study in DPN (primary endpoint: nerve conduction studies, 420 subjects, 24 weeks) was conducted, but the development was discontinued in 2010.

CXB909

CXB909 (CeNeRx BioPharma) is a small molecule drug that enhances the effect of NGF downstream. Unlike rhNGF, it can be orally administered and penetrate the blood-brain barrier (BBB), with a long half-life. In preclinical studies, it enhanced the effect of NGF by approximately 7 times *in vitro* and improved chemically-induced neurodegenerative diseases *in vivo*. It is therefore expected to be effective for neurodegenerative diseases including chemically-induced or diabetic peripheral neuropathy, Alzheimer's disease, and Huntington's disease by increasing NGF production.

In the first phase I study, it was tolerated with an increase in the blood NGF concentration after single administration. As of 2012, preparation is underway for a phase I/II study.

Neuroimmunophilin Ligands Analog

In the 1990s, it was found that immunosuppressants such as ciclosporin A and tacrolimus (FK-506) have a neuroprotective effect. More recently, it was shown that only FK-506 has a neuroregenerative effect, demonstrating that the mechanism of neuroregeneration is independent from that of inhibition of calcineurin, which is necessary for immunosuppression. It was also shown that FK-506 has immunosuppressive and neuroregenerative effects through different receptors: FK-506-binding proteins (FKBPs) for immunosuppression and FKBP-12 and FKBP-52 for neuroregeneration. These findings led to the development of so-called neuroimmunophilin ligands (NILs), drugs that have neurotrophic effects such as stimulation of repair and growth of nerve cells, but no immunosuppressive effects (no inhibition of calcineurin) [153]. NILs stimulate the production of multiple nerve growth factors such as NGF, NT-3, and BDNF without binding to neurotrophic receptors such as trk. NILs have been shown to repair and regenerate damaged nerves without affecting normal nerves in animal models of Parkinson's disease and acute/chronic neurological diseases [154, 155]. In 2001, a phase II study of NIL-A (Guilford) in Parkinson's disease failed to demonstrate the efficacy for motor symptoms. Development of NILs for the treatment of DPN is awaited.

In preclinical studies, Timcodar dimesilate (VX-853; Vertex) has been reported to be effective in improving nerve function and neuroprotective in models of DPN, toxin-induced neuropathy, and Parkinson's disease. A phase II study in DPN was completed in 1999, but the development was discontinued. TAK-428 (Takeda Pharmaceutical) repairs and regenerates the peripheral nervous tissue by stimulating the production of neurotrophic factors in DPN. Currently, a phase II study is ongoing in Europe and the US.

DIABETIC RETINOPATHY

Metabolic disorders associated with persistent hyperglycemia induce a network of many cytokines and chemokines, which causes microvascular damage to the retina and thus various secondary ocular fundus lesions such as increased vascular permeability, microvascular occlusion, and angiogenesis. This is called diabetic retinopathy (DR), and is one of the main causes of blindness. The clinical condition is correlated with the duration of diabetes and poor blood glucose control, and it is classified into nonproliferative diabetic retinopathy (simple retinopathy), pre-proliferative diabetic retinopathy, and proliferative diabetic retinopathy (PDR), according to the degree of progression. PDR may lead to vitreous hemorrhage, retinal detachment, and further to blindness. Diabetic macular edema (DME), which involves edema and/or vascular occlusion in the retinal macular region, also causes loss of vision.

Photocoagulation and vitreous surgery are performed to inhibit the progression of DR. Especially for PDR-related tractional retinal detachment and massive vitreous hemorrhage, vitreous surgery is the only option to prevent blindness. Based on the finding that neovascularization is greatly involved in the development of DR and DME, intravitreal treatment with anti-VEGF antibodies is being established. In addition, steroids are intraocularly administered to inhibit edema. With no other options available, however, there is a need for drugs with a new mechanism of action. In particular, those that can be used in eye drops or through the oral route are highly needed.

Anti-Angiogenic Drugs

Anti-VEGF Antibodies

Bevacizumab

Bevacizumab (Avastin®; Genentech/Roche/Chugai Pharmaceutical), which is the world's first recombinant humanized monoclonal IgG1 antibody (149 kD) against human VEGF intended for intravenous infusion, was approved for the treatment of colorectal cancer in the US in 2004 and the EU in 2005. VEGF is involved in division of vascular endothelial cells and increased vascular permeability. While it is essential to normal angiogenesis, it is overexpressed in various cancer cells. In cancer cells, it promotes angiogenesis to supply more nutrition and oxygen, contributing to cancer cell proliferation and metastasis. Bevacizumab is an angiogenesis inhibitor that exerts antitumor effects by binding specifically to VEGF to inhibit its biological activity.

In ophthalmology, bevacizumab is intravitreally administered to treat AMD and DME on an off-label basis. In the Bevacizumab or Laser Therapy (BOLT) study, it was reported that 2-year treatment with bevacizumab significantly improved the primary endpoint, defined as the difference in ETDRS best-corrected visual acuity (BCVA), and reduced the central macular thickness in 80 patients with DME [156].

Ranibizumab

Ranibizumab (Lucentis[®]; Genentech/Roche/Novartis Pharma) is a recombinant humanized monoclonal IgG1 κ-isotype antibody fragment (Fab, 48 kD) designed for intraocular use. Indications of ranibizumab were originally wet AMD and macular edema following retinal vein occlusion (RVO), and DME was additionally approved in the EU (2011) and the USA (2012).

Results of three phase III clinical trials for DME have been reported. In RESTORE study (primary endpoint: mean average change in BCVA from baseline, 345 subjects, 12 months), ranibizumab monotherapy and combined with laser treatment provided superior visual acuity gain over standard laser treatment in patients with visual impairment due to DME [157]. In RISE study and RIDE study (primary endpoint: proportion of patients gaining ≥15 letters in BCVA from baseline, 377 and 382 subjects, 24 months), significantly more ranibizumab-treated patients gained ≥15 letters, and improvements in macular edema were noted on OCT [158].

Anti-VEGF Aptamers

Pegaptanib

Pegaptanib sodium injection (Macugen[®]; Eyetech/Valeant Pharmaceuticals/ Pfizer) is a pegylated anti-VEGF aptamer (a single-stranded oligonucleotides constructed to bind to targets with high affinity and selectivity) that binds to $VEGF_{165}$ isomer and prevents the VEGF signal transduction. In 2004, Pegaptanib sodium was approved by the FDA for the treatment of wet AMD. Every 6 weeks intraocular treatment of pegaptanib sodium reduced the risk of moderate vision loss in patients with wet AMD [159].

In a phase II/III clinical trial for the treatment of DME (Macugen 1013; primary endpoint: proportion gaining $\geq$ 10 letters of visual acuity (VA) from baseline, 467 subjects, 1 or 2 years), 36.8% subjects from a pegaptanib group and 19.7% from a sham group experienced VA improvement of $\geq$ 10 letters at week 54 compared with baseline. The pegaptanib treatment group showed significantly better results on the National Eye Institute-Visual Functioning Questionnaire (NEI-VFQ 25) than sham for subscales important in this population [160, 161].

VEGF Recombinant Fusion Proteins

Aflibercept

Aflibercept, or VEGF trap-eye (Eylea®; Regeneron Pharmaceuticals/Bayer/Santen Pharmaceutical), is a soluble recombinant fusion protein, consisting of VEGF-binding extracellular domains of human VEGF receptors 1 and 2 fused to the Fc region of human IgG1, binds to all VEGF-A isoforms and PlGF. Results of two phase III studies, in the USA (2011) and the EU (2012), demonstrated that the intravitreal treatment of aflibercept is effective for AMD [162], and aflibercept is approved for the treatment of wet AMD.

A phase II clinical trial (DA-VINCI study) assessed the efficacy and safety of intravitreal aflibercept *vs.* laser photocoagulation in the treatment of DME (primary endpoint: change in BCVA at 24 weeks, 221 subjects, 1 year). At 24 weeks, the mean change in BCVA for aflibercept arms ranged from +8.5 to +11.4 letters, compared to the mean change of +2.5 letters in the laser-treated eyes. Mean change in central retinal thickness ranged from −127.3 μm to −194.5 μm in aflibercept arms, compared to −67.9 μm in laser-treated eyes at 24 weeks [163]. The effects were maintained or improved at week 52 [164]. Two phase III studies (VIVID-DME and VISTA-DME) are ongoing.

VEGF Inhibitors

MP0112

DARPins® (Designed Ankyrin Repeat Proteins) is a novel class of proteins selected for specific, high-affinity binding to a target protein. MP0112 (Allergan/Molecular Partners) is a DARPin that specifically binds to and antagonizes all relevant subtypes of VEGF-A. Because of its high potency (IC_{50} < 10 pM) and long half-life in the eye (> 6 days), the frequency of intravitreal injections needed is reduced compared to the current standard drugs. DARPins have shown efficacy when applied as eye drops. MP-0112 is in phase II clinical studies for the treatment of DME [165] and wet AMD.

Hypoxia-Induced Factor-1α (HIF-1α) Expression Inhibitors

CLT-003

CLT-003 (Charlesson) is a small molecule that inhibits the HIF-1α, a primary transcription factor responses to hypoxic stress, which retinas in DME and wet-AMD patients are exposed to. Because HIF-1α regulates VEGF downstream in its pathway, CLT-003 also suppresses VEGF and ICAM-1 expression in retina under hypoxic conditions. CLT-003 significantly reduced retinal vascular leakage, endothelial cell proliferation, and inflammation in preclinical studies. CLT-003 has been formulated into nanoparticles (DD-001) to confer sustained efficacy through controlled and long-lasting drug release.

c-Raf Kinase Antisenses

iCo-007

The MAPK/ERK signal transduction pathway, downstream of VEGF signaling, is of interest as a new approach against increased vascular permeability and angiogenesis. c-Raf (Raf-1) kinase participates in a protein kinase cascade in the MAPK/ERK signal transduction pathway and is involved in cell proliferation. iCo-007 (iCo Therapeutics/ISIS Pharmaceuticals), an antisense drug targeting c-Raf kinase, is under development for the treatment of DME and DR [166]. A phase I study was completed in 2010. In an ongoing phase II 12-month study in DME (iDEAL study, primary endpoint: change in visual acuity (VA) from baseline to month 8, 208 subjects, intravitreal injection), iCo-007 is administered alone or in combination with ranibizumab or laser photocoagulation.

Insulin Receptor Substrate-1 (IRS-1) Expression Inhibitors

Aganirsen®

IRS-1, a protein involved in angiogenesis, contributes to growth of new blood vessels. Aganirsen® (GS-101; Gene Signal), which is an antisense oligonucleotide that inhibits the expression of IRS-1, has been shown to inhibit the angiogenic process in preclinical studies [167, 168]. Currently, clinical studies are under way not only for the treatment of orphan diseases such as corneal graft rejection (eye drops, phase III) [169], neovascular glaucoma (ointment, phase II), and retinopathy of prematurity (ointment or intravitreal, phase I), but also for the treatment of AMD (ointment or intravitreal, phase II) and DR (ointment or intravitreal, phase I).

Growth Hormon Receptor (GHr) Expression Inhibitors

<u>ATL1103</u>

ATL1103 (Antisense Therapeutics) is an antisense aimed to inhibit GHr expression, and reduces insulin-like growth factor-I (IGF-I) level in the blood. Diseases associated with excessive growth hormone and IGF-I action include acromegaly, an abnormal growth disorder, DR, and some forms of cancer. Reduction of IGF-I to normal levels is the primary marker of an effective drug treatment for acromegaly.

In the case of DR, reduction in IGF-I levels retarded the progression of the disease and improved vision in patients. ATL 227446, an antisense to the GHr, suppressed blood IGF-I levels in mice [170] and inhibited retinopathy in a mouse retinopathy model [171]. ATL1103 has also prevented retinopathy, by significantly reducing retinal neovascularisation in preclinical studies. A phase I study of ATL1103 has been completed.

Chemokine CXCR4 Antagonists

Stromal cell-derived factor-1 (SDF-1/CXCL12) is a chemokine that regulates homing and retention of hematopoietic stem cells in the bone marrow. SDF-1 binds with high affinity to the chemokine receptors CXCR4 and CXCR7. The SDF-1/CXCR4/CXCR7 axis has been shown to play a role in stem cell mobilization, vasculogenesis, tumor growth, and metastasis. Inhibition of SDF-1 increases sensitivity of tumor cells to chemotherapy and prevents invasion and metastasis of some solid tumors, suggesting that antagonism of CXCR could be beneficial in the treatment of a broad range of cancers. Also, anti-vasculogenesis effect is expected in eyes.

TG-0054

TG-0054 (TaiGen Biotechnology) is a chemokine receptor CXCR4 antagonist currently undergoing phase II clinical trials for use in stem cell transplantation in cancer patients [172]. TG-0054 rapidly mobilizes stem cells and progenitor cells from the bone marrow into peripheral circulation. Therefore, TG-0054 is being developed for the treatment of stem cell transplantation in multiple myeloma, non-Hodgkin's lymphoma, Hodgkin's lymphoma, and myocardial infarction. Preclinical studies are also ongoing for the treatment of DR and AMD caused by neovascularization inside eyes. In preclinical disease models, topical application of TG-0054 was able to seal leaky vessels and block ocular neovascularization.

NOX-A12

NOX-A12 (Noxxon Pharma) is a potent Spiegelmer® (oligonucleotides made from the *L*-stereosiomer of RNA) that specifically antagonized SDF-1 in a phase II trial for the treatment of chronic lymphocytic leukemia and multiple myeloma [172]. NOX-A12 is in development for the treatment of AMD and for use in autologous stem cell transplantation in hematological disorders. Noxxon Pharma is preclinically evaluating the potential of NOX-A12 in the treatment of DR.

Anti-CD105 Antibodies

TRC-105

TRC-105 (Roswell Park Cancer Institute/Tracon Pharmaceuticals) is a chimeric monoclonal anti-CD105 antibody in early clinical trials for the treatment of solid tumors. Phase II trials are also ongoing for the treatment of recurrent glioblastoma after anti-VEGF or other anti-angiogenic therapy. By binding to CD105 (endoglin), a receptor over-expressed on proliferating endothelium that is required for angiogenesis, TRC-105 inhibits tumor growth. Preclinical data suggest that CD105 expression is increased after treatment of anti-VEGF antibody in an experimental angiogenesis model [173]. Therefore, TRC-105 may complement treatment with anti-VEGF antibodies. The compound is also being evaluated preclinically for the treatment of AMD.

RTP-801 Expression Inhibitors

PF-655

PF-655 (Quark Pharmaceuticals/Pfizer) is a synthetic siRNA that inhibits the expression of the hypoxia-inducible gene RTP-801 [174]. Since expression of the RTP-801 gene is rapidly up-regulated in the presence of ischemia, hypoxia, and/or oxidative stress, it is hypothesized that it may regulate hypoxia-induced pathogenesis by a mechanism that is independent of VEGF. In addition, attenuation of apoptosis in the retina and reduction of retinal neovascularization were observed in an RTP801 knockout mouse model of oxygen-induced retinopathy [175]. PF-655 is currently being evaluated in phase II studies for DME and wet AMD.

Anti-Integrin Oligopeptides

ALG-1001

By targeting VEGF production and VEGF effects at different levels of the affected tissues, anti-integrin oligopeptide ALG-1001 (Allegro Ophthalmics) has effect on tissues such as the retina and retinal pigment epithelium. The primary mechanism of ALG-1001 is mediation of the integrin α5-β1: the peptide works in the extracellular matrix to inhibit activation of the fibronectin ligand as part of the assembly of new blood vessels. ALG-1001 also targets tyrosine kinase, which plays a key role in VEGF signal transduction, thereby inhibiting angiogenesis and decreasing inflammation of the endothelial cells. In addition, by causing posterior vitreous detachments and liquefying the vitreous humor, ALG-1001 may inhibit angiogenesis or macular edema through enhanced turnover of VEGF.

In a phase I study, 15 end-stage DME patients received three monthly intravitreal injections of ALG-1001. While the study was primarily a safety study, efficacy data shows that eight patients improved by three to five lines on the eye chart. Also, 30-80% reduction in the thickness of macula was observed in half of the patients. These results have been achieved despite the fact that these first patients were end stage, several failed to respond to bevacizumab, and some had PDR. Phase I/II clinical trials for the treatment of wet AMD are currently enrolling subjects.

Tie2 Activators

Tie2 signaling is responsible for stabilizing blood vessels and maintaining vascular integrity. Tie2 is the receptor for the angiopoietin family of growth factors, and angiopoietin-1 (Ang1) and angiopoietin-2 (Ang2) are the natural Tie2 agonist and antagonist, respectively. Ang2 expressions are known to increase in a disease states involving vascular leakage and pathologic angiogenesis such as PDR [176]. By binding to the Tie2 receptor, Ang2 inhibits Tie2 signaling, which in turn compromises vascular integrity and promotes vascular leakage and pathologic angiogenesis.

AKB-9778

AKB-9778 (Aerpio Therapeutics) acts by inhibiting HPTPβ, a negative regulator of Tie2 receptor, to remove this negative regulation on Tie2 receptor. Therefore, AKN-9778 stabilizes blood vessels to prevent vascular leakage and pathologic angiogenesis by restoring Tie2 signaling, and enhances the effect of Ang2. AKB-9778 is initially being developed for the treatment of DME (phase I).

PI3K/Akt/mTOR Inhibitors

The phosphoinositide 3-kinase (PI3K)/Akt/mammalian target of rapamycin (mTOR) pathway integrates signals from multiple receptor kinases to regulate cellular growth and metabolism. Mechanisms for pathway activation include dysfunction of PTEN (a tumor suppressor), amplification or mutation of PI3K and Akt, activation of growth factor receptors, and exposure to carcinogens. Activation of the PI3K/Akt/mTOR pathway enhances the angiogenesis in diseases such as cancer and in ophthalmic diseases.

P529

Palomid 529 (P529; Paloma Pharmaceuticals) is a nonsteroidal, synthetic, small-molecule drug that inhibits the PI3K/Akt/mTOR signal transduction pathway as an allosteric dual dissociative inhibitor of both the mTOR/raptor (TORC1) and the mTOR/protor-1 (TORC2) complexes. P529 is in phase I/II clinical trials for the treatment of advanced neovascular AMD. Preclinical studies are also ongoing for the treatment of DR and proliferative vitreoretinopathy [177].

Sirolimus (rapamycin)

Sirolimus (rapamycin) is an immunosuppressant drug isolated from *Streptomyces hygroscopicus* in soil samples from Easter Island. Sirolimus binds to the cytosolic protein FK-binding protein 12 (FKBP12), and sirolimus-FKBP12 complex inhibits mTOR pathway by directly binding to TORC1. Rapamune® (Pfizer), an immunosuppressive agent used in renal transplant patients, and the Cypher® (Cordis), a sirolimus-eluting coronary stent for symptomatic ischemic disease, have been approved in the USA and the EU.

Sirolimus inhibits the production, signaling, and activity of many growth factors relevant to the development of DR. In an open-label phase I study of Perceiva™ (DE-109; MacuSight/Santen Pharmaceutical), subconjunctival (SCJ) and intravitreal (IVT) injection of sirolimus in DME eyes were well-tolerated, and the median increase in BCVA was 4.0 letters after 90 days. At day 45, median decreases in retinal thickness were -23.7 μm (SCJ) and -52.0 μm (ITV) [178]. A phase II clinical trial is ongoing for treatment of DME (DIAMOND trial) in parallel with trials for treatment of wet AMD (phase II) and of non-infectious uveitis (phase III).

Recombinant Pentraxin-2 (rPTX2)

Pentraxin-2 is a natural human protein that expresses on damaged tissue and stimulates monocytes to differentiate into regulatory macrophages rather than pro-fibrotic macrophages and fibrocytes, thereby reversing inflammatory, fibrotic, and neovascular processes and promoting normal healing. In eyes, fibrosis, inflammation, and neovascularization cause retinal damage.

PRM-167

PRM-167 (Promedior) is an rPTX-2 variant for intravitreal injection under development for the treatment of fibroproliferative retinal diseases such as AMD, DR, and proliferative vitreoretinopathy (PVR, retinal detachment). In a preclinical study, intraocular injections of rPTX-2 prevented vascular leak, neovascularization, and collagen deposition, with reduced monocyte/macrophage numbers in the retina. These effects of PRM-167 are mediated through similar inductions in IL-10, an anti-inflammatory cytokine produced by regulatory macrophage. By these mechanisms, rPTX-2 may become a new treatment for debilitating choroidal and retinal neovascular diseases in addition to that currently provided by VEGF antagonists.

Tubulin Polymerization Inhibitors

Tubulin is a protein dimer composed of α and β subunits, which form microtubules that are essential to cellular functions such as mitosis, transport, and cytokinesis, and also helps form the cell cytoskeleton. Tubulin inhibitors such as vinca alkaloids, taxols, colchicine, and combretastatin are used as anticancer drugs. Of these, combretastatin is being investigated for the treatment of AMD. Most tubulin inhibitors, however, have significant toxicity, giving them a narrow therapeutic window and causing adverse effect on normal cells.

OC-10X

OC-10X (OcuCure Therapeutics) is a selective tubulin inhibitor currently under development as a topical eye drop treatment for AMD. In animal studies, OC-10X achieved therapeutic concentrations at the vitreous humor, retinal, and choroidal levels after instillation into eyes. In a preclinical study, OC-10X significantly inhibited ocular neovascularization in laser-induced choroidal neovascularization (CNV) primate model. OC-10X works independently of growth factors, and thus could provide improved long-term visual outcome and overall efficacy. OC-10X is also being evaluated for the treatment of DR.

Pharmacologic Vitreolysis

Vitreous is composed of collagen bound to hyaluronan in a network with very high elasticity. Without this scaffolding, there is no structure for blood vessel growth and the growth factors diffuse out into the vitreous humor, away from the retina. Partial posterior vitreous detachment (PVD) with vitreoretinal traction is a major risk factor for progression of PDR, but complete PVD protects against the progression of PDR [179]. The vitreous humor also has been implicated as a cause of DME by several mechanical and physiological mechanisms, such as macular traction and concentration of vasopermeable factors in the macular region [180]. Therefore, pharmacologic vitreolysis might avoid ocular complications such as PDR and DME [181].

Hyaluronidase

Hyaluronan is a major macromolecule of vitreous humor and forms a proteoglycan that plays a pivotal role in stabilizing the vitreous gel. Hyaluronidase cleaves glycosidic bonds in the connective tissue. Dissolution of the hyaluronic acid and collagen complex results in decreased viscosity of the extracellular matrix. In two phase III clinical trials, intravitreal hyaluronidase injection (Vitrase®; ISTA Pharmaceuticals/Bausch & Lomb) has shown evidence of safety and efficacy in reducing vitreous hemorrhage secondary to different etiologies, including PDR, although it has not been approved by the FDA for this indication [182, 183].

Vitreosolve®

Vitreosolve® (VitreoRetinal Technologies/Innovations In Sight) is a urea-derived molecule (non-enzymatic agent) developed for the treatment of DR. Vitreosolve® was injected once or twice during one-month clinical trials, and demonstrated no significant adverse events. In phase I and II trials, 83% of patients treated with Vitreosolve® showed a total PVD and have reported either a reduction or an elimination of floaters. The compound is now in phase III clinical trials.

Microplasmin (Ocriplasmin)

Plasmin is an enzyme in blood responsible for dissolving fibrin clots. Intravitreal injection of autologous plasmin has been reported to induce PVD, reduce macular thickening due to DME in cases that fail to respond to conventional laser photocoagulation, and improve visual acuity [184, 185]. Although plasmin has the capability of both PVD induction and liquefaction, clinical use is not available due to its instability. Jetrea® intravitreal injection (ThromboGenics) is a truncated form of human plasmin that retains intact serine protease activity, and was approved by the FDA in 2012 for the treatment of symptomatic vitreomacular adhesion (VMA). In the phase III trial, intravitreal injection of ocriplasmin (125 µg) induced PVD and resolved vitreomacular traction and closed macular holes significantly compared with a placebo group [186-188]. A phase IIa clinical trial for treatment of patients with DME (MIVI II-DME study) has also been successfully completed (primary endpoint: PVD induction at 14 day post-injection, 51 eyes of 51 patients).

PKCβ Inhibitors

Ruboxistaurin

Activation of DAG/PKC pathway plays an important role in the development and progression of diabetic microvascular complications, including DR. A selective PKCβ inhibitor, ruboxistaurin mesylate (Arxxant[®]; Eli Lilly) ameliorated the retinal circulation in diabetic rats [71, 189]. In the phase III clinical trials, the PKC-Diabetic Retinopathy Study (PKC-DRS) reported that ruboxistaurin mesylate reduced the risk of visual loss, but did not prevent progression of DR (252 subjects, 36-46 months) [190, 191]. The PKC-DRS2 reported that ruboxistaurin mesylate reduced vision loss, need for laser treatment, and DME progression, while increasing occurrence of visual improvement in patients with non-proliferative retinopathy (514 subjects, 36 months) [191, 192]. In the PKC-DME Study (PKC-DMES), ruboxistaurin mesylate delayed progression of DME to a sight-threatening stage, although this was a secondary endpoint (686 subjects, 30 months) [193]. Ruboxistaurin mesylate has not received approval from the FDA. JTT-010 (Japan Tobacco) also showed efficacy on early DR in a preclinical study [194], but the development was discontinued.

AGE Inhibitors

GLY-230

GLY-230 (Glycadia Pharmaceuticals) is an AGE inhibitor that inhibits the formation of Amadori-modified proteins *in vitro*. In preclinical studies, GLY-230 decreases microalbuminuria, ameliorates abnormalities in molecular mediators, and attenuates glomerulosclerosis and the kidney insufficiency. In patients with diabetes, GLY-230 lowers glycated albumin, and the decrease is associated with a reduction in urine albumin excretion [195]. In preclinical studies, GLY-230 restored the balance between proangiogenic and anti-angiogenic factors in ocular fluid, and reduced products of oxidative stress. A phase I clinical study for DR has been completed.

Lipoprotein Associated Phospholipase A2 (Lp-PLA2) Inhibitors

Darapladib

Lp-PLA2 is prominently involved in macrophage infiltration, pro-inflammatory cytokine expression, and adhesion molecule upregulation in macrovascular disease. Darapladib (SB-480848; GlaxoSmithKline), a Lp-PLA2 inhibitor [196], is currently ongoing phase III clinical trials for the treatment of chronic coronary heart disease and atherosclerosis [197].

Since DME is recognized to have an inflammation-linked etiology, and Lp-LPA2 was localized to the retinal vascular endothelium retinal capillaries, it was hypothesized that inhibition of Lp-PLA2 could protect blood retinal barrier (BRB) from breakdown. In a preclinical model, darapladib prevented inner BRB breakdown caused by hyperglycemia. Therefore, Lp-PLA2 may be a useful target for preventing DME, and phase II clinical trials are ongoing.

Nrf2 Activators

Nrf2 is a transcription factor involved in the expression of antioxidant and detoxification genes. Activation of Nrf2 results in inhibition of the increase in intracellular antioxidants and inflammatory signaling pathways, leading to protection of tissues from inflammation. It is said that chronic tissue inflammation contributes to the development of complications of cardiovascular disease, type 2 diabetes, diabetic retinopathy, and chronic renal failure.

OT-551

OT-551 (Othera Pharmaceuticals/Colby Pharmaceuticals) is a small molecule that acts on oxidative stress and inflammation through Nrf2 activation. OT-551 is under development in phase II clinical trials to arrest the progression of or prevent cataracts in patients after vitrectomy surgery. In collaboration with the National Eye Institute, additional phase II trials for the treatment of patients with geographic atrophy (GA), a sign of dry or atrophic AMD, are ongoing. In this phase II study, topical 0.45% administration of OT-551 three times per day for 2 years significantly improved the primary efficacy outcome measure (change in best-corrected visual acuity at 24 months, 10 subjects) [198].

OT-551 is metabolized to its active metabolite, TEMPOL-H (TP-H). TP-H protects cells from free radical damage. Inside the eye, TP-H permeates tissues at both the lens and retina due to the high ocular bioavailability of the compound, providing antioxidant protection against both cataract and AMD.

Somatostatin Analogs

Somatoprim

Somatoprim (DG-3173; Aspireo Pharmaceuticals) is a somatostatin analog (SSA) peptide with a backbone ring structure to stabilize the peptide. By its original profiles, somatoprim has demonstrated a less side effect compared to other SSAs. Furthermore, growth hormone secretion in somatotroph adenoma treated with somatoprim was significantly increased the response rate in acromegalic patients to SSA therapy [199]. A phase II clinical trial for the treatment of acromegaly is ongoing. Preliminary data with Somatoprim indicate that the compound inhibits incipient DR by significantly reducing pericyte loss and reversing formation of acellular capillaries. A phase I clinical study with the compound for the treatment of DR is also under way.

Anti-Inflammatory Drugs

Steroids are not suitable for systemic administration, because they reduce edema but worsen diabetes. Therefore, long-acting steroids are injected directly into the eye. The effect persists for several months and injection is repeated as needed. Many attempts have been made to design prodrugs and improve the DDS for prolongation of effect, as well as to develop eye drop formulations.

Steroids

Triamcinolone

Triamcinolone acetonide (Kenalog®/Kenacort®/Kenacort A®; Bristol-Myers Squibb) is a derivative of prednisolone, a synthetic adrenocortical hormone, and various formulations are used in different disciplines. In ophthalmology, it is used on an off-label basis, but the efficacy of intravitreal triamcinolone acetonide (IVTA) for DME is widely recognized [200]. A measurable concentration in the vitreous humor persisted for 3 months [201].

It is also used to clearly visualize the vitreous body during vitreous surgery based on the chemical characteristic that its water-insoluble white crystalline powder sticks to gel-like material in water [202]. In Japan, MaQaid® intravitreal injection (Wakamoto Pharmaceutical) was approved in 2010 as a vitreous visualizing agent containing no additives harmful to the eye tissue, and additionally approved in 2012 for the treatment of DME.

Dexamethasone

Dexamethasone intravitreal implant (Ozurdex®; Allergan/Sanwa Kagaku Kenkyusyo) is a biodegradable implant that delivers an extended release (6 months) of the corticosteroid dexamethasone *via* intravitreal injection with Novadur® solid polymer delivery system. Ozurdex® is approved by the FDA (2009) and the EMA (2010) (phase II/III in Japan as SK-0503) for the treatment of macular edema following branch retinal vein occlusion (BRVO) or central retinal vein occlusion (CRVO) [203] and noninfectious uveitis [204]. Clinical phase III trial treatment for DME is ongoing (primary endpoint: the change in visual acuity (number of ETDRS letters) and the change in central foveal thickness (microns on high resolution OCT)).

Cortiject® (Novagali Pharma) is a preservative-free and solvent-free ophthalmic injectable emulsion that contains a tissue-activated dexamethasone palmitate prodrug, and is in phase I/II clinical development for the treatment of DME and DR in the USA. Based on Eyeject® technology, sustained release of the prodrug over 6-9 months is expected. The prodrug is converted into the drug by enzymes that are present in the retina and choroid. This specific distribution of the enzymes could enable avoidance of common corticosteroid side effects such as the increase of the intraocular pressure and the opacification of the lens. Additional advantages conferred by Eyeject® technology are ease of administration for the practitioner as well as prolonged effect and safety for the patient.

Fluocinolone Acetonide

Iluvien® (pSivida/Alimera Sciences) is an injectable intravitreal insert delivering a low dose of corticosteroid (fluocinolone acetonide) to the retina for up to 3 years as a treatment for DME. Using a proprietary 25-gauge injector system, a practitioner injects the Iluvien® insert into the vitreous humor through a minimally invasive procedure in an out-patient setting [205, 206].

In two separate clinical studies (FAME; Fluocinolone Acetonide in DME), 956 patients with DME were treated with Iluvien® for 3 years. The primary endpoint was the percentage of patients who had an improvement in best-corrected visual acuity (BCVA) of ≥15 letters from baseline on the Early Treatment Diabetic Retinopathy Study (ETDRS) eye chart at month 24. In these two studies, 29-31% of patients receiving Iluvien[®] and 15-19% of control patients had an improvement in BCVA of ≥15 letters over baseline through month 24 to 36.

From these data, European countries (Austria, France, Portugal, Germany, and the UK) have approved Iluvien[®] for treating DME (2012). An application for approval of Iluvien[®] has been submitted to the FDA (2012). In addition to the FAME study, a phase II clinical trial with Iluvien[®] in wet and dry AMD and retinal vein occlusion (RVO) is ongoing.

Betamethasone

Chroniject[TM] platform is a polymer microsphere system for the sustained release of injectable drugs developed by Oakwood Laboratories. DE-102 (Santen Pharmaceutical) is a betamethasone microsphere product for a sustained-release injection using this technology. In Japan, phase II/III clinical trials are ongoing for the treatment of DME and ME following branch retinal vein occlusion (primary endpoint: change from baseline of BCVA in ETDRS letter score). Preclinical data suggest that ocular injection of DE-102 shows sustained efficacy in DME.

Difluprednate

Difluprednate (Durezol[®]; Sirion Therapeutics/Alcon Laboratories) is a corticosteroid used for the treatment of inflammation and pain associated with eye surgery. Difluprednate rapidly penetrates the corneal epithelium and is metabolized by deacetylates to difluoroprednisolone butyrate (DFPB), an active metabolite. DFPB has strong corticosteroid receptor agonist activity, more potent than betamethasone and prednisolone [207]. In Japan, difluprednate is now in a phase II clinical trial for treating DME, as SJE-2079 (Mitsubishi Tanabe Pharma/Senjyu Pharmaceutical).

Non-Steroidal Anti-Inflammatory Drugs (NSAIDs)

Nepafenac

Nepafenac (Nevanac[®]; Alcon Laboratories) is a NSAID for the management of pain and inflammation associated with cataract surgery. After topical ocular dosing, nepafenac is converted by ocular tissue hydrolases to amfenac after penetrating the cornea. Amfenac inhibits cyclooxygenase (COX), an enzyme that produces prostaglandin, which causes pain and swelling in the eye.

In a phase III, double-blind, comparative study using fluorometholone eye drops as the control in cataract surgery patients, the incidence of cystoid macular edema (CME) in the postoperative 5-week period was 81.5% in the fluorometholone eye drops 0.1% group and 14.3% in the nepafenac group, showing that nepafenac significantly prevented CME. In a rabbit model of anterior chamber paracentesis-induced vascular permeability, nepafenac 0.1% eye drops exerted anti-inflammatory effects, as shown by 61% suppression of protein influx to aqueous humor and inhibition of PGE2 accumulation associated with tissue injury [208].

In 2011, the Committee for Medicinal Products for Human Use (CHMP) of the European Medicines Agency recommended the use of nepafenac to reduce the risk of cataract surgery-related postoperative DME in diabetic patients.

Retinal Ganglion Cell (RGC) Protectants

DR has been recognized as a vascular disease for a long time. However, the recent researches have shown that neuronal cells of the retina are also affected by hyperglycemia, resulting in dysfunction and even degeneration of some neuronal cells. Although RGCs are the major retinal neurons with respect to the effect of diabetes, the relationship between RGC damages and visual dysfunction in diabetes has not yet been revealed. Nevertheless, therapies protecting RGCs are a putative novel approach to treatment of DR.

Calpain Inhibitors

<u>SNJ-1945</u>

Calpains are a family of Ca^{2+}-regulated, intracellular cysteine proteases, which modulate cellular functions by limited, specific proteolysis, and excess activation of calpain leads cells to apoptosis. Calpain is intimately related to neurodegenerative diseases such as Alzheimer's disease. Activation of calpain and caspase cascades leads to apoptotic death of RGCs, as observed in glaucoma [209].

Orally available calpain inhibitor SNJ-1945 (Senjyu Pharmaceutical) is being developed as drug for testing in preventing progression of glaucomatous retinal degeneration [210]. SNJ-1945 has a neuroprotective effect against apoposis of RGC in glaucoma and diabetic optic neuropathy [211]. Because metabolic-induced RGC degeneration is caused by hyperglycemia and oxidative stress, antioxidant and calpain inhibition may become a new neuroprotective treatment against RGC death in various metabolic stress-induced diseases, including DR [212]. Furthermore, SNJ-1945 inhibited VEGF-mediated angiogenesis in retinal endothelial cells *in vitro* [213]. Therefore, inhibition of calpain activity might provide both neuroprotective and anti-angiogenic effects for management of PDR and AMD.

DIABETIC NEPHROPATHY

Diabetic nephropathy (DN) is a major cause of end-stage renal failure. Since the number of patients with DN is steadily increasing with the increasing number of diabetic patients, it is a medical and social imperative to clarify and overcome its mechanisms of development and progression, which remain unknown. DN is characterized by impaired renal function due to glomerular damage associated with disease progression, and may progress to nephrotic syndrome and chronic renal failure and end in dependence on dialysis as the only therapeutic option.

The main treatment strategy for diabetes is correction of hyperglycemia and hypertension. While a variety of hypoglycemic and hypotensive agents have become available in recent years, DN cannot be managed effectively, as indicated by the fact that dialysis is still unavoidable despite strict control of blood glucose and blood pressure. In conclusion, treatment of DN is far from satisfactory, and there is an urgent need for new effective drugs.

Renin-Angiotensin System (RAS) Inhibitors

It is well known that the RAS plays important roles in the regulation of blood pressure and electrolytes, development of cardic/renal/vascular lesions, and remodeling, and is deeply involved in the development and progression of hypertension and organopathy. Many clinical studies have demonstrated that RAS inhibitors effectively inhibit the progression of DN, and some are approved for the treatment of this disease.

Angiotensin-I Converting Enzyme (ACE) inhibitors

<u>Trandolapril</u>

The BENEDICT study was conducted to determine whether trandolapril could inhibit the progression to microalbuminuria in hypertensive diabetic patients with normal renal function [214]. Trandolapril (Odrik®/Gopten®; Abbott Laboratories, Mavik®; AbbVie) significantly inhibited the progression to microalbuminuria by 53% compared with placebo. There was no difference in the glomerular filtration rate.

<u>Ramipril</u>

Ramipril (Altace®/Triatec®/Delix®/Tritace®; Sanofi, Ramace®; AstraZeneca) is approved for the treatment of DN in Britain. The MICRO-HOPE study was among the studies of ramipril [215]. This was a randomized, double-blind, comparative study designed to verify the renoprotective effect of ramipril in diabetic patients with a history of cardiovascular disease or any cardiovascular risk factor (hyperlipidemia, hypertension, microalbuminuria, smoking). Ramipril inhibited the incidences of cardiovascular events by 25% and overt nephropathy by 24%, compared with placebo. While reduction in blood pressure was significantly greater in the ramipril group, analysis of data corrected for blood pressure also showed that ramipril significantly decreased the relative risk of nephropathy by 25%.

<u>Lisinopril</u>

Lisinopril (Zestril®; AstraZeneca, Carace®/Prinivil®; Bristol-Myers Squibb/Merck, Longes®; Shionogi) is approved for the treatment of DN in Italy, New Zealand, and Britain. The Italian Microalbuminuria Study was one of the studies of lisinopril [216]. This study was conducted to evaluate the inhibition of progression to overt proteinuria in normotensive type 1 diabetic patients with microalbuminuria. Lisinopril decreased the risk of progression to overt proteinuria by 49.1%. In addition, the proportion of subjects with at least 50% increase in urinary albumin excretion (UAE) from baseline over 1 year was significantly lower in the lisinopril group.

Enalapril

Ahmad *et al.* reported the results of a randomized, placebo-controlled, single-blind study of enalapril (Renitec®/Xanef®/Enapren®/Renivace®/Innovace®; Merck) [217]. This study was conducted to evaluate the prevention of progression of DN in normotensive type 2 diabetic patients. The duration of follow-up was 5 years. Mean UAE decreased from 55 μg/min at baseline to 20 μg/min after treatment in the enalapril group. In the placebo group, on the other hand, it increased from 53 μg/min at baseline to 85 μg/min, with 12 subjects (23.5%) with a UAE of ≥200 μg/min, compared with 4 subjects (7.7%) in the enalapril group. During the follow-up, there were no changes in blood pressure, glomerular filtration rate (GFR)/renal plasma flow (RPF), electrolytes, or blood lipids.

Captopril

The Microalbuminuria Captopril Study was among the studies of captopril (Capoten®; Bristol-Myers Squibb, Captril®; Daiichi Sankyo) [218]. This was a randomized, double-blind, comparative study designed to evaluate the inhibition of progression to overt albuminuria in type 1 diabetic patients with microalbuminuria. The proportion of subjects with progression to overt albuminuria was 21.9% in the placebo group and 7.2% in the captopril group. Captopril reduced the risk by 69.2%, and a similar outcome was obtained after correction for blood pressure. While UAE significantly decreased in the captopril group at 3 months after the start of the study, creatinine clearance significantly increased in the placebo group at 18 months after the start of the study. These results showed that captopril had a renoprotective effect independent from the hypotensive effect in type 1 diabetic patients.

Imidapril

Imidapril hydrochloride (Tanatril®; Mitsubishi Tanabe Pharma) is approved for the treatment of DN in type 1 diabetic patients in Japan. JAPAN-IDDM was one of the studies of imidapril hydrochloride [219]. This was a randomized, double-blind, comparative study designed to evaluate the renoprotective effect of imidapril hydrochloride in type 1 diabetic patients with albuminuria. UAE increased by 57% in the placebo group, but decreased by 27% in the imidapril group. Blood pressure was not significantly different between the two groups, but tended to be higher in the placebo group.

Angiotensin-II Receptor Blockers (ARBs)

Losartan

Losartan potassium (Cozaar®/Nu-Lotan®; Merck) is approved for the treatment of DN in hypertensive type 2 diabetic patients in the US and Japan. RENAAL was one of the studies of losartan [220]. This was a randomized, double-blind, comparative study designed to evaluate the renoprotective effect of losartan potassium in type 2 diabetic patients with nephropathy. Losartan potassium reduced the risk of doubling of serum creatinine by 25% and the risk of end-stage renal failure (dialysis/renal transplant) by 28%. It also decreased urinary protein excretion by 35%. Risk reduction in the losartan group under blood pressure control comparable to the placebo group indicated that the renoprotective effect of losartan potassium was independent from the hypotensive effect.

Irbesartan

Irbesartan (Karvea®/Avapro®; Bristol-Myers Squibb/Sanofi/Dainippon Sumitomo Pharma, Irbetan®; Shionogi) is approved for the treatment of DN in hypertensive type 2 diabetic patients in Canada, the EU, and the US. The IDNT study was one of the studies of irbesartan [221]. This was a randomized, double-blind, comparative study designed to verify the renoprotective effect of irbesartan and amlodipine, a Ca^{2+} channel blocker, in hypertensive patients with type 2 DN. Patients were randomized to receive irbesartan, amlodipine, or placebo. During the study period, mean blood pressure was not significantly different between the irbesartan and amlodipine groups. In the placebo group, mean blood pressure was significantly higher than in the two active drug groups. Irbesartan significantly reduced the relative risk of doubling of serum creatinine level by 33% and 37% compared with placebo and amlodipine, respectively. Irbesartan reduced the relative risk of progression to end-stage renal failure by 23% compared with placebo and amlodipine. No significant differences were observed in the relative risk of doubling of serum creatinine level or progression to end-stage renal failure between the placebo and amlodipine groups. A similar outcome was obtained after correction for the baseline characteristics of patients or blood pressure in the study period. These results indicated that the renoprotective effect of irbesartan was independent from the degree of reduction in blood pressure.

Valsartan

The MARVAL study was among the studies of valsartan (Diovan®; Novartis Pharma) [222]. This was a randomized, double-blind, comparative study designed to determine whether the efficacy of valsartan for microalbuminuria was dependent on blood pressure or not in type 2 diabetic patients with microalbuminuria. In the study, subjects received valsartan or amlodipine for 24 weeks. While valsartan and amlodipine reduced final blood pressure to a similar level with a similar hypotensive effect, only valsartan decreased UAE. These results indicated that valsartan inhibited proteinuria, irrespective of blood pressure.

Telmisartan

The INNOVATION study was a randomized, double-blind, comparative study designed to verify the inhibitory effect of telmisartan (Micardis®; Boehringer Ingelheim/Astellas Pharma, Pritor®; GlaxoSmithKline, Kinzar®; Bayer) on progression of nephropathy in normotensive or hypertensive type 2 diabetic patients with microalbuminuria [223]. Patients were randomized to receive telmisartan at a dose of 80 or 40 mg, or placebo. The proportion of subjects with progression to overt nephropathy was significantly lower in the telmisartan groups (80 mg group: 16.7%, 40 mg group: 22.6%, placebo group: 49.9%). The proportion of subjects who recovered from microalbuminuria was significantly higher in the telmisartan groups (80 mg group: 21.2%, 40 mg group: 12.8%, placebo group: 1.2%). In normotensive patients as well, the proportion of subjects with progression to overt nephropathy was significantly lower in the telmisartan groups (80 mg group: 11.0%, 40 mg group: 21.0%, placebo group: 44.2%). These results showed that telmisartan had a renoprotective effect in not only hypertensive patients, but also normotensive patients.

Olmesartan

The ROADMAP study was one of the studies of olmesartan medoxomil (Olmetec®/Benicar®; Daiichi Sankyo/Kowa) [224]. This was a randomized, double-blind, comparative study designed to determine whether olmesartan medoxomil could inhibit the development of microalbuminuria in type 2 diabetic patients with normal albuminuria. The proportion of subjects who achieved target blood pressure was 80% in the olmesartan medoxomil group and 71% in the placebo group, and casual blood pressure was lower by 3.1/1.9 mmHg in the olmesartan medoxomil group than in the placebo group. The incidence of microalbuminuria was 8.2% (178/2160 subjects) in the olmesartan medoxomil group and 9.8% (210/2139) in the placebo group, and olmesartan medoxomil increased the time to development of microalbuminuria by 23%. In this study, however, the incidence of fatal cardiovascular events was 0.7% in the olmesartan medoxomil group, which was significantly higher than in the placebo group (0.1%).

Direct Renin Inhibitors (DRIs)

Aliskiren

The AVOID study was among the studies of aliskiren hemifumarate (Rasilez®/Tekturna®; Novartis Pharma) [225]. This was a randomized, double-blind, comparative study designed to verify the renoprotective effect of aliskiren hemifumarate added to losartan at the the maximum recommended dose in type 2 diabetic patients with hypertension and renal disorder. Aliskiren hemifumarate added to losartan decreased UAE by another 20% compared with losartan alone. Blood pressure was minimally affected in both groups, indicating that the renoprotective effect of aliskiren hemifumarate was independent from the hypotensive effect. The ALTITUDE study was another study of aliskiren hemifumarate [226]. This study was conducted to evaluate the cardiovascular protective effect of aliskiren added to conventional therapy with ACE inhibitor or ARB in type 2 diabetic patients with renal impairment. The study was prematurely discontinued, because the incidence of adverse events increased in the aliskiren hemifumarate add-on therapy group, compared with the placebo group. After discontinuation, analysis revealed that aliskiren hemifumarate add-on therapy reduced final blood pressure and UAE compared with placebo, but had no cardiovascular protective effect.

Other Drugs Under Development With New Mechanisms of Action

Nrf2 Activators

Bardoxolone

Bardoxolone methyl (RTA402; Reata Pharmaceuticals/Abbott Laboratories/Kyowa Hakko Kirin) is a low-molecular compound that activates Nrf2. In 2011, the results of a phase IIb clinical study in type 2 diabetic patients with chronic kidney disease were reported [227]. At Week 24 of treatment, bardoxolone methyl improved renal function as shown by a significant increase in estimated glomerular filtration rate (eGFR) compared with the placebo group. Significant improvement in eGFR was maintained at Week 52. In 2012, however, it was announced that an ongoing phase III study was discontinued due to safety concerns.

NADPH Oxidase (Nox) Inhibitors

Nox is a membrane-bound enzyme complex and a synthase for reactive oxygen species (ROS). ROS is necessary to eliminate microorganisms and other foreign substances that invade the body, but is harmful to the body if overproduced, because it also damages normal cells. Nox has seven isoforms; namely, Nox1, Nox2, Nox3, Nox4, Nox5, Duox1, and Duox2. Despite differences in tissue expression, intracellular location, alignment, structure, and mechanism of activation, all of these isoforms have one thing in common: they produce ROS. It has been reported that the renal Nox4 activity and ROS production are elevated in the kidney in an animal model of diabetes [228]. It has also been reported that antisense oligonucleotide inhibits diabetic glomerular injury by inhibiting renal Nox4 expression [229]. These findings suggest that Nox (especially Nox4) may serve as a molecular target for DN therapy.

GKT137831

GKT137831 (Genkyotex) is a low-molecular compound that inhibits Nox1 and Nox4. Currently, a clinical study in DN patients is being designed.

NOX-E36

NOX-E36 (Noxxon Pharma) is a Spiegelmer® targeting MCP-1. It is an RNA aptamer that binds to and inhibits MCP-1. Currently, a clinical study is under way.

Growth Factor Modulators

It is suggested that various growth factors are involved in the development and progression of DN. TGF-β, which is induced by hyperglycemia and potently enhances extracellular matrix production and fibrosis, may play a central role in the development and progression of DN. It is known to potently induce the connective tissue growth factor (CTGF) [230]. CTGF stimulates fibroblast proliferation and extracellular matrix production, and it has been reported that TGF-β-stimulated fibrosis is partly mediated by CTGF [231]. These findings suggest that TGF-β and CTGF may serve as a promising target for anti-fibrotic therapy.

LY2382770

LY2382770 (Eli Lilly) is a human monoclonal antibody against TGF-β. Currently, a clinical study is under way, with the primary endpoint defined as the change in serum creatinine at month 12 of treatment.

FG-3019

FG-3019 (FibroGen) is a human monoclonal antibody against CTGF. Currently, a clinical study is under way.

MCP-1/CCR2 System Inhibitors

CCR2 (C-C chemokine receptor type 2) is a seven-transmembrane receptor for which MCP-1 (Monocyte Chemoattractant Protein-1) serves as a ligand. MCP-1 is produced in macrophages, fibroblasts, and vascular endothelial cells, and potently stimulates migration of monocytes and T lymphocytes. It may also be involved in fibrosis by enhancing TGF-β-mediated type 1 collagen gene expression in fibroblasts. In a streptozotocin (STZ)-induced diabetes model, renal disorder was less severe with less increase in UAE and plasma creatinine in MCP-1 knockout mice, compared with wild-type mice [232]. Similar results were reported in CCR2 knockout mice and with CCR2 antagonist [233]. These findings suggest that the blockade of MCP-1/CCR2 signaling may effectively inhibit the development and progression of DN.

CCX140

CCX140 (ChemoCentryx) is a CCR2 antagonist for which a clinical study was conducted in patients with diabetes, and another is being conducted in patients with DN. In diabetic patients, it reduced fasting blood glucose and HbA1c after 4-week treatment. The clinical study in DN patients is under way, and results are awaited.

PF-04634817

PF-04634817 (Pfizer) is a CCR2/5 antagonist. In 2012, phase II trials for the treatment of diabetic neuropathy commenced.

Janus Kinase (JAK) Inhibitors

JAKs are nonreceptor tyrosine kinases. JAKs are classified as JAK1, JAK2, JAK3, or TYK2, according to the function and genomic location, and many of them are involved in cell proliferation, survial, and differentiation and play important roles especially in immune and blood cells. The signal transduction is mediated by the transcription factor STAT (Signal Transducers and Activator of Transcription). Activation of the JAK/STAT pathway in the kidney is reported in DN patients [234], suggesting that this may contribute to the progression of renal disorder by inducing inflammatory cytokine production and cell differentiation/outgrowth.

Baricitinib

Baricitinib (Incyte/Eli Lilly) is a low-molecular compound that inhibits JAK1 and JAK2. Currently, clinical studies are under way for the treatment of rheumatism, psoriasis, and DN. In the ongoing clinical study in DN, the primary endpoint is the decrease in UAE.

AGE Inhibitors

Pyridoxamine

Pyridoxamine dihydrochloride (Pyridorin™; BioStratum/Kowa/NephroGenex) is a compund that classified into vitamine B6, and blocks the multiple pathways of

AGE formation. In the preclinical study, pyridoxamine prevents increased UAE in daibeic animal models [235, 236]. hyperglycemia-induced renal dysfunction. Currentry, phase II clinical studies are ongoing for treatment of DN [237]. The FDA has granted fast track status to pyridoxamine for the treatment of DN.

GLY-230

GLY-230 (Glycadia Pharmaceuticals) is an AGE inhibitor (Maillard's reaction inhibitor). In clinical studies in patients with diabetes, GLY-230 significantly reduced abnormal urine levels of markers indicating damage to the kidney filtration barrier [195]. Currently, phase II clinical study is under way.

Endothelin A (ETA) Receptor Antagonists

Endothelin is a highly vasoconstrictive, physiologically active substance that was isolated and purified from the culture supernatant of porcine aortic endothelial cells, with the gene identified [238]. There are three isoforms (ET-1, ET-2, and ET-3) of endothelin, which exhibit their actions by binding to specific receptors (ETA and ETB). The ETA receptor is highly expressed in the heart and lung and also found in the brain, kidney, and intestinal tract. In blood vessel walls, the ETA receptor is expressed in the smooth muscle and involved in vascular contraction and cell proliferation. In a STZ-induced diabetes model, an ETA antagonist significantly decreased urinary proteins [239]. It has also been reported that increased expression of ET-1 and ETA was observed in the kidney in DN patients and correlated with urinary proteins [240].

Atrasentan

Atrasentan (Xinlay[TM]/ABT-627; Abbott Laboratories) is a selective antagonist for ETA. In 2011, the results of a phase II study in DN patients were reported [241]. After 8-week treatment with atrasentan at doses of 0.25, 0.75, and 1.75 mg in type 2 DN patients receiving an RAS inhibitor, the urinary albumin-creatinine ratio was significantly decreased at doses of 0.75 and 1.75 mg. However, the incidence of adverse drug reactions, especially edema (1.75 mg group: 46%, placebo group: 9%), significantly increased at a dose of 1.75 mg. Currently, clinical studies are under way for the treatment of DN and renal cell carcinoma.

Avosentan

Avosentan (SPP301; Roche/Speedel Pharma) is a second-generation antagonist for ETA. In a phase II clinical trial, 12 weeks treatment of avosentan (5-50 mg) decreased mean relative UAE (-16.3 to -29.9%) compared with placebo (35.5%) [242]. However, phase III clinical trial was terminated based on a recommendation from the Data Safety Monitoring Board, following a significant imbalance in fluid retention in patients amongst the study arms.

CONCLUSION

A number of pre-clinical and clinical trials have demonstrated various approaches to prevent/ameliorate diabetic microvascular complications. Although the development of these new drugs requires an extraordinary effort and time to be approved by authorities, a large number of patients need to receive benefits. Drugs with new mechanisms to reduce blood glucose levels, such as DPP-IV inhibitors, GLP-1 analogues, and SGLT2 inhibitors, have been launched these years, and the options for treating hyperglycemia have been widely extended. However, the development of new hypoglycemic agents seems to be quite difficult. Development of new drugs for diabetic microvascular complications is an ongoing pharmacological approach for all diabetics suffering threat of serious complications.

ACKNOWLEDGEMENTS

Declared None.

CONFLICT OF INTEREST

TS wrote diabetic peripheral neuropathy and diabetic retinopathy sections. RS wrote diabetic nephropathy section. All the authors have no competing interests.

REFERENCES

[1]   International Diabetes Federation. IDF Diabetes Atlas. 5th ed. Brussels, Belgium: International Diabetes Federation 2011.
[2]   Ohkubo Y, Kishikawa H, Araki E, Miyata T, Isami S, Motoyoshi S, *et al.* Intensive insulin therapy prevents the progression of diabetic microvascular complications in Japanese

patients with non-insulin-dependent diabetes mellitus: a randomized prospective 6-year study. Diabetes Res Clin Pract. 1995 May;28(2):103-17.

[3]   The Diabetes Control and Complications Trial Research Group. The effect of intensive treatment of diabetes on the development and progression of long-term complications in insulin-dependent diabetes mellitus. N Engl J Med. 1993 Sep 30;329(14):977-86.

[4]   UK Prospective Diabetes Study (UKPDS) Group. Intensive blood-glucose control with sulphonylureas or insulin compared with conventional treatment and risk of complications in patients with type 2 diabetes (UKPDS 33). Lancet. 1998 Sep 12;352(9131):837-53.

[5]   Rosenstock J, Tuchman M, LaMoreaux L, Sharma U. Pregabalin for the treatment of painful diabetic peripheral neuropathy: a double-blind, placebo-controlled trial. Pain. 2004 Aug;110(3):628-38.

[6]   Gee NS, Brown JP, Dissanayake VU, Offord J, Thurlow R, Woodruff GN. The novel anticonvulsant drug, gabapentin (Neurontin), binds to the alpha2delta subunit of a calcium channel. J Biol Chem. 1996 Mar 8;271(10):5768-76.

[7]   Cundy KC, Annamalai T, Bu L, De Vera J, Estrela J, Luo W, *et al.* XP13512 [(+/-)-1-([(alpha-isobutanoyloxyethoxy) carbonyl] aminomethyl)-1-cyclohexane acetic acid], a novel gabapentin prodrug: II. Improved oral bioavailability, dose proportionality, and colonic absorption compared with gabapentin in rats and monkeys. J Pharmacol Exp Ther. 2004 Oct;311(1):324-33.

[8]   Backonja MM, Canafax DM, Cundy KC. Efficacy of gabapentin enacarbil *vs.* placebo in patients with postherpetic neuralgia and a pharmacokinetic comparison with oral gabapentin. Pain Med. 2011 Jul;12(7):1098-108.

[9]   Kushida CA, Becker PM, Ellenbogen AL, Canafax DM, Barrett RW. Randomized, double-blind, placebo-controlled study of XP13512/GSK1838262 in patients with RLS. Neurology. 2009 Feb 3;72(5):439-46.

[10]   Ellenbogen AL, Thein SG, Winslow DH, Becker PM, Tolson JM, Lassauzet ML, *et al.* A 52-week study of gabapentin enacarbil in restless legs syndrome. Clin Neuropharmacol. 2011 Jan-Feb;34(1):8-16.

[11]   Kochar DK, Jain N, Agarwal RP, Srivastava T, Agarwal P, Gupta S. Sodium valproate in the management of painful neuropathy in type 2 diabetes - a randomized placebo controlled study. Acta Neurol Scand. 2002 Nov;106(5):248-52.

[12]   Kochar DK, Rawat N, Agrawal RP, Vyas A, Beniwal R, Kochar SK, *et al.* Sodium valproate for painful diabetic neuropathy: a randomized double-blind placebo-controlled study. QJM. 2004 Jan;97(1):33-8.

[13]   Rull JA, Quibrera R, Gonzalez-Millan H, Lozano Castaneda O. Symptomatic treatment of peripheral diabetic neuropathy with carbamazepine (Tegretol): double blind crossover trial. Diabetologia. 1969 Aug;5(4):215-8.

[14]   Raskin P, Donofrio PD, Rosenthal NR, Hewitt DJ, Jordan DM, Xiang J, *et al.* Topiramate *vs* placebo in painful diabetic neuropathy: analgesic and metabolic effects. Neurology. 2004 Sep 14;63(5):865-73.

[15]   Sindrup SH, Jensen TS. Efficacy of pharmacological treatments of neuropathic pain: an update and effect related to mechanism of drug action. Pain. 1999 Dec;83(3):389-400.

[16]   Wong MC, Chung JW, Wong TK. Effects of treatments for symptoms of painful diabetic neuropathy: systematic review. BMJ. 2007 Jul 14;335(7610):87.

[17]   Dworkin RH, O'Connor AB, Backonja M, Farrar JT, Finnerup NB, Jensen TS, *et al.* Pharmacologic management of neuropathic pain: evidence-based recommendations. Pain. 2007 Dec 5;132(3):237-51.

[18]   Richelson E, Pfenning M. Blockade by antidepressants and related compounds of biogenic amine uptake into rat brain synaptosomes: most antidepressants selectively block norepinephrine uptake. Eur J Pharmacol. 1984 Sep 17;104(3-4):277-86.

[19]   Davis JL, Lewis SB, Gerich JE, Kaplan RA, Schultz TA, Wallin JD. Peripheral diabetic neuropathy treated with amitriptyline and fluphenazine. JAMA. 1977 Nov 21;238(21):2291-2.

[20]   Horioka T, Nakamura H, Masuda Y, Murai K, Kaneko H. [Pharmacological studies of a new antidepressant nortriptyline (Noritren)]. Igaku Kenkyu. 1970 Jun;40(3):289-304.

[21]   Gomez-Perez FJ, Rull JA, Dies H, Rodriquez-Rivera JG, Gonzalez-Barranco J, Lozano-Castaneda O. Nortriptyline and fluphenazine in the symptomatic treatment of diabetic neuropathy. A double-blind cross-over study. Pain. 1985 Dec;23(4):395-400.

[22]   Cohen H, Cohen MR. Nortriptyline and neuropathic pain. Nurse Pract. 1999 Apr;24(4):124.

[23]   Pritchett YL, McCarberg BH, Watkin JG, Robinson MJ. Duloxetine for the management of diabetic peripheral neuropathic pain: response profile. Pain Med. 2007 Jul-Aug;8(5):397-409.

[24]   Skljarevski V, Desaiah D, Zhang Q, Chappell AS, Detke MJ, Gross JL, *et al.* Evaluating the maintenance of effect of duloxetine in patients with diabetic peripheral neuropathic pain. Diabetes Metab Res Rev. 2009 Oct;25(7):623-31.

[25]   Kaur H, Hota D, Bhansali A, Dutta P, Bansal D, Chakrabarti A. A comparative evaluation of amitriptyline and duloxetine in painful diabetic neuropathy: a randomized, double-blind, cross-over clinical trial. Diabetes Care. 2011 Apr;34(4):818-22.

[26]   Basile AS, Janowsky A, Golembiowska K, Kowalska M, Tam E, Benveniste M, *et al.* Characterization of the antinociceptive actions of bicifadine in models of acute, persistent, and chronic pain. J Pharmacol Exp Ther. 2007 Jun;321(3):1208-25.

[27]   Wang RI, Johnson RP, Lee JC, Waite EM. The oral analgesic efficacy of bicifadine hydrochloride in postoperative pain. J Clin Pharmacol. 1982 Apr;22(4):160-4.

[28]   Karchewski LA, Bloechlinger S, Woolf CJ. Axonal injury-dependent induction of the peripheral benzodiazepine receptor in small-diameter adult rat primary sensory neurons. Eur J Neurosci. 2004 Aug;20(3):671-83.

[29]   Ferzaz B, Brault E, Bourliaud G, Robert JP, Poughon G, Claustre Y, *et al.* SSR180575 (7-chloro-N,N,5-trimethyl- 4-oxo-3-phenyl-3,5-dihydro-4H-pyridazino[4,5-b]indole-1 -acetamide), a peripheral benzodiazepine receptor ligand, promotes neuronal survival and repair. J Pharmacol Exp Ther. 2002 Jun;301(3):1067-78.

[30]   Gimbel JS, Richards P, Portenoy RK. Controlled-release oxycodone for pain in diabetic neuropathy: a randomized controlled trial. Neurology. 2003 Mar 25;60(6):927-34.

[31]   Harati Y, Gooch C, Swenson M, Edelman S, Greene D, Raskin P, *et al.* Double-blind randomized trial of tramadol for the treatment of the pain of diabetic neuropathy. Neurology. 1998 Jun;50(6):1842-6.

[32]   Harati Y, Gooch C, Swenson M, Edelman SV, Greene D, Raskin P, *et al.* Maintenance of the long-term effectiveness of tramadol in treatment of the pain of diabetic neuropathy. J Diabetes Complications. 2000 Mar-Apr;14(2):65-70.

[33]   Freeman R, Raskin P, Hewitt DJ, Vorsanger GJ, Jordan DM, Xiang J, *et al.* Randomized study of tramadol/acetaminophen *vs.* placebo in painful diabetic peripheral neuropathy. Curr Med Res Opin. 2007 Jan;23(1):147-61.

[34] Christoph T, De Vry J, Tzschentke TM. Tapentadol, but not morphine, selectively inhibits disease-related thermal hyperalgesia in a mouse model of diabetic neuropathic pain. Neurosci Lett. 2010 Feb 12;470(2):91-4.

[35] Schwartz S, Etropolski M, Shapiro DY, Okamoto A, Lange R, Haeussler J, *et al.* Safety and efficacy of tapentadol ER in patients with painful diabetic peripheral neuropathy: results of a randomized-withdrawal, placebo-controlled trial. Curr Med Res Opin. 2011 Jan;27(1):151-62.

[36] Amir R, Argoff CE, Bennett GJ, Cummins TR, Durieux ME, Gerner P, *et al.* The role of sodium channels in chronic inflammatory and neuropathic pain. J Pain. 2006 May;7(5 Suppl 3):S1-29.

[37] Rogers M, Tang L, Madge DJ, Stevens EB. The role of sodium channels in neuropathic pain. Semin Cell Dev Biol. 2006 Oct;17(5):571-81.

[38] Sun W, Miao B, Wang XC, Duan JH, Wang WT, Kuang F, *et al.* Reduced conduction failure of the main axon of polymodal nociceptive C-fibres contributes to painful diabetic neuropathy in rats. Brain. 2012 Feb;135(Pt 2):359-75.

[39] Wang N, Orr-Urtreger A, Korczyn AD. The role of neuronal nicotinic acetylcholine receptor subunits in autonomic ganglia: lessons from knockout mice. Prog Neurobiol. 2002 Dec;68(5):341-60.

[40] Bannon AW, Decker MW, Kim DJ, Campbell JE, Arneric SP. ABT-594, a novel cholinergic channel modulator, is efficacious in nerve ligation and diabetic neuropathy models of neuropathic pain. Brain Res. 1998 Aug 10;801(1-2):158-63.

[41] Rowbotham MC, Arslanian A, Nothaft W, Duan WR, Best AE, Pritchett Y, *et al.* Efficacy and safety of the alpha4beta2 neuronal nicotinic receptor agonist ABT-894 in patients with diabetic peripheral neuropathic pain. Pain. 2012 Apr;153(4):862-8.

[42] Bitar MS, Bajic KT, Farook T, Thomas MI, Pilcher CW. Spinal cord noradrenergic dynamics in diabetic and hypercortisolaemic states. Brain Res. 1999 May 29;830(1):1-9.

[43] Guo TZ, Davies MF, Kingery WS, Patterson AJ, Limbird LE, Maze M. Nitrous oxide produces antinociceptive response *via* alpha2B and/or alpha2C adrenoceptor subtypes in mice. Anesthesiology. 1999 Feb;90(2):470-6.

[44] Courteix C, Bardin M, Chantelauze C, Lavarenne J, Eschalier A. Study of the sensitivity of the diabetes-induced pain model in rats to a range of analgesics. Pain. 1994 May;57(2):153-60.

[45] Byas-Smith MG, Max MB, Muir J, Kingman A. Transdermal clonidine compared to placebo in painful diabetic neuropathy using a two-stage 'enriched enrollment' design. Pain. 1995 Mar;60(3):267-74.

[46] Campbell CM, Kipnes MS, Stouch BC, Brady KL, Kelly M, Schmidt WK, *et al.* Randomized control trial of topical clonidine for treatment of painful diabetic neuropathy. Pain. 2012 Sep;153(9):1815-23.

[47] Caterina MJ, Schumacher MA, Tominaga M, Rosen TA, Levine JD, Julius D. The capsaicin receptor: a heat-activated ion channel in the pain pathway. Nature. 1997 Oct 23;389(6653):816-24.

[48] Bandell M, Macpherson LJ, Patapoutian A. From chills to chilis: mechanisms for thermosensation and chemesthesis *via* thermoTRPs. Curr Opin Neurobiol. 2007 Aug;17(4):490-7.

[49] Nelson KA, Park KM, Robinovitz E, Tsigos C, Max MB. High-dose oral dextromethorphan *vs.* placebo in painful diabetic neuropathy and postherpetic neuralgia. Neurology. 1997 May;48(5):1212-8.

[50]   Criner TM, Perdun CS. Dextromethorphan and diabetic neuropathy. Ann Pharmacother. 1999 Nov;33(11):1221-3.

[51]   Shaibani AI, Pope LE, Thisted R, Hepner A. Efficacy and safety of dextromethorphan/quinidine at two dosage levels for diabetic neuropathic pain: a double-blind, placebo-controlled, multicenter study. Pain Med. 2012 Feb;13(2):243-54.

[52]   Beyreuther B, Callizot N, Stohr T. Antinociceptive efficacy of lacosamide in a rat model for painful diabetic neuropathy. Eur J Pharmacol. 2006 Jun 6;539(1-2):64-70.

[53]   Ziegler D, Hidvegi T, Gurieva I, Bongardt S, Freynhagen R, Sen D, *et al.* Efficacy and safety of lacosamide in painful diabetic neuropathy. Diabetes Care. 2010 Apr;33(4):839-41.

[54]   Oskarsson P, Ljunggren JG, Lins PE. Efficacy and safety of mexiletine in the treatment of painful diabetic neuropathy. The Mexiletine Study Group. Diabetes Care. 1997 Oct;20(10):1594-7.

[55]   Huang Z, Liu S, Zhang L, Salem M, Greig GM, Chan CC, *et al.* Preferential inhibition of human phosphodiesterase 4 by ibudilast. Life Sci. 2006 May 1;78(23):2663-8.

[56]   Sone H, Okuda Y, Asakura Y, Asano M, Mizutani M, Bannai C, *et al.* Efficacy of Ibudilast on lower limb circulation of diabetic patients with minimally impaired baseline flow: a study using color Doppler ultrasonography and laser Doppler flowmetry. Angiology. 1995 Aug;46(8):699-703.

[57]   Yoshioka A, Shimizu Y, Hirose G, Kitasato H, Pleasure D. Cyclic AMP-elevating agents prevent oligodendroglial excitotoxicity. J Neurochem. 1998 Jun;70(6):2416-23.

[58]   Lane NE, Schnitzer TJ, Birbara CA, Mokhtarani M, Shelton DL, Smith MD, *et al.* Tanezumab for the treatment of pain from osteoarthritis of the knee. N Engl J Med. 2010 Oct 14;363(16):1521-31.

[59]   Brown MT, Murphy FT, Radin DM, Davignon I, Smith MD, West CR. Tanezumab reduces osteoarthritic knee pain: results of a randomized, double-blind, placebo-controlled phase III trial. J Pain. 2012 Aug;13(8):790-8.

[60]   Kuninaka T, Senga Y, Senga H, Weiner M. Nature of enhanced mitochondrial oxidative metabolism by a calf blood extract. J Cell Physiol. 1991 Jan;146(1):148-55.

[61]   Obermaier-Kusser B, Muhlbacher C, Mushack J, Seffer E, Ermel B, Machicao F, *et al.* Further evidence for a two-step model of glucose-transport regulation. Inositol phosphate-oligosaccharides regulate glucose-carrier activity. Biochem J. 1989 Aug 1;261(3):699-705.

[62]   Jacob S, Dietze GJ, Machicao F, Kuntz G, Augustin HJ. Improvement of glucose metabolism in patients with type II diabetes after treatment with a hemodialysate. Arzneimittelforschung. 1996 Mar;46(3):269-72.

[63]   Ziegler D, Movsesyan L, Mankovsky B, Gurieva I, Abylaiuly Z, Strokov I. Treatment of symptomatic polyneuropathy with actovegin in type 2 diabetic patients. Diabetes Care. 2009 Aug;32(8):1479-84.

[64]   Kamiya H, Nakamura J, Hamada Y, Nakashima E, Naruse K, Kato K, *et al.* Polyol pathway and protein kinase C activity of rat Schwannoma cells. Diabetes Metab Res Rev. 2003 Mar-Apr;19(2):131-9.

[65]   Kasuya Y, Nakamura J, Hamada Y, Nakayama M, Sasaki H, Komori T, *et al.* An aldose reductase inhibitor prevents the glucose-induced increase in PDGF-beta receptor in cultured rat aortic smooth muscle cells. Biochem Biophys Res Commun. 1999 Aug 11;261(3):853-8.

[66]   Kikkawa R, Hatanaka I, Yasuda H, Kobayashi N, Shigeta Y, Terashima H, *et al.* Effect of a new aldose reductase inhibitor, (E)-3-carboxymethyl-5-[(2E)-methyl-3-

phenylpropenylidene]rhodanine (ONO-2235) on peripheral nerve disorders in streptozotocin-diabetic rats. Diabetologia. 1983 Apr;24(4):290-2.

[67]   Hotta N, Akanuma Y, Kawamori R, Matsuoka K, Oka Y, Shichiri M, *et al*. Long-term clinical effects of epalrestat, an aldose reductase inhibitor, on diabetic peripheral neuropathy: the 3-year, multicenter, comparative Aldose Reductase Inhibitor-Diabetes Complications Trial. Diabetes Care. 2006 Jul;29(7):1538-44.

[68]   Hotta N, Kawamori R, Atsumi Y, Baba M, Kishikawa H, Nakamura J, *et al*. Stratified analyses for selecting appropriate target patients with diabetic peripheral neuropathy for long-term treatment with an aldose reductase inhibitor, epalrestat. Diabet Med. 2008 Jul;25(7):818-25.

[69]   Bril V, Buchanan RA. Aldose reductase inhibition by AS-3201 in sural nerve from patients with diabetic sensorimotor polyneuropathy. Diabetes Care. 2004 Oct;27(10):2369-75.

[70]   Bril V, Hirose T, Tomioka S, Buchanan R. Ranirestat for the management of diabetic sensorimotor polyneuropathy. Diabetes Care. 2009 Jul;32(7):1256-60.

[71]   Ishii H, Jirousek MR, Koya D, Takagi C, Xia P, Clermont A, *et al*. Amelioration of vascular dysfunctions in diabetic rats by an oral PKC beta inhibitor. Science. 1996 May 3;272(5262):728-31.

[72]   Nakamura J, Kato K, Hamada Y, Nakayama M, Chaya S, Nakashima E, *et al*. A protein kinase C-beta-selective inhibitor ameliorates neural dysfunction in streptozotocin-induced diabetic rats. Diabetes. 1999 Oct;48(10):2090-5.

[73]   Casellini CM, Barlow PM, Rice AL, Casey M, Simmons K, Pittenger G, *et al*. A 6-month, randomized, double-masked, placebo-controlled study evaluating the effects of the protein kinase C-beta inhibitor ruboxistaurin on skin microvascular blood flow and other measures of diabetic peripheral neuropathy. Diabetes Care. 2007 Apr;30(4):896-902.

[74]   Vinik AI, Bril V, Kempler P, Litchy WJ, Tesfaye S, Price KL, *et al*. Treatment of symptomatic diabetic peripheral neuropathy with the protein kinase C beta-inhibitor ruboxistaurin mesylate during a 1-year, randomized, placebo-controlled, double-blind clinical trial. Clin Ther. 2005 Aug;27(8):1164-80.

[75]   Sasase T, Yamada H, Sakoda K, Imagawa N, Abe T, Ito M, *et al*. Novel protein kinase C-beta isoform selective inhibitor JTT-010 ameliorates both hyper- and hypoalgesia in streptozotocin- induced diabetic rats. Diabetes Obes Metab. 2005 Sep;7(5):586-94.

[76]   Wada R, Yagihashi S. Role of advanced glycation end products and their receptors in development of diabetic neuropathy. Ann N Y Acad Sci. 2005 Jun;1043:598-604.

[77]   Toth C, Rong LL, Yang C, Martinez J, Song F, Ramji N, *et al*. Receptor for advanced glycation end products (RAGEs) and experimental diabetic neuropathy. Diabetes. 2008 Apr;57(4):1002-17.

[78]   Yagihashi S, Kamijo M, Baba M, Yagihashi N, Nagai K. Effect of aminoguanidine on functional and structural abnormalities in peripheral nerve of STZ-induced diabetic rats. Diabetes. 1992 Jan;41(1):47-52.

[79]   Kihara M, Schmelzer JD, Poduslo JF, Curran GL, Nickander KK, Low PA. Aminoguanidine effects on nerve blood flow, vascular permeability, electrophysiology, and oxygen free radicals. Proc Natl Acad Sci U S A. 1991 Jul 15;88(14):6107-11.

[80]   Freedman BI, Wuerth JP, Cartwright K, Bain RP, Dippe S, Hershon K, *et al*. Design and baseline characteristics for the aminoguanidine Clinical Trial in Overt Type 2 Diabetic Nephropathy (ACTION II). Control Clin Trials. 1999 Oct;20(5):493-510.

[81]   Bolton WK, Cattran DC, Williams ME, Adler SG, Appel GB, Cartwright K, *et al.* Randomized trial of an inhibitor of formation of advanced glycation end products in diabetic nephropathy. Am J Nephrol. 2004 Jan-Feb;24(1):32-40.

[82]   Nakamura S, Makita Z, Ishikawa S, Yasumura K, Fujii W, Yanagisawa K, *et al.* Progression of nephropathy in spontaneous diabetic rats is prevented by OPB-9195, a novel inhibitor of advanced glycation. Diabetes. 1997 May;46(5):895-9.

[83]   Wada R, Nishizawa Y, Yagihashi N, Takeuchi M, Ishikawa Y, Yasumura K, *et al.* Effects of OPB-9195, anti-glycation agent, on experimental diabetic neuropathy. Eur J Clin Invest. 2001 Jun;31(6):513-20.

[84]   Forbes JM, Thallas V, Thomas MC, Founds HW, Burns WC, Jerums G, *et al.* The breakdown of preexisting advanced glycation end products is associated with reduced renal fibrosis in experimental diabetes. FASEB J. 2003 Sep;17(12):1762-4.

[85]   Kim JB, Song BW, Park S, Hwang KC, Cha BS, Jang Y, *et al.* Alagebrium chloride, a novel advanced glycation end-product cross linkage breaker, inhibits neointimal proliferation in a diabetic rat carotid balloon injury model. Korean Circ J. 2010 Oct;40(10):520-6.

[86]   Yonekura H, Yamamoto Y, Sakurai S, Petrova RG, Abedin MJ, Li H, *et al.* Novel splice variants of the receptor for advanced glycation end-products expressed in human vascular endothelial cells and pericytes, and their putative roles in diabetes-induced vascular injury. Biochem J. 2003 Mar 15;370(Pt 3):1097-109.

[87]   Humpert PM, Papadopoulos G, Schaefer K, Djuric Z, Konrade I, Morcos M, *et al.* sRAGE and esRAGE are not associated with peripheral or autonomic neuropathy in type 2 diabetes. Horm Metab Res. 2007 Dec;39(12):899-902.

[88]   Nagamatsu M, Nickander KK, Schmelzer JD, Raya A, Wittrock DA, Tritschler H, *et al.* Lipoic acid improves nerve blood flow, reduces oxidative stress, and improves distal nerve conduction in experimental diabetic neuropathy. Diabetes Care. 1995 Aug;18(8):1160-7.

[89]   Ziegler D, Nowak H, Kempler P, Vargha P, Low PA. Treatment of symptomatic diabetic polyneuropathy with the antioxidant alpha-lipoic acid: a meta-analysis. Diabet Med. 2004 Feb;21(2):114-21.

[90]   Nickander KK, Schmelzer JD, Rohwer DA, Low PA. Effect of alpha-tocopherol deficiency on indices of oxidative stress in normal and diabetic peripheral nerve. J Neurol Sci. 1994 Oct;126(1):6-14.

[91]   Tutuncu NB, Bayraktar M, Varli K. Reversal of defective nerve conduction with vitamin E supplementation in type 2 diabetes: a preliminary study. Diabetes Care. 1998 Nov;21(11):1915-8.

[92]   Eliasson MJ, Sampei K, Mandir AS, Hurn PD, Traystman RJ, Bao J, *et al.* Poly(ADP-ribose) polymerase gene disruption renders mice resistant to cerebral ischemia. Nat Med. 1997 Oct;3(10):1089-95.

[93]   Lo EH, Bosque-Hamilton P, Meng W. Inhibition of poly(ADP-ribose) polymerase: reduction of ischemic injury and attenuation of N-methyl-D-aspartate-induced neurotransmitter dysregulation. Stroke. 1998 Apr;29(4):830-6.

[94]   Li F, Szabo C, Pacher P, Southan GJ, Abatan OI, Charniauskaya T, *et al.* Evaluation of orally active poly(ADP-ribose) polymerase inhibitor in streptozotocin-diabetic rat model of early peripheral neuropathy. Diabetologia. 2004 Apr;47(4):710-7.

[95]   Drel VR, Pacher P, Stavniichuk R, Xu W, Zhang J, Kuchmerovska TM, *et al.* Poly(ADP-ribose)polymerase inhibition counteracts renal hypertrophy and multiple manifestations of peripheral neuropathy in diabetic Akita mice. Int J Mol Med. 2011 Oct;28(4):629-35.

[96]   Stracke H, Hammes HP, Werkmann D, Mavrakis K, Bitsch I, Netzel M, *et al.* Efficacy of benfotiamine *vs.* thiamine on function and glycation products of peripheral nerves in diabetic rats. Exp Clin Endocrinol Diabetes. 2001;109(6):330-6.

[97]   Haupt E, Ledermann H, Kopcke W. Benfotiamine in the treatment of diabetic polyneuropathy--a three-week randomized, controlled pilot study (BEDIP study). Int J Clin Pharmacol Ther. 2005 Feb;43(2):71-7.

[98]   Stracke H, Gaus W, Achenbach U, Federlin K, Bretzel RG. Benfotiamine in diabetic polyneuropathy (BENDIP): results of a randomised, double blind, placebo-controlled clinical study. Exp Clin Endocrinol Diabetes. 2008 Nov;116(10):600-5.

[99]   Leung T, Manser E, Tan L, Lim L. A novel serine/threonine kinase binding the Ras-related RhoA GTPase which translocates the kinase to peripheral membranes. J Biol Chem. 1995 Dec 8;270(49):29051-4.

[100] Ishizaki T, Maekawa M, Fujisawa K, Okawa K, Iwamatsu A, Fujita A, *et al.* The small GTP-binding protein Rho binds to and activates a 160 kDa Ser/Thr protein kinase homologous to myotonic dystrophy kinase. EMBO J. 1996 Apr 15;15(8):1885-93.

[101] Matsui T, Amano M, Yamamoto T, Chihara K, Nakafuku M, Ito M, *et al.* Rho-associated kinase, a novel serine/threonine kinase, as a putative target for small GTP binding protein Rho. EMBO J. 1996 May 1;15(9):2208-16.

[102] Shimokawa H, Takeshita A. Rho-kinase is an important therapeutic target in cardiovascular medicine. Arterioscler Thromb Vasc Biol. 2005 Sep;25(9):1767-75.

[103] Begum N, Sandu OA, Ito M, Lohmann SM, Smolenski A. Active Rho kinase (ROK-alpha) associates with insulin receptor substrate-1 and inhibits insulin signaling in vascular smooth muscle cells. J Biol Chem. 2002 Feb 22;277(8):6214-22.

[104] Kanda T, Wakino S, Homma K, Yoshioka K, Tatematsu S, Hasegawa K, *et al.* Rho-kinase as a molecular target for insulin resistance and hypertension. FASEB J. 2006 Jan;20(1):169-71.

[105] Melendez-Vasquez CV, Einheber S, Salzer JL. Rho kinase regulates schwann cell myelination and formation of associated axonal domains. J Neurosci. 2004 Apr 21;24(16):3953-63.

[106] Ohsawa M, Aasato M, Hayashi SS, Kamei J. RhoA/Rho kinase pathway contributes to the pathogenesis of thermal hyperalgesia in diabetic mice. Pain. 2011 Jan;152(1):114-22.

[107] Chitaley K, Wingard CJ, Clinton Webb R, Branam H, Stopper VS, Lewis RW, *et al.* Antagonism of Rho-kinase stimulates rat penile erection *via* a nitric oxide-independent pathway. Nat Med. 2001 Jan;7(1):119-22.

[108] Lohn M, Plettenburg O, Ivashchenko Y, Kannt A, Hofmeister A, Kadereit D, *et al.* Pharmacological characterization of SAR407899, a novel rho-kinase inhibitor. Hypertension. 2009 Sep;54(3):676-83.

[109] Guagnini F, Ferazzini M, Grasso M, Blanco S, Croci T. Erectile properties of the Rho-kinase inhibitor SAR407899 in diabetic animals and human isolated corpora cavernosa. J Transl Med. 2012;10:59.

[110] Wahren J, Johansson BL, Wallberg-Henriksson H. Does C-peptide have a physiological role? Diabetologia. 1994 Sep;37 Suppl 2:S99-107.

[111] Sima AA, Zhang W, Sugimoto K, Henry D, Li Z, Wahren J, *et al.* C-peptide prevents and improves chronic Type I diabetic polyneuropathy in the BB/Wor rat. Diabetologia. 2001 Jul;44(7):889-97.

[112] Cotter MA, Ekberg K, Wahren J, Cameron NE. Effects of proinsulin C-peptide in experimental diabetic neuropathy: vascular actions and modulation by nitric oxide synthase inhibition. Diabetes. 2003 Jul;52(7):1812-7.

[113] Ido Y, Vindigni A, Chang K, Stramm L, Chance R, Heath WF, *et al.* Prevention of vascular and neural dysfunction in diabetic rats by C-peptide. Science. 1997 Jul 25;277(5325):563-6.

[114] Kamiya H, Zhang W, Ekberg K, Wahren J, Sima AA. C-Peptide reverses nociceptive neuropathy in type 1 diabetes. Diabetes. 2006 Dec;55(12):3581-7.

[115] Hills CE, Brunskill NJ, Squires PE. C-peptide as a therapeutic tool in diabetic nephropathy. Am J Nephrol. 2010;31(5):389-97.

[116] Ekberg K, Brismar T, Johansson BL, Lindstrom P, Juntti-Berggren L, Norrby A, *et al.* C-Peptide replacement therapy and sensory nerve function in type 1 diabetic neuropathy. Diabetes Care. 2007 Jan;30(1):71-6.

[117] Pacifici L, Bellucci A, Piovesan P, Maccari F, Gorio A, Ramacci MT. Counteraction on experimentally induced diabetic neuropathy by levocarnitine acetyl. Int J Clin Pharmacol Res. 1992;12(5-6):231-6.

[118] Sima AA, Ristic H, Merry A, Kamijo M, Lattimer SA, Stevens MJ, *et al.* Primary preventive and secondary interventionary effects of acetyl-L-carnitine on diabetic neuropathy in the bio-breeding Worcester rat. J Clin Invest. 1996 Apr 15;97(8):1900-7.

[119] Sima AA. Acetyl-L-carnitine in diabetic polyneuropathy: experimental and clinical data. CNS Drugs. 2007;21 Suppl 1:13-23; discussion 45-6.

[120] Evans JD, Jacobs TF, Evans EW. Role of acetyl-L-carnitine in the treatment of diabetic peripheral neuropathy. Ann Pharmacother. 2008 Nov;42(11):1686-91.

[121] De Grandis D, Minardi C. Acetyl-L-carnitine (levacecarnine) in the treatment of diabetic neuropathy. A long-term, randomised, double-blind, placebo-controlled study. Drugs R D. 2002;3(4):223-31.

[122] Sima AA, Calvani M, Mehra M, Amato A. Acetyl-L-carnitine improves pain, nerve regeneration, and vibratory perception in patients with chronic diabetic neuropathy: an analysis of two randomized placebo-controlled trials. Diabetes Care. 2005 Jan;28(1):89-94.

[123] Schratzberger P, Walter DH, Rittig K, Bahlmann FH, Pola R, Curry C, *et al.* Reversal of experimental diabetic neuropathy by VEGF gene transfer. J Clin Invest. 2001 May;107(9):1083-92.

[124] Koike H, Morishita R, Iguchi S, Aoki M, Matsumoto K, Nakamura T, *et al.* Enhanced angiogenesis and improvement of neuropathy by cotransfection of human hepatocyte growth factor and prostacyclin synthase gene. FASEB J. 2003 Apr;17(6):779-81.

[125] Nakae M, Kamiya H, Naruse K, Horio N, Ito Y, Mizubayashi R, *et al.* Effects of basic fibroblast growth factor on experimental diabetic neuropathy in rats. Diabetes. 2006 May;55(5):1470-7.

[126] Hanft JR, Pollak RA, Barbul A, van Gils C, Kwon PS, Gray SM, *et al.* Phase I trial on the safety of topical rhVEGF on chronic neuropathic diabetic foot ulcers. J Wound Care. 2008 Jan;17(1):30-2, 4-7.

[127] Price SA, Dent C, Duran-Jimenez B, Liang Y, Zhang L, Rebar EJ, *et al.* Gene transfer of an engineered transcription factor promoting expression of VEGF-A protects against experimental diabetic neuropathy. Diabetes. 2006 Jun;55(6):1847-54.

[128] Nakamura Y, Morishita R, Higaki J, Kida I, Aoki M, Moriguchi A, *et al.* Hepatocyte growth factor is a novel member of the endothelium-specific growth factors: additive

stimulatory effect of hepatocyte growth factor with basic fibroblast growth factor but not with vascular endothelial growth factor. J Hypertens. 1996 Sep;14(9):1067-72.

[129] Kato N, Nemoto K, Nakanishi K, Morishita R, Kaneda Y, Uenoyama M, *et al.* Nonviral gene transfer of human hepatocyte growth factor improves streptozotocin-induced diabetic neuropathy in rats. Diabetes. 2005 Mar;54(3):846-54.

[130] Akita S, Akino K, Imaizumi T, Hirano A. A basic fibroblast growth factor improved the quality of skin grafting in burn patients. Burns. 2005 Nov;31(7):855-8.

[131] Wieman TJ, Smiell JM, Su Y. Efficacy and safety of a topical gel formulation of recombinant human platelet-derived growth factor-BB (becaplermin) in patients with chronic neuropathic diabetic ulcers. A phase III randomized placebo-controlled double-blind study. Diabetes Care. 1998 May;21(5):822-7.

[132] Papanas N, Maltezos E. Benefit-risk assessment of becaplermin in the treatment of diabetic foot ulcers. Drug Saf. 2010 Jun 1;33(6):455-61.

[133] Kajiguchi M, Kondo T, Izawa H, Kobayashi M, Yamamoto K, Shintani S, *et al.* Safety and efficacy of autologous progenitor cell transplantation for therapeutic angiogenesis in patients with critical limb ischemia. Circ J. 2007 Feb;71(2):196-201.

[134] Naruse K, Hamada Y, Nakashima E, Kato K, Mizubayashi R, Kamiya H, *et al.* Therapeutic neovascularization using cord blood-derived endothelial progenitor cells for diabetic neuropathy. Diabetes. 2005 Jun;54(6):1823-8.

[135] Hasegawa T, Kosaki A, Shimizu K, Matsubara H, Mori Y, Masaki H, *et al.* Amelioration of diabetic peripheral neuropathy by implantation of hematopoietic mononuclear cells in streptozotocin-induced diabetic rats. Exp Neurol. 2006 Jun;199(2):274-80.

[136] Shibata T, Naruse K, Kamiya H, Kozakae M, Kondo M, Yasuda Y, *et al.* Transplantation of bone marrow-derived mesenchymal stem cells improves diabetic polyneuropathy in rats. Diabetes. 2008 Nov;57(11):3099-107.

[137] Andersson LO, Barrowcliffe TW, Holmer E, Johnson EA, Sims GE. Anticoagulant properties of heparin fractionated by affinity chromatography on matrix-bound antithrombin iii and by gel filtration. Thromb Res. 1976 Dec;9(6):575-83.

[138] Rosenberg RD. Biochemistry and pharmacology of low molecular weight heparin. Semin Hematol. 1997 Oct;34(4 Suppl 4):2-8.

[139] Jorneskog G, Brismar K, Fagrell B. Low molecular weight heparin seems to improve local capillary circulation and healing of chronic foot ulcers in diabetic patients. Vasa. 1993;22(2):137-42.

[140] Kalani M, Apelqvist J, Blomback M, Brismar K, Eliasson B, Eriksson JW, *et al.* Effect of dalteparin on healing of chronic foot ulcers in diabetic patients with peripheral arterial occlusive disease: a prospective, randomized, double-blind, placebo-controlled study. Diabetes Care. 2003 Sep;26(9):2575-80.

[141] Kalani M, Silveira A, Blomback M, Apelqvist J, Eliasson B, Eriksson JW, *et al.* Beneficial effects of dalteparin on haemostatic function and local tissue oxygenation in patients with diabetes, severe vascular disease and foot ulcers. Thromb Res. 2007;120(5):653-61.

[142] Planes A. Review of bemiparin sodium--a new second-generation low molecular weight heparin and its applications in venous thromboembolism. Expert Opin Pharmacother. 2003 Sep;4(9):1551-61.

[143] Chapman TM, Goa KL. Bemiparin: a review of its use in the prevention of venous thromboembolism and treatment of deep vein thrombosis. Drugs. 2003;63(21):2357-77.

[144] Rullan M, Cerda L, Frontera G, Masmiquel L, Llobera J. Treatment of chronic diabetic foot ulcers with bemiparin: a randomized, triple-blind, placebo-controlled, clinical trial. Diabet Med. 2008 Sep;25(9):1090-5.

[145] Fernyhough P, Diemel LT, Hardy J, Brewster WJ, Mohiuddin L, Tomlinson DR. Human recombinant nerve growth factor replaces deficient neurotrophic support in the diabetic rat. Eur J Neurosci. 1995 May 1;7(5):1107-10.

[146] Tomlinson DR, Fernyhough P, Diemel LT. Role of neurotrophins in diabetic neuropathy and treatment with nerve growth factors. Diabetes. 1997 Sep;46 Suppl 2:S43-9.

[147] Karamoysoyli E, Burnand RC, Tomlinson DR, Gardiner NJ. Neuritin mediates nerve growth factor-induced axonal regeneration and is deficient in experimental diabetic neuropathy. Diabetes. 2008 Jan;57(1):181-9.

[148] Mizisin AP, Vu Y, Shuff M, Calcutt NA. Ciliary neurotrophic factor improves nerve conduction and ameliorates regeneration deficits in diabetic rats. Diabetes. 2004 Jul;53(7):1807-12.

[149] Anitha M, Gondha C, Sutliff R, Parsadanian A, Mwangi S, Sitaraman SV, *et al.* GDNF rescues hyperglycemia- induced diabetic enteric neuropathy through activation of the PI3K/Akt pathway. J Clin Invest. 2006 Feb;116(2):344-56.

[150] Freeman R. Human studies of recombinant human nerve growth factor and diabetic peripheral neuropathy. Eur Neurol. 1999;41 Suppl 1:20-6.

[151] Apfel SC, Schwartz S, Adornato BT, Freeman R, Biton V, Rendell M, *et al.* Efficacy and safety of recombinant human nerve growth factor in patients with diabetic polyneuropathy: A randomized controlled trial. rhNGF Clinical Investigator Group. JAMA. 2000 Nov 1;284(17):2215-21.

[152] Kakinoki B, Sekimoto S, Yuki S, Ohgami T, Sejima M, Yamagami K, *et al.* Orally active neurotrophin-enhancing agent protects against dysfunctions of the peripheral nerves in hyperglycemic animals. Diabetes. 2006 Mar;55(3):616-21.

[153] Gold BG. Neuroimmunophilin ligands: evaluation of their therapeutic potential for the treatment of neurological disorders. Expert Opin Investig Drugs. 2000 Oct;9(10):2331-42.

[154] Sharkey J, Butcher SP. Immunophilins mediate the neuroprotective effects of FK506 in focal cerebral ischaemia. Nature. 1994 Sep 22;371(6495):336-9.

[155] Steiner JP, Hamilton GS, Ross DT, Valentine HL, Guo H, Connolly MA, *et al.* Neurotrophic immunophilin ligands stimulate structural and functional recovery in neurodegenerative animal models. Proc Natl Acad Sci U S A. 1997 Mar 4;94(5):2019-24.

[156] Rajendram R, Fraser-Bell S, Kaines A, Michaelides M, Hamilton RD, Esposti SD, *et al.* A 2-year prospective randomized controlled trial of intravitreal bevacizumab or laser therapy (BOLT) in the management of diabetic macular edema: 24-month data: report 3. Arch Ophthalmol. 2012 Aug 1;130(8):972-9.

[157] Mitchell P, Bandello F, Schmidt-Erfurth U, Lang GE, Massin P, Schlingemann RO, *et al.* The RESTORE study: ranibizumab monotherapy or combined with laser *vs.* laser monotherapy for diabetic macular edema. Ophthalmology. 2011 Apr;118(4):615-25.

[158] Nguyen QD, Brown DM, Marcus DM, Boyer DS, Patel S, Feiner L, *et al.* Ranibizumab for diabetic macular edema: results from 2 phase III randomized trials: RISE and RIDE. Ophthalmology. 2012 Apr;119(4):789-801.

[159] Chakravarthy U, Adamis AP, Cunningham ET, Jr., Goldbaum M, Guyer DR, Katz B, *et al.* Year 2 efficacy results of 2 randomized controlled clinical trials of pegaptanib for

neovascular age-related macular degeneration. Ophthalmology. 2006 Sep;113(9):1508 e1-25.

[160] Sultan MB, Zhou D, Loftus J, Dombi T, Ice KS. A phase 2/3, multicenter, randomized, double-masked, 2-year trial of pegaptanib sodium for the treatment of diabetic macular edema. Ophthalmology. 2011 Jun;118(6):1107-18.

[161] Loftus JV, Sultan MB, Pleil AM. Changes in vision- and health-related quality of life in patients with diabetic macular edema treated with pegaptanib sodium or sham. Invest Ophthalmol Vis Sci. 2011 Sep;52(10):7498-505.

[162] Heier JS, Brown DM, Chong V, Korobelnik JF, Kaiser PK, Nguyen QD, *et al.* Intravitreal aflibercept (VEGF trap-eye) in wet age-related macular degeneration. Ophthalmology. 2012 Dec;119(12):2537-48.

[163] Do DV, Schmidt-Erfurth U, Gonzalez VH, Gordon CM, Tolentino M, Berliner AJ, *et al.* The DA VINCI Study: phase 2 primary results of VEGF Trap-Eye in patients with diabetic macular edema. Ophthalmology. 2011 Sep;118(9):1819-26.

[164] Do DV, Nguyen QD, Boyer D, Schmidt-Erfurth U, Brown DM, Vitti R, *et al.* One-year outcomes of the DA VINCI Study of VEGF Trap-Eye in eyes with diabetic macular edema. Ophthalmology. 2012 Aug;119(8):1658-65.

[165] Campochiaro PA, Channa R, Berger BB, Heier JS, Brown DM, Fiedler U, *et al.* Treatment of Diabetic Macular Edema With a Designed Ankyrin Repeat Protein That Binds Vascular Endothelial Growth Factor: A Phase 1/2 Study. Am J Ophthalmol. 2012 Dec 3.

[166] Hnik P, Boyer DS, Grillone LR, Clement JG, Henry SP, Green EA. Antisense oligonucleotide therapy in diabetic retinopathy. J Diabetes Sci Technol. 2009 Jul;3(4):924-30.

[167] Hos D, Regenfuss B, Bock F, Onderka J, Cursiefen C. Blockade of insulin receptor substrate-1 inhibits corneal lymphangiogenesis. Invest Ophthalmol Vis Sci. 2011 Jul;52(8):5778-85.

[168] Cloutier F, Lawrence M, Goody R, Lamoureux S, Al-Mahmood S, Colin S, *et al.* Antiangiogenic activity of aganirsen in nonhuman primate and rodent models of retinal neovascular disease after topical administration. Invest Ophthalmol Vis Sci. 2012 Mar;53(3):1195-203.

[169] Cursiefen C, Bock F, Horn FK, Kruse FE, Seitz B, Borderie V, *et al.* GS-101 antisense oligonucleotide eye drops inhibit corneal neovascularization: interim results of a randomized phase II trial. Ophthalmology. 2009 Sep;116(9):1630-7.

[170] Tachas G, Lofthouse S, Wraight CJ, Baker BF, Sioufi NB, Jarres RA, *et al.* A GH receptor antisense oligonucleotide inhibits hepatic GH receptor expression, IGF-I production and body weight gain in normal mice. J Endocrinol. 2006 Apr;189(1):147-54.

[171] Wilkinson-Berka JL, Lofthouse S, Jaworski K, Ninkovic S, Tachas G, Wraight CJ. An antisense oligonucleotide targeting the growth hormone receptor inhibits neovascularization in a mouse model of retinopathy. Mol Vis. 2007;13:1529-38.

[172] de Nigris F, Schiano C, Infante T, Napoli C. CXCR4 inhibitors: tumor vasculature and therapeutic challenges. Recent Pat Anticancer Drug Discov. 2012 Sep;7(3):251-64.

[173] MacMillan CJ, Furlong SJ, Doucette CD, Chen PL, Hoskin DW, Easton AS. Bevacizumab diminishes experimental autoimmune encephalomyelitis by inhibiting spinal cord angiogenesis and reducing peripheral T-cell responses. J Neuropathol Exp Neurol. 2012 Nov;71(11):983-99.

[174] Shoshani T, Faerman A, Mett I, Zelin E, Tenne T, Gorodin S, *et al.* Identification of a novel hypoxia-inducible factor 1-responsive gene, RTP801, involved in apoptosis. Mol Cell Biol. 2002 Apr;22(7):2283-93.

[175] Brafman A, Mett I, Shafir M, Gottlieb H, Damari G, Gozlan-Kelner S, *et al.* Inhibition of oxygen-induced retinopathy in RTP801-deficient mice. Invest Ophthalmol Vis Sci. 2004 Oct;45(10):3796-805.

[176] Watanabe D, Suzuma K, Suzuma I, Ohashi H, Ojima T, Kurimoto M, *et al.* Vitreous levels of angiopoietin 2 and vascular endothelial growth factor in patients with proliferative diabetic retinopathy. Am J Ophthalmol. 2005 Mar;139(3):476-81.

[177] Lewis GP, Chapin EA, Byun J, Luna G, Sherris D, Fisher SK. Muller cell reactivity and photoreceptor cell death are reduced after experimental retinal detachment using an inhibitor of the Akt/mTOR pathway. Invest Ophthalmol Vis Sci. 2009 Sep;50(9):4429-35.

[178] Dugel PU, Blumenkranz MS, Haller JA, Williams GA, Solley WA, Kleinman DM, *et al.* A randomized, dose-escalation study of subconjunctival and intravitreal injections of sirolimus in patients with diabetic macular edema. Ophthalmology. 2012 Jan;119(1):124-31.

[179] Jalkh A, Takahashi M, Topilow HW, Trempe CL, McMeel JW. Prognostic value of vitreous findings in diabetic retinopathy. Arch Ophthalmol. 1982 Mar;100(3):432-4.

[180] Harbour JW, Smiddy WE, Flynn HW, Jr., Rubsamen PE. Vitrectomy for diabetic macular edema associated with a thickened and taut posterior hyaloid membrane. Am J Ophthalmol. 1996 Apr;121(4):405-13.

[181] Gandorfer A. Enzymatic vitreous disruption. Eye (Lond). 2008 Oct;22(10):1273-7.

[182] Kuppermann BD, Thomas EL, de Smet MD, Grillone LR. Pooled efficacy results from two multinational randomized controlled clinical trials of a single intravitreous injection of highly purified ovine hyaluronidase (Vitrase) for the management of vitreous hemorrhage. Am J Ophthalmol. 2005 Oct;140(4):573-84.

[183] Kuppermann BD, Thomas EL, de Smet MD, Grillone LR. Safety results of two phase III trials of an intravitreous injection of highly purified ovine hyaluronidase (Vitrase) for the management of vitreous hemorrhage. Am J Ophthalmol. 2005 Oct;140(4):585-97.

[184] Azzolini C, D'Angelo A, Maestranzi G, Codenotti M, Della Valle P, Prati M, *et al.* Intrasurgical plasmin enzyme in diabetic macular edema. Am J Ophthalmol. 2004 Oct;138(4):560-6.

[185] Diaz-Llopis M, Udaondo P, Arevalo F, Salom D, Garcia-Delpech S, Quijada A, *et al.* Intravitreal plasmin without associated vitrectomy as a treatment for refractory diabetic macular edema. J Ocul Pharmacol Ther. 2009 Aug;25(4):379-84.

[186] DeCroos FC, Toth CA, Folgar FA, Pakola S, Stinnett SS, Heydary CS, *et al.* Characterization of vitreoretinal interface disorders using OCT in the interventional phase 3 trials of ocriplasmin. Invest Ophthalmol Vis Sci. 2012 Sep;53(10):6504-11.

[187] Stalmans P, Benz MS, Gandorfer A, Kampik A, Girach A, Pakola S, *et al.* Enzymatic vitreolysis with ocriplasmin for vitreomacular traction and macular holes. N Engl J Med. 2012 Aug 16;367(7):606-15.

[188] Folgar FA, Toth CA, DeCroos FC, Girach A, Pakola S, Jaffe GJ. Assessment of retinal morphology with spectral and time domain OCT in the phase III trials of enzymatic vitreolysis. Invest Ophthalmol Vis Sci. 2012;53(11):7395-401.

[189] Nonaka A, Kiryu J, Tsujikawa A, Yamashiro K, Miyamoto K, Nishiwaki H, *et al.* PKC-beta inhibitor (LY333531) attenuates leukocyte entrapment in retinal microcirculation of diabetic rats. Invest Ophthalmol Vis Sci. 2000 Aug;41(9):2702-6.

[190] PKC-DRS Study Group. The effect of ruboxistaurin on visual loss in patients with moderately severe to very severe nonproliferative diabetic retinopathy: initial results of the Protein Kinase C beta Inhibitor Diabetic Retinopathy Study (PKC-DRS) multicenter randomized clinical trial. Diabetes. 2005 Jul;54(7):2188-97.

[191] Aiello LP, Vignati L, Sheetz MJ, Zhi X, Girach A, Davis MD, *et al*. Oral protein kinase c beta inhibition using ruboxistaurin: efficacy, safety, and causes of vision loss among 813 patients (1,392 eyes) with diabetic retinopathy in the Protein Kinase C beta Inhibitor-Diabetic Retinopathy Study and the Protein Kinase C beta Inhibitor-Diabetic Retinopathy Study 2. Retina. 2011 Nov;31(10):2084-94.

[192] Aiello LP, Davis MD, Girach A, Kles KA, Milton RC, Sheetz MJ, *et al*. Effect of ruboxistaurin on visual loss in patients with diabetic retinopathy. Ophthalmology. 2006 Dec;113(12):2221-30.

[193] PKC-DMES Study Group. Effect of ruboxistaurin in patients with diabetic macular edema: thirty-month results of the randomized PKC-DMES clinical trial. Arch Ophthalmol. 2007 Mar;125(3):318-24.

[194] Sasase T, Morinaga H, Abe T, Miyajima K, Ohta T, Shinohara M, *et al*. Protein kinase C beta inhibitor prevents diabetic peripheral neuropathy, but not histopathological abnormalities of retina in Spontaneously Diabetic Torii rat. Diabetes Obes Metab. 2009 Nov;11(11):1084-7.

[195] Kennedy L, Solano MP, Meneghini L, Lo M, Cohen MP. Anti-glycation and anti-albuminuric effects of GLY-230 in human diabetes. Am J Nephrol. 2010;31(2):110-6.

[196] Blackie JA, Bloomer JC, Brown MJ, Cheng HY, Hammond B, Hickey DM, *et al*. The identification of clinical candidate SB-480848: a potent inhibitor of lipoprotein-associated phospholipase A2. Bioorg Med Chem Lett. 2003 Mar 24;13(6):1067-70.

[197] O'Donoghue ML, Braunwald E, White HD, Serruys P, Steg PG, Hochman J, *et al*. Study design and rationale for the Stabilization of pLaques usIng Darapladib-Thrombolysis in Myocardial Infarction (SOLID-TIMI 52) trial in patients after an acute coronary syndrome. Am Heart J. 2011 Oct;162(4):613-9 e1.

[198] Wong WT, Kam W, Cunningham D, Harrington M, Hammel K, Meyerle CB, *et al*. Treatment of geographic atrophy by the topical administration of OT-551: results of a phase II clinical trial. Invest Ophthalmol Vis Sci. 2010 Dec;51(12):6131-9.

[199] Plockinger U, Hoffmann U, Geese M, Lupp A, Buchfelder M, Flitsch J, *et al*. DG3173 (somatoprim), a unique somatostatin receptor subtypes 2-, 4- and 5-selective analogue, effectively reduces GH secretion in human GH-secreting pituitary adenomas even in Octreotide non-responsive tumours. Eur J Endocrinol. 2012 Feb;166(2):223-34.

[200] Martidis A, Duker JS, Greenberg PB, Rogers AH, Puliafito CA, Reichel E, *et al*. Intravitreal triamcinolone for refractory diabetic macular edema. Ophthalmology. 2002 May;109(5):920-7.

[201] Beer PM, Bakri SJ, Singh RJ, Liu W, Peters GB, 3rd, Miller M. Intraocular concentration and pharmacokinetics of triamcinolone acetonide after a single intravitreal injection. Ophthalmology. 2003 Apr;110(4):681-6.

[202] Sakamoto T, Miyazaki M, Hisatomi T, Nakamura T, Ueno A, Itaya K, *et al*. Triamcinolone-assisted pars plana vitrectomy improves the surgical procedures and decreases the postoperative blood-ocular barrier breakdown. Graefes Arch Clin Exp Ophthalmol. 2002 Jun;240(6):423-9.

[203] Haller JA, Bandello F, Belfort R, Jr., Blumenkranz MS, Gillies M, Heier J, *et al.* Dexamethasone intravitreal implant in patients with macular edema related to branch or central retinal vein occlusion twelve-month study results. Ophthalmology. 2011 Dec;118(12):2453-60.

[204] Lowder C, Belfort R, Jr., Lightman S, Foster CS, Robinson MR, Schiffman RM, *et al.* Dexamethasone intravitreal implant for noninfectious intermediate or posterior uveitis. Arch Ophthalmol. 2011 May;129(5):545-53.

[205] Kane FE, Burdan J, Cutino A, Green KE. Iluvien: a new sustained delivery technology for posterior eye disease. Expert Opin Drug Deliv. 2008 Sep;5(9):1039-46.

[206] Campochiaro PA, Hafiz G, Shah SM, Bloom S, Brown DM, Busquets M, *et al.* Sustained ocular delivery of fluocinolone acetonide by an intravitreal insert. Ophthalmology. 2010 Jul;117(7):1393-9 e3.

[207] Foster CS, Davanzo R, Flynn TE, McLeod K, Vogel R, Crockett RS. Durezol (Difluprednate Ophthalmic Emulsion 0.05%) compared with Pred Forte 1% ophthalmic suspension in the treatment of endogenous anterior uveitis. J Ocul Pharmacol Ther. 2010 Oct;26(5):475-83.

[208] Gamache DA, Graff G, Brady MT, Spellman JM, Yanni JM. Nepafenac, a unique nonsteroidal prodrug with potential utility in the treatment of trauma-induced ocular inflammation: I. Assessment of anti-inflammatory efficacy. Inflammation. 2000 Aug;24(4):357-70.

[209] Das A, Garner DP, Del Re AM, Woodward JJ, Kumar DM, Agarwal N, *et al.* Calpeptin provides functional neuroprotection to rat retinal ganglion cells following Ca2+ influx. Brain Res. 2006 Apr 21;1084(1):146-57.

[210] Oka T, Walkup RD, Tamada Y, Nakajima E, Tochigi A, Shearer TR, *et al.* Amelioration of retinal degeneration and proteolysis in acute ocular hypertensive rats by calpain inhibitor ((1S)-1-((((1S)-1-benzyl-3-cyclopropylamino-   2,3-di-oxopropyl)amino)carbonyl)-3-me thylbutyl)carbamic acid 5-methoxy-3-oxapentyl ester. Neuroscience. 2006 Sep 15;141(4):2139-45.

[211] Ryu M, Yasuda M, Shi D, Shanab AY, Watanabe R, Himori N, *et al.* Critical role of calpain in axonal damage-induced retinal ganglion cell death. J Neurosci Res. 2012 Apr;90(4):802-15.

[212] Shanab AY, Nakazawa T, Ryu M, Tanaka Y, Himori N, Taguchi K, *et al.* Metabolic stress response implicated in diabetic retinopathy: the role of calpain, and the therapeutic impact of calpain inhibitor. Neurobiol Dis. 2012 Dec;48(3):556-67.

[213] Ma H, Tochigi A, Shearer TR, Azuma M. Calpain inhibitor SNJ-1945 attenuates events prior to angiogenesis in cultured human retinal endothelial cells. J Ocul Pharmacol Ther. 2009 Oct;25(5):409-14.

[214] Ruggenenti P, Fassi A, Ilieva AP, Bruno S, Iliev IP, Brusegan V, *et al.* Preventing microalbuminuria in type 2 diabetes. N Engl J Med. 2004 Nov 4;351(19):1941-51.

[215] Heart Outcomes Prevention Evaluation Study Investigators. Effects of ramipril on cardiovascular and microvascular outcomes in people with diabetes mellitus: results of the HOPE study and MICRO-HOPE substudy. Lancet. 2000 Jan 22;355(9200):253-9.

[216] Crepaldi G, Carta Q, Deferrari G, Mangili R, Navalesi R, Santeusanio F, *et al.* Effects of lisinopril and nifedipine on the progression to overt albuminuria in IDDM patients with incipient nephropathy and normal blood pressure. The Italian Microalbuminuria Study Group in IDDM. Diabetes Care. 1998 Jan;21(1):104-10.

[217] Ahmad J, Siddiqui MA, Ahmad H. Effective postponement of diabetic nephropathy with enalapril in normotensive type 2 diabetic patients with microalbuminuria. Diabetes Care. 1997 Oct;20(10):1576-81.

[218] The Microalbuminuria Captopril Study Group. Captopril reduces the risk of nephropathy in IDDM patients with microalbuminuria. Diabetologia. 1996 May;39(5):587-93.

[219] Katayama S, Kikkawa R, Isogai S, Sasaki N, Matsuura N, Tajima N, *et al.* Effect of captopril or imidapril on the progression of diabetic nephropathy in Japanese with type 1 diabetes mellitus: a randomized controlled study (JAPAN-IDDM). Diabetes Res Clin Pract. 2002 Feb;55(2):113-21.

[220] Brenner BM, Cooper ME, de Zeeuw D, Keane WF, Mitch WE, Parving HH, *et al.* Effects of losartan on renal and cardiovascular outcomes in patients with type 2 diabetes and nephropathy. N Engl J Med. 2001 Sep 20;345(12):861-9.

[221] Lewis EJ, Hunsicker LG, Clarke WR, Berl T, Pohl MA, Lewis JB, *et al.* Renoprotective effect of the angiotensin-receptor antagonist irbesartan in patients with nephropathy due to type 2 diabetes. N Engl J Med. 2001 Sep 20;345(12):851-60.

[222] Viberti G, Wheeldon NM. Microalbuminuria reduction with valsartan in patients with type 2 diabetes mellitus: a blood pressure-independent effect. Circulation. 2002 Aug 6;106(6):672-8.

[223] Makino H, Haneda M, Babazono T, Moriya T, Ito S, Iwamoto Y, *et al.* Prevention of transition from incipient to overt nephropathy with telmisartan in patients with type 2 diabetes. Diabetes Care. 2007 Jun;30(6):1577-8.

[224] Haller H, Ito S, Izzo JL, Jr., Januszewicz A, Katayama S, Menne J, *et al.* Olmesartan for the delay or prevention of microalbuminuria in type 2 diabetes. N Engl J Med. 2011 Mar 10;364(10):907-17.

[225] Parving HH, Persson F, Lewis JB, Lewis EJ, Hollenberg NK. Aliskiren combined with losartan in type 2 diabetes and nephropathy. N Engl J Med. 2008 Jun 5;358(23):2433-46.

[226] Parving HH, Brenner BM, McMurray JJ, de Zeeuw D, Haffner SM, Solomon SD, *et al.* Cardiorenal end points in a trial of aliskiren for type 2 diabetes. N Engl J Med. 2012 Dec 6;367(23):2204-13.

[227] Pergola PE, Raskin P, Toto RD, Meyer CJ, Huff JW, Grossman EB, *et al.* Bardoxolone methyl and kidney function in CKD with type 2 diabetes. N Engl J Med. 2011 Jul 28;365(4):327-36.

[228] Sedeek M, Callera G, Montezano A, Gutsol A, Heitz F, Szyndralewiez C, *et al.* Critical role of Nox4-based NADPH oxidase in glucose-induced oxidative stress in the kidney: implications in type 2 diabetic nephropathy. Am J Physiol Renal Physiol. 2010 Dec;299(6):F1348-58.

[229] Gorin Y, Block K, Hernandez J, Bhandari B, Wagner B, Barnes JL, *et al.* Nox4 NAD(P)H oxidase mediates hypertrophy and fibronectin expression in the diabetic kidney. J Biol Chem. 2005 Nov 25;280(47):39616-26.

[230] Igarashi A, Okochi H, Bradham DM, Grotendorst GR. Regulation of connective tissue growth factor gene expression in human skin fibroblasts and during wound repair. Mol Biol Cell. 1993 Jun;4(6):637-45.

[231] Kothapalli D, Frazier KS, Welply A, Segarini PR, Grotendorst GR. Transforming growth factor beta induces anchorage-independent growth of NRK fibroblasts *via* a connective tissue growth factor-dependent signaling pathway. Cell Growth Differ. 1997 Jan;8(1):61-8.

[232] Chow FY, Nikolic-Paterson DJ, Ozols E, Atkins RC, Rollin BJ, Tesch GH. Monocyte chemoattractant protein-1 promotes the development of diabetic renal injury in streptozotocin-treated mice. Kidney Int. 2006 Jan;69(1):73-80.

[233] Awad AS, Kinsey GR, Khutsishvili K, Gao T, Bolton WK, Okusa MD. Monocyte/macrophage chemokine receptor CCR2 mediates diabetic renal injury. Am J Physiol Renal Physiol. 2011 Dec;301(6):F1358-66.

[234] Berthier CC, Zhang H, Schin M, Henger A, Nelson RG, Yee B, *et al.* Enhanced expression of Janus kinase-signal transducer and activator of transcription pathway members in human diabetic nephropathy. Diabetes. 2009 Feb;58(2):469-77.

[235] Degenhardt TP, Alderson NL, Arrington DD, Beattie RJ, Basgen JM, Steffes MW, *et al.* Pyridoxamine inhibits early renal disease and dyslipidemia in the streptozotocin-diabetic rat. Kidney Int. 2002 Mar;61(3):939-50.

[236] Alderson NL, Chachich ME, Youssef NN, Beattie RJ, Nachtigal M, Thorpe SR, *et al.* The AGE inhibitor pyridoxamine inhibits lipemia and development of renal and vascular disease in Zucker obese rats. Kidney Int. 2003 Jun;63(6):2123-33.

[237] Williams ME, Bolton WK, Khalifah RG, Degenhardt TP, Schotzinger RJ, McGill JB. Effects of pyridoxamine in combined phase 2 studies of patients with type 1 and type 2 diabetes and overt nephropathy. Am J Nephrol. 2007;27(6):605-14.

[238] Yanagisawa M, Kurihara H, Kimura S, Tomobe Y, Kobayashi M, Mitsui Y, *et al.* A novel potent vasoconstrictor peptide produced by vascular endothelial cells. Nature. 1988 Mar 31;332(6163):411-5.

[239] Saleh MA, Pollock JS, Pollock DM. Distinct actions of endothelin A-selective *vs.* combined endothelin A/B receptor antagonists in early diabetic kidney disease. J Pharmacol Exp Ther. 2011 Jul;338(1):263-70.

[240] Zanatta CM, Veronese FV, Loreto Mda S, Sortica DA, Carpio VN, Eldeweiss MI, *et al.* Endothelin-1 and endothelin a receptor immunoreactivity is increased in patients with diabetic nephropathy. Ren Fail. 2012;34(3):308-15.

[241] Kohan DE, Pritchett Y, Molitch M, Wen S, Garimella T, Audhya P, *et al.* Addition of atrasentan to renin-angiotensin system blockade reduces albuminuria in diabetic nephropathy. J Am Soc Nephrol. 2011 Apr;22(4):763-72.

[242] Wenzel RR, Littke T, Kuranoff S, Jurgens C, Bruck H, Ritz E, *et al.* Avosentan reduces albumin excretion in diabetics with macroalbuminuria. J Am Soc Nephrol. 2009 Mar;20(3):655-64.

Index

A

ABCA1   127, 169-70

Absorption of gabapentin   214

Acarbose   69-70

Accord-lipid   179-80, 185

ACE inhibitors   3-5, 7, 11, 13, 15-19, 53, 211, 267

ACEI/ARB   9, 11, 21

ACEIs   6, 8-9, 15, 20

Acetyl-L-carnitine   237-8

Acromegalic patients   258

Acute kidney injury (AKI)   8, 13

Adipocyte differentiation   131, 136

Adipocytes   37, 130-2, 135, 161, 164, 171-2, 198

Adipogenesis   131, 134-5, 168, 172

Adipose   121, 129, 131

Adipose tissue   35-7, 130-2, 137, 172, 192

Adiposity   33, 36-7, 138

Adjunctive treatment   225

Adrenaline receptors   215, 223

Advanced glycation endproducts (AGE)   33, 36-52, 54-5, 57, 232-4

AGE accumulation   33, 43, 46, 50, 52-4

AGE formation   33, 37-8, 40, 42, 47-8, 50-1, 271

AGE-induced cross-linking of type IV collagen   40

AGE production inhibitor   232-3

AGE-rich diet   37

AKB-9778   252

Alagebrium   52-3, 233

Albumin   83, 109, 175, 188, 190

Albumin creatinine ratios (ACR)   11

Albumin excretion rate (AER)   11

Albuminuria   11, 13, 45, 53, 187, 265

Aldose reductase (AR)   230-1

I

www.ingramcontent.com/pod-product-compliance
Lightning Source LLC
Chambersburg PA
CBHW050810220326
41598CB00006B/166